KU-408-851

BRITISH GEOLOGICAL SURVEY

British Regional Geology

The Grampian Highlands

FOURTH EDITION

By D Stephenson, PhD and
D Gould, PhD

Contributors:

G C Clark PhD
D J Fettes PhD
T P Fletcher PhD
M J Gallagher PhD
G S Johnstone BSc
R M Key PhD
D I J Mallick PhD
J R Mendum PhD
J W Merritt BSc
R M W Musson PhD
W Mykura DSc
J D Peacock PhD
D I Smith PhD

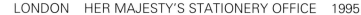

LONDON   HER MAJESTY'S STATIONERY OFFICE   1995

HER MAJESTY'S STATIONERY OFFICE

HMSO publications are available from:

HMSO Publications Centre
(Mail, fax and telephone orders only)
PO Box 276, London SW8 5DT
Telephone orders 0171-873 9090
General enquiries 0171-873 0011
Queuing system in operation for both numbers
Fax orders 0171-873 8200

HMSO Bookshops
49 High Holborn, London WC1V 6HB
(counter service only)
0171-873 0011   Fax 0171-831 1326
68–69 Bull Street, Birmingham B4 6AD
0121-236 9696   Fax 0121-236 9699
33 Wine Street, Bristol BS1 2BQ
0117-9264306   Fax 0117-9294515
9 Princess Street, Manchester M60 8AS
0161-834 7201   Fax 0161-833 0634
16 Arthur Street, Belfast BT1 4GD
01232-238451   Fax 01232-235401
71 Lothian Road, Edinburgh EH3 9AZ
0131-228 4181   Fax 0131-229 2734
HMSO Oriel Bookshop, The Friary,
Cardiff CF1 4AA
01222-395548   Fax 01222-384347

HMSO's Accredited Agents
(see Yellow Pages)

And through good booksellers

BRITISH GEOLOGICAL SURVEY

Keyworth, Nottingham NG12 5GG
0115-936 3100

Murchison House, West Mains Road,
Edinburgh, EH9 3LA   0131-667 1000

London Information Office, Natural
History Museum, Earth Galleries, Exhibition Road, London SW7 2DE
0171-589 4090

The full range of Survey publications is available through the Sales Desks at Keyworth and at Murchison House, Edinburgh, and in the BGS London Information Office in the Natural History Museum (Earth Galleries). The adjacent bookshop stocks the more popular books for sale over the counter. Most BGS books and reports can be bought from HMSO and through HMSO agents and retailers. Maps are listed in the BGS Map Catalogue, and can be bought together with books and reports through BGS-approved stockists and agents as well as direct from BGS.

The British Geological Survey carries out the geological survey of Great Britain and Northern Ireland (the latter as an agency service for the government of Northern Ireland), and of the surrounding continental shelf, as well as its basic research projects. It also undertakes programmes of British technical aid in geology in developing countries as arranged by the Overseas Development Administration.

The British Geological Survey is a component body of the Natural Environment Research Council.

Bibliographic reference
STEPHENSON, D, and GOULD, D.   1995.
British regional geology: the Grampian Highlands (4th edition). (London: HMSO for the British Geological Survey.)

Maps and diagrams in this book use topography based on Ordnance Survey

© *NERC copyright 1995*
First published 1935
Second edition 1948
Third edition 1966

ISBN 0 11 884521 7

Contents

Figures

Plates

Numbers in brackets refer to photographs in the BGS collections.

Tables

Foreword to the fourth edition

The first edition of the Grampian Highlands regional guide was published in 1935, with subsequent editions being issued in 1948 and 1966. Throughout the period since the appearance of the third edition there has been a considerable amount of work in the Grampian Highlands region (which here is taken to include the South-west Highlands), both systematic mapping by the British Geological Survey and research by the universities. The result of this work has been substantial improvement to our understanding of the geology. Even now, however, that understanding is incomplete and there is still controversy over several issues of fundamental significance. The present edition of the guide presents a comprehensive summary of our existing knowledge and reviews the on-going debates, particularly those relating to the Precambrian geology.

The Grampian Highlands region covers a large area and understanding its complex geology calls for a wide variety of geological expertise. In consequence, this fourth edition of the guide is a cooperative product from the BGS Highlands and Islands Group, (including several former members of staff), and other parts of BGS.

The Grampians region is mostly one of uplands and mountains, and includes the highest peak in Britain, Ben Nevis (1343 m). It is an area much admired for its wild and unspoiled rugged topography and the deep and beautiful valleys in which run some of the most renowned of salmon and trout rivers such as the Spey and the Dee. In the north of the region, however, the countryside along the southern side of the Moray Firth is rather different, with more gently undulating hills, and a flat coastal plain is largely cultivated.

These and other topographical differences reflect the underlying geology which provides an insight into the evolution of the area over 700 million years. The rocks of the uplands mostly record the development of the root zone of the Caledonian mountain belt in late Precambrian and early Palaeozoic times; subsequently many kilometres of rock were eroded to expose the rocks we see at present.

It is only at the margins of the region that there are large developments of younger rocks, most notably the Devonian of the southern margin of the Moray Firth basin and the widespread but thin cover of Quaternary glacial and periglacial deposits.

Understanding the nature of the rocks, the geological history of the area and the processes involved to produce the present-day landscapes is essential if we are to properly understand and conserve the beautiful country and wild scenery which is the very heart of the region. It is also essential if we are to understand the distribution of natural resources and possible hazards which

x

may result from any manmade developments. The resources include gold,
baryte, small amounts of copper, lead, zinc and manganese and large quantities of both hard rock for aggregate, and sand and gravel.

I am sure that this fourth edition of the regional guide to the Grampian
Highlands will prove as popular as its predecessors with geologists, planners,
environmentalists, students and tourists and all those whose wish to have a
better geological understanding of this magnificent region of Britain.

Peter J Cook, DSc, C Geol, FGS
Director
British Geological Survey
Kingsley Dunham Centre
Keyworth
Nottingham NG12 5GG

October 1995

ACKNOWLEDGEMENTS

This fourth edition of the Grampian Highlands regional guide is mostly a
cooperative product from BGS Highlands and Islands Group, with input also
from former members of its staff. The major part of the work, Chapters 4, 5
and 6 on the metamorphic rocks, was written by Dr D Stephenson, with contributions from Dr R M Key, Dr J R Mendum and Mr G S Johnstone. The other
major chapter, on the Caledonian igneous rocks, was written by Dr
D Gould who also contributed to Chapters 11 and 12. Other authors and contributors were Dr G C Clark on Chapter 1, Dr D I J Mallick on Chapters 2 and
3, Dr D J Fettes, Chapters 7 and 9; the Devonian, Carboniferous and Mesozoic
chapters, 10, 11 and 12, were written by the late Dr W Mykura and by Dr T P
Fletcher. The chapter on the Quaternary is by Dr J D Peacock and Mr J W
Merritt, the latter also contributing the short chapter on the Neogene. Dr D I
Smith was responsible for the chapters on the minor intrusions (Chapter 13)
and late brittle faulting (Chapter 16); Dr R M W Musson contributed the seismicity section in the latter. The final chapter, 17, on the economic geology
was prepared by the late Dr M J Gallagher and by Mr J W Merritt. The guide
was compiled and edited by Dr D I J Mallick and Dr G C Clark.

We are grateful to the following copyright holders who have given permission to reproduce or modify their illustrative material: Blackie for Figures 5, 6
and 8; the Geological Society of London for Figures 19 and 25; the Royal
Society of Edinburgh for Figure 20 and the Edinburgh Geological Society
and the Geological Society of Glasgow for Figures, 11, 12, 13, 14, 15, 16, 17,
18, 28 and 29.

1 Introduction

The boundaries of the region described in this regional guide are set by well-defined natural features. On the north the limit is the deep trench of Glen Mor—the Great Glen of Scotland—which traverses the whole of the Highlands from the Moray Firth to the Firth of Lorn and whose floor never rises more than 40 m above OD. On the south the margin is the Highland Border, a line running from Stonehaven in the north-east to Arran and the lower estuary of the Clyde in the south-west, along which the relatively subdued topography of the Midland Valley of Scotland abruptly gives place northwards to the more rugged and elevated hills of the Scottish Highlands. Both these major topographical features follow, or largely follow, lines of faults of major structural importance—the Great Glen Fault and Highland Boundary Fault respectively. The area is thus clearly determined geologically as well as topographically (Figure 1).

The name Grampian Highlands or 'Grampians' is strictly applicable only to the central mountainous part of the region but for this guide the name is also applied to the fiord coastal region in the south-west and the low-lying ground inland from the Moray Firth and Aberdeenshire coast.

In early days the 'high land' was known as 'the Mounth' (Gaelic—Monadh) and this name was in use (at least for certain parts) as late as the cattle-droving times of the late 18th and early 19th centuries. In 1526 Hector Boece, the Scottish historian, identified the Mounth with the Mons Graupius (or Grampius) in whose south-eastern foothills an important battle was fought in AD 86 at the time of a Roman foray into what is now Scotland. In Blaeu's 'Atlas' (1662) a reference is made to 'Grampios Montes', showing that the name was understood in at least some quarters by that time, but just when the term 'Grampians' came into general use is not clear. The areas covered by the geographical terms used in this account are shown on Figure 2. They are largely based on the old Scottish 'Lordships' and are still in fairly common use, if not too clearly defined. Many have been used in geological descriptions of the Grampian Highlands and are familiar to most geologists.

PHYSICAL FEATURES

Although the region described in this guide is by no means all mountainous (Figure 1) it contains the highest land in the British Isles. Ben Nevis reaches 1344 m (4406 ft) above OD, with three attendant summits exceeding 1220 m (4000 ft), while in the Cairngorms massif four mountains reach or exceed 1200 m above OD with a considerable area of hill-top over 1050 m (c.3500 ft). Of the 276 separate mountains recognised by hill-walkers as 'Munros'

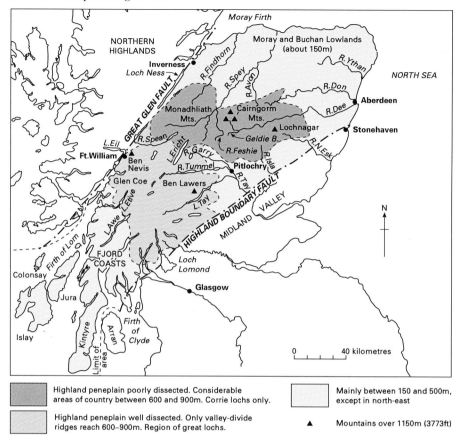

Figure 1 Physiography of the Grampian Highlands.

(mountains exceeding 3000 ft (914.5 m) above OD), 164 are found in the Grampian Highlands.

Viewed from a distance, or from the top of any commanding height within them, the Grampians show a general tendency for ridges and mountain tops to reach up to a more or less uniform level. The impression given by the skyline is one of a plateau-like surface showing very little variation in overall elevation. The Grampians appear to have been carved by the agents of denudation from a single elevated planar surface.

That this view is too simple has long been appreciated as several workers have noted groupings of summit levels which suggest the existence of successively lower erosion surfaces. To some, the highest surface is a plane of marine erosion with the lower summit levels formed in periods of still-stand during intermittent uplift; to others, it is an exhumed feature modified by drainage instituted on a former flat cover of Mesozoic sediments, with the lower level surfaces formed by subaerial erosion as the land was intermittently elevated.

George (1966) gives a detailed analysis of the development of the topography based on the hypothesis that the high surface was an uplifted plane of

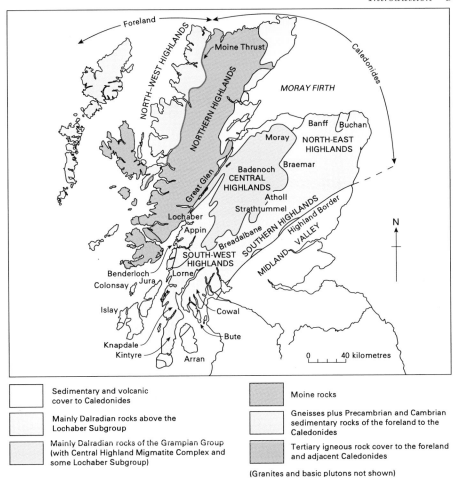

Figure 2 Main geological divisions of the Scottish Highlands and geographical areas within the Grampian Highlands.

marine erosion. Sissons (1967) advocates the upwarp of the Highland block with streams originating on a widespread cover of Mesozoic rocks. His book includes a review of the extensive literature on the subject, including the ideas of George, and should be read by anyone interested in this specialist subject.

It is clear, however, that the planation and subsequent uplift of the Grampian Highlands is later than the outpourings of the lavas of the Tertiary (Palaeogene) Volcanic Province of the western seaboard of Scotland (see BGS regional guide—the Tertiary Volcanic Districts) since feeder dykes to former lava fields are found east of the present lava outcrop and transecting the metamorphic rocks of the Highland block up to the general level of the mountain tops. These dykes must, of course, have consolidated under at least moderate cover. A considerable thickness of lavas (and some underlying strata) has thus been eroded to form the present high Highland surface, which extends alike over Caledonian basement and Palaeogene (formerly called Tertiary) rocks.

Watson (1984) discusses in detail the tectonics of block and extension faulting which, since Palaeozoic times, has provided the control on Highland physiographic evolution. She notes that since the end of the Caledonian Orogeny the Highland block of mainland Scotland has been a stable region, with vertical movements of any great extent occurring only in the marginal areas. As a consequence, the present Highland surface is not far removed from that of the eroded Caledonian mountain belt, which was the surface on which the Palaeozoic Old Red Sandstone sediments were deposited and which is now exposed at moderately high levels throughout the Highlands. Watson considers that this old surface has been exhumed from a thin cover of Palaeozoic and Mesozoic rocks following a Palaeogene uplift of 0.5–1.5 km, the first major uplift since Devonian times. The Highland surface extends over the Tertiary Volcanic Province and was probably the product of subaerial erosion which exhumed the old, uplifted Caledonian basement, removed several hundred metres of lavas and planated the Tertiary rocks.

Whatever the origins of the Neogene planation, it seems probable that the maximum elevation of the uplifted high Grampian Highland surface was located in the general area of the present day north–south divide, which Sissons (1967) considers to be a continuation of that in the Northern Highlands. The original consequent drainage that developed, probably instituted on a smooth, albeit thin, cover of Mesozoic rocks, included long eastward-flowing rivers. Relics of the original river drainage are preserved in the courses of some present day rivers, notably in the Don, the Dee and the Tummel–Tay river system. As the overlying cover rocks were removed by denudation the drainage pattern was then strongly modified by the NE–SW-trending 'grain' of the emerging metamorphic rocks. Several examples of capture of the early drainage by rivers with courses controlled by differential erosion of the heterogeneous metamorphic rocks can be identified, most notably the capture of the headwaters of the original consequent River Dee (the Geldie Burn) by the River Feshie, and the capture of the headwaters of the Don by the River Avon (Figure 1).

The pattern of denudation of the Grampians has also been strongly influenced by the presence of many faults, with accompanying belts of shattered, weakened rock, which trend in a general north-easterly, or northerly, direction. These narrow zones of weakened rock have been actively eroded to form deep elongated hollows.

North-east of a line roughly joining Pitlochry and Spean Bridge the high Highland surface is poorly dissected, with broad upland areas lying between heights of 600 and 900 m above OD. South-west of this line only narrow valley-divide ridges reach these altitudes. This variation in physiography is probably partly due to the heterogeneity of the Dalradian rocks which make up much of the latter area, in contrast to the relatively uniform lithologies of the Grampian Group, and the huge Caledonian plutons that make up much of the north-eastern portion. In addition, variation in intensity of glacial erosion also produced variation in physiography. In Pleistocene times ice accumulated in the areas of maximum elevation in the west. The Moor of Rannoch formed a huge 'ice cauldron' from which ice debouched more or less radially to begin with, but became confined to the old valley lines as time progressed. This ice eroded deeply, encroaching on the valley sides to reduce the interfluve areas to narrow ridges. Less intense valley erosion took place in the more open, comparatively shallow valleys of the north-east and, although deep trenches were gouged locally, large areas of upland remained, smoothed off but otherwise little affected.

It was this concentration of ice in the western valleys which promoted overdeepening to such an extent that, on glacial retreat, many large lochs were left occupying the valley floors of the area. Glacial erosion was especially active along shatter belts which commonly accompany the major transcurrent faults cutting the Grampian Highlands. An example is Loch Ericht, which is 23 km in length but does not exceed 0.8 km in width, with its greatest depth of 156 m occurring where ice was concentrated in the narrowest part of the valley; the loch overlies the Ericht–Laidon Fault. Another example is Loch Ness on the Great Glen Fault.

The north-eastern and south-western sea margins of the Grampians area are totally different in character. As the sea level rose following the melting of the Pleistocene ice the overdeepened valleys on the south-western seaboard were inundated, despite isostatic uplift on removal of the weight of the ice cap, resulting in the 'drowned valley' coastline of the area. Several of these 'fjords' penetrate deeply into the Highland massif. In the north-east, although overdeepened valleys do enter the North Sea and Moray Firth, they are choked with glaciofluvial debris and later deposits; their original form is thus concealed.

The lowland areas of Moray and Buchan are the result of removal from the underlying metamorphic basement surface of the soft Mesozoic rocks (and Old Red Sandstone) which originally covered it. The rectilinear coast of Moray and Buchan trends parallel to the southern margin of the Mesozoic sedimentary basin occupying the Moray Firth Graben, a branch of the extensive fault system of the North Sea.

SCENERY

The two major boundaries of the Grampian Highlands—the Great Glen and the Highland Border—are both spectacular scenic features. Differential erosion between the softer rocks of the Midland Valley and the more resistant metamorphic rocks of the Highlands on opposite sides of the Highland Boundary Fault has resulted in an abrupt fault-line scarp which can be traced from Stonehaven in the north-east to Loch Lomond in the south-west. The feature is especially obvious when viewed across the low-lying ground between Stirling and Loch Lomond; it is the 'Highland Line' of history.

The huge straight-running depression of the Great Glen is a very obvious physical feature (Plate 1), best appreciated from some way up from its valley floor, such as from the A82 road north of Spean Bridge or north of Drumnadrochit. Aerial photographs provide particularly dramatic views of this spectacular linear feature, which extends across the width of the Scottish mainland.

Two contrasted types of mountain scenery are presented by the Grampian Highlands. In the Cairngorm area and around Glen Clova great relics of the original planar surface remain as broad, level moorlands cut into by deep glens and scarred by gigantic corrie-cliffs. Towards the south-west this type of mountain scenery passes into a more highly dissected type with rugged

Plate 1   The Great Glen, looking north-east from above Banavie.

The fault-controlled glen contains the glacially eroded basin of Loch Lochy and, in the foreground, a meandering river system with prominent gravel bars, and the Caledonian Canal (D 2128).

pinnacles, crests and ridges. The detail of this latter scenery depends upon the geological character and structure of the rocks. Resistant beds, such as quartzites, grits or massive gneisses, rise into linear ridges and summits. If no marked guiding planes are present, conical forms result, as in the quartzite mountains of Schiehallion, Beinn a'Ghlo and Paps of Jura. Between these resistant quartzites, grits and gneisses, belts of weaker strata such as slates, limestones and phyllites have been excavated into valleys. The tors, scree-slopes and bold corries of the Cairngorms and Lochnagar exemplify the mode of denudation of the granites which form these mountain groups. The bold cliff scenery of Glen Coe, Ben Nevis and adjacent regions is carved out of volcanic rocks or larger granitic masses.

2 Summary of the geology

The Grampian Highlands are mostly made up of metamorphic and igneous rocks, part of the eroded root zone of the Caledonian mountain belt, which developed in late Precambrian to early Palaeozoic times (Table 1). The name 'Caledonides' was given by E Suess to this mountain belt which extends from the eastern seaboard of North America to Scandinavia and Greenland; in Britian and Ireland its width is from north-west Scotland to central Wales. The Grampian Highlands portion of the Caledonides belt is very well defined by two major dislocations, the Great Glen and Highland Boundary faults (Figure 3). Late Palaeozoic and Mesozoic rocks are now found in large basins of deposition to the north and south of the Grampians and in small internal basins, but may formerly have extended over much of the region. Since Devonian times, however, the area has been mainly one of erosion. Apart from some Carboniferous rocks along the Highland Border and a small area of Permian to Jurassic rocks near Lossiemouth, the only significant post-Devonian deposits are the widespread Quaternary glacial deposits.

The only pre-Caledonian basement rocks exposed appear to be the metamorphosed acid and basic plutonic rocks of the Rhinns of Islay and Colonsay, dated as about 1800 Ma (million years old) (Table 1); they are not correlatives of the Lewisian gneisses of the foreland to the north-west of the Great Glen Fault. The Rhinns rocks are overlain by the low-grade Colonsay Group metasedimentary rocks, whose correlation across a splay of the Great Glen Fault with the rocks on the mainland of Scotland is still uncertain, although a late Precambrian age for the Colonsay Group seems most likely.

Late Precambrian metamorphosed sedimentary rocks of the Dalradian Supergroup, and a wide variety of igneous rocks intruded into them, form most of the Grampian Highlands. The Dalradian sedimentary rocks, many kilometres thick, are mostly of shallow-water origin, although deeper-water turbidite deposition became predominant in late Dalradian times. The oldest of the metasedimentary rocks are now largely migmatised (Central Highland Migmatite Complex) and there is still debate as to whether they are stratigraphically part of the Grampian Group of the Dalradian or constitute an earlier basement; this handbook will take the former view. The protoliths of the migmatites and of the Grampian Group were mainly sandy sediments (the main rock types now are various psammites); most were shallow-marine shelf deposits but there are indications of penecontemporaneous faulting which led to development of small basins in which more muddy turbiditic sandstones accumulated. Stable shallow-marine deposition followed, resulting in deposition of quartzite-shale-limestone sequences of the Appin Group, in which some stratigraphical units maintain a constancy of character across the Grampians from north-east to south-west, and even into Ireland. Deposition

Table 1   Geological sequence and events in the Grampian Highlands.

Age	(Ma)	Sedimentation	Deformation and metamorphism	Igneous events
QUATERNARY	0 Ma	Holocene deposits; glacial erosion and deposition		
NEOGENE	2 Ma	Deep weathering and erosion		
PALAEOGENE				Basaltic dyke swarms
CRETACEOUS	100 Ma	?Cretaceous sedimentation over the Grampian Highlands		
JURASSIC	200 Ma	Moray Firth basin margin clastic sediments		
TRIASSIC				
PERMIAN				Alkaline lamprophyre dykes
				Quartz-dolerite dykes
CARBONIFEROUS	300 Ma	Limestone Coal Formation to Coal Measures deposition		
DEVONIAN	400 Ma	Old Red Sandstone terrestrial deposits	Renewed movement on the Great Glen Fault	
SILURIAN			Major and rapid uplift	Post-tectonic granitic and volcanic rocks
			D_4 Late open folding and wrench faulting	Late-tectonic granitic rocks
ORDOVICIAN	500 Ma	Highland Border Complex sediments	Emplacement of Highland Border Complex	
			D_2–D_3 Metamorphic maximum; folding*	
CAMBRIAN				Syn- to late-tectonic basic and ultramafic intrusions
	600 Ma	Southern Highland Group	D_1 Deformation and increasing metamorphic grade in Southern Highlands	Ben Vurich Granite
				Tayvallich Volcanic Formation
NEOPROTEROZOIC	700 Ma	Dalradian (and Colonsay Group?) deposition — Argyll Group / Appin Group	Early movements on basement faults controlling sedimentary basins	Pre-tectonic basic intrusions
	800 Ma	Grampian Group (base not seen)	D_2 Folding and thrusting with migmatite development in the Central Highlands	
	1000 Ma		?Juxtaposition of Islay and Lewisian terranes along the Great Glen Fault?	
MESOPROTEROZOIC	2000 Ma			Alkaline plutonic basement of Islay and Colonsay (Rhinns Complex)

Caledonian Igneous Suite

CALEDONIAN OROGENY

* This Lower Ordovician event has been termed the 'Grampian Event' or 'Grampian Orgeny' by some authors. The term is not used in this handbook as

of shallow-marine deposits continued and formed the Argyll Group, but growing instability on the shelf led to development of relatively small fault-bounded basins. The resultant deposits generally show poor stratigraphical continuity (although one barium-rich horizon extends intermittently for about 90 km) and increasing influence of turbidite deposition. Further indication of increasing instability and crustal extension is provided by the basic volcanic rocks which appear in the upper parts of the Argyll Group, and continue into the Southern Highland Group, particularly in the South-west Highlands. The dominant rocks of the Southern Highland Group are, however, turbiditic metagreywacke sandstones and siltstones.

There are still considerable disputes about the detailed structure of the Grampian Highlands, and particularly about the timing of the various events within the Caledonian Orogeny; radiometric ages conflict to some extent with the stratigraphical evidence. Nevertheless, the overall general structure is fairly well understood. It involved early recumbent nappe folding, probably directed towards the north-west overall but with at least one major SE-facing fold, the Tay Nappe, which dominates the Southern Highlands. The nappe folds were deformed by subsequent folds and thrusts of generally similar trend and form, and by a series of later, generally upright folds which trend in a variety of directions; they include a major monoform or downbend along the Highland Border. This downbend was probably associated with uplift prior to the obduction on to the margin of the Highlands of the Cambro-Ordovician rocks and ophiolite suite which together constitute the Highland Border Complex and which are preserved as a number of separate slivers within the Highland Boundary Fault Zone.

Generally, the metamorphism reflects the stratigraphy with the highest grade (upper amphibolite facies) being in the older rocks of the north-west and the lowest grade, greenschist facies, in the youngest Southern Highland Group rocks along the Highland Border. This general pattern is, however, modified in the North-east Highlands where there is extensive development of migmatite in the Argyll Group rocks of Angus and southern Aberdeenshire, and in northern Aberdeenshire where there are lower pressure–high temperature andalusite- and sillimanite-bearing assemblages. These are probably the result of the high local heat flow which also resulted in the intrusion of large volumes of basic and acid plutonic igneous rocks there. A narrow zone of greenschist facies rocks at a low structural level is preserved adjacent to the Great Glen Fault Zone.

Early magmatic activity, already noted in the upper parts of the Dalradian sequence, is also found in small amounts at lower levels in the Appin and Grampian groups, and even in the migmatite complex, as small basic intrusions, now mainly represented by amphibolites. Magmatism culminated, however, during and after the late stages of the Caledonian deformation and metamorphism; in the Ordovician the layered basic masses and granites of the north-east referred to above, and then large plutons of granitic rocks and associated volcanics and dyke swarms in the late Silurian and early Devonian (Table 1) were intruded. All of these igneous rocks constitute the Caledonian Igneous Suite.

It is evident that the post-orogenic plutonic rocks were intruded during a phase of very active uplift and erosion so that by early Devonian time the area of the Grampian Highlands was one of considerable relief in which the full range of Dalradian metasedimentary rocks and Caledonian igneous rocks was exposed at the surface. Late Silurian to Devonian deposition (Old Red Sandstone facies) was mainly to the north and south of the Grampian Highlands, in the Midland Valley and Orcadian basins respectively, but sediments also accumulated in mainly fault-controlled basins within the mountain belt (e.g. those of Tomintoul, Cabrach and Rhynie) or on the periphery (e.g. those of Oban and Kintyre). The rocks are mostly clastic— scree breccias, conglomerates, sandstones and siltstones. Around Oban there is also an extensive development of andesitic lavas within the Lower Old Red Sandstone. The Moray–Buchan coastal area represents the southern margin of the large, continental Orcadian Basin. The basal units of the Old Red Sandstone here are typically scree and fan breccias and conglomerates which pass upwards and northwards into a sequence of mainly lacustrine fine sandstones with fish-bearing limestone and calcareous sandstone beds.

Along the Highland Border near Loch Lomond, Old Red Sandstone facies rocks of Devonian age pass up generally conformably into cornstone-bearing sandstones of Lower Carboniferous age. Apart from two other small occurrences in the South-west Highlands the only other Carboniferous sedimentary rocks occur in the Machrihanish Coalfield on the Kintyre peninsula. Although north-west of the Highland Boundary Fault, this has stratigraphical affinities with the Midland Valley, and includes a condensed sequence of Carboniferous strata extending up to the Coal Measures.

The late Palaeozoic is also represented by two suites of minor intrusions. Thick dykes of quartz-dolerite, the northern edge of a swarm which is centred in the Midland Valley, occur in the Southern Highlands from Loch Awe to Aberdeenshire and are of late Carboniferous to earliest Permian age. Numerous thin dykes and small vents of alkaline lamprophyre occur in parts of the South-west and Central Highlands; they are younger than the quartz-dolerites, but have yielded some older radiometric ages.

North of the Central Highlands, the fault-controlled Moray Firth Basin (an embayment of the North Sea) developed from the southern part of the Orcadian Basin with more or less continuous deposition from Late Devonian times to the present. Most of this deposition remains offshore but on the Moray coast around Lossiemouth there are small outcrops of Permo-Triassic and Jurassic rocks, mainly continental facies sandstones which have yielded a rich reptile fauna.

There is no unequivocal evidence of Cretaceous rocks in situ in the Grampian Highlands but they occur offshore to the north and east and may have encroached on land; indeed the present river system has been interpreted as having developed initially on a Cretaceous cover, and angular flints are common in some Quaternary gravels.

The Palaeogene and Neogene appear to have been mainly times of erosion in the Grampians; little evidence of deposits remain apart from some deep weathering profiles and gravels of probable fluviatile origin in the north-east which are interpreted as being of Neogene age. In the south-west, however, there are several swarms of dykes, mainly basaltic, associated with the volcanic centres of the British Tertiary (Palaeogene) Volcanic Province, notably those of Mull, Arran and the submarine Blackstones Complex.

During Quaternary times it is probable that an ice cap was centred on the Northern Highlands and South-west Highlands during several glacial episodes. Over much of the area there is evidence of considerable glacial erosion, but deposits are mostly confined to those of the last (Late Devensian) glaciation; these include tills, various fluvioglacial deposits and, near Inverness, glaciomarine sands and gravels. Inland from the Moray Firth and North Sea coasts, however, there are various isolated occurrences of older interglacial peats and glacial tills and gravels.

3 Basement of the Grampian Highlands

Our knowledge of the basement to the Dalradian rocks of the Grampian Highlands is still very limited and in the previous edition of this handbook the analogy drawn by the original surveyors between the pre-Dalradian of Islay and Colonsay and the Lewisian and Torridonian of North-west Scotland was generally accepted. Recent work, however, has indicated that the exposed basement rocks are not Lewisian but represent new, 1800 Ma-old, Proterozoic crust, and that the overlying Colonsay Group is not equivalent to the Torridonian. The affinity of the Bowmore Sandstone of Islay likewise remains uncertain. These findings have major implications for the pre-Dalradian history of the area, and suggest that the Great Glen Fault may be a very old structure.

Seismic profiling indicates two crustal layers beneath the surface metamorphic rocks of the Grampians, a high-grade metamorphic upper crustal basement with P-wave velocities greater than 6.4 km/sec at depths of 6 to 14 km and a lower crust of mafic/ultramafic rocks with a P-wave velocity of approximately 7 km/sec (Figure 4; Bamford, 1979). The precise nature of these crustal materials is not known, but they may be partly represented by

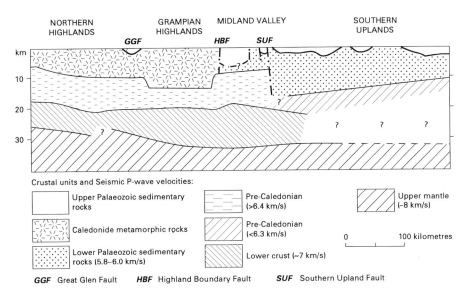

Figure 4   Schematic cross-section through the crust and uppermost mantle of northern Britain (modified after Bamford, 1979).

the quartzofeldspathic gneisses and basic granulites respectively occurring as xenoliths in volcanic rocks of the Midland Valley and western Grampians. The xenoliths also indicate that the mantle under at least the western part of the Grampians consists of harzburgite, lherzolite and cumulate ultramafites (Upton et al., 1983).

SUB-DALRADIAN OF ISLAY AND COLONSAY

The eastern part of the island of Islay is made up of undoubted Dalradian rocks, a continuation of those found on neighbouring Jura and the mainland, and also of probable Dalradian rocks—the Bowmore Sandstone. These are separated from the older rocks of western Islay and Colonsay by the Loch Gruinart Fault, a splay from the Great Glen Fault (Figure 3). These older rocks consist of a gneissic basement overlain by a 5.5 to 6 km-thick suite of low-grade metasedimentary rocks, the Colonsay Group.

The basement rocks, called the *Rhinns Complex* by Muir et al. (1992), crop out over an area of about 20 km² on the Rhinns of Islay and as a very small inlier at the north end of Colonsay (Figure 5). Similar rocks occur 50 km south-west of Islay on Inishtrahull Island near the northern coast of Ireland (Dickin and Bowes, 1991).

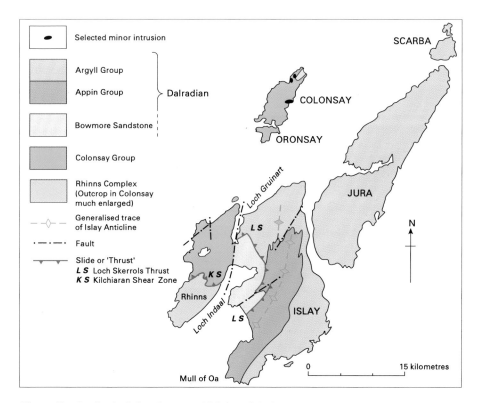

Figure 5   Geological sketch map of Islay and Colonsay.

The Rhinns Complex on Islay consists of two main components, respectively acid and basic gneisses (Wilkinson, 1907). These constitute an alkali igneous complex which has suffered intense multiple deformation and has been intersliced with the overlying Colonsay Group rocks (Muir et al., 1992). Examination of a low-strain zone on the south-west side of the Rhinns has shown that the acid gneisses were derived from a coarse, pink alkali-feldspar syenite with minor bodies of granite, and the basic gneisses from coarse gabbros. The original igneous rock was subject to pervasive ductile deformation soon after its formation, resulting in development of the gneissic banding. Wilkinson (1907) noted that the rocks had suffered considerable crushing, mylonitisation and metamorphic downgrading; the basic rocks, for example, are now represented mainly by amphibolites. He also noted that the intensity of the cataclastic and mylonitic effects increased as the contact with the overlying Colonsay Group is approached.

The basement inlier on north Colonsay covers only about 0.3 km^2 and is largely obscured by blown sand. The exposures are of quartzofeldspathic gneisses, much of it coarse grey pegmatite, with dark knots, streaks and layers which are amphibolitic and all are now brecciated and sheared (Craig et al., 1911). The contact with the Colonsay Group is rarely seen but, where exposed, it is marked by a high-strain zone of phyllonitisation and mylonitisation several centimetres thick, originally interpreted as a thrust plane.

Isotope studies of rocks from the Rhinns of Islay have shown that the syenites and gabbros were emplaced during the early Proterozoic (about 1800 Ma) and that they are juvenile, mantle-derived material, not reworked Archaean crust (Marcantonio, 1988; Muir et al., 1992); similar ages have been obtained from acid and basic material from Inishtrahull (Dickin and Bowes, 1991). These isotope studies have effectively ruled out any direct correlation between the Rhinns rocks and the Lewisian Complex of north-west Scotland, in which there is no evidence for major crustal addition of mantle material at about 1800 Ma, although the Laxfordian tectonothermal cycle resulted in reworking of the Lewisian at about that time (Fettes et al., 1992). The Islay–Colonsay basement, an early Proterozoic mantle-derived terrane devoid of Archaean material, has analogies with the Ketilidian belt of South Greenland and the Svecofennian of Scandinavia (Marcantonio et al., 1988) and may form a link between these two segments of the early Proterozoic mobile belt around the southern margin of Laurentia (Muir et al., 1989; 1992). Dickin and Bowes (1991) interpreted the isotopic data on Grampian Highlands granites as indicating that the Rhinns Complex terrane extended at depth to the North-east Highlands and to western Ireland, forming a block measuring at least 600 × 100 km.

These observations have raised the question of the relationship between the Rhinns rocks and the Lewisian. The fact that true Lewisian occurs on Iona, only 20 km north-west of Colonsay, raises the possibility that their present close juxtaposition is not original. Bentley et al. (1988) suggest that the Rhinns Complex might represent an allochthonous terrane transported to its present position by strike-slip movements on the Great Glen Fault. This is not yet proven. What is clear, however, is that the Rhinns terrane was in its present position relative to the Dalradian by about 620 Ma, the age of intrusions that were foliated by events which affected the Dalradian.

COLONSAY GROUP

The Colonsay Group consists of a 5.5 to 6 km-thick sequence of low-grade, strongly deformed metasandstones and phyllites, with minor calcareous beds, overlying the basement rocks and exposed on Islay to the north of the Rhinns, and on Colonsay. The overall strike of the beds is north-easterly so that the succession exposed, younging generally towards the north-west, presents an oblique section through the basin of accumulation (Bentley, 1988). Stewart (1962) and Stewart and Hackman (1973) divided the sequence into 18 lithostratigraphical units, the lower ten on Islay and the upper eight on Colonsay (Figure 6); it is probable that part of the succession is covered by sea between Islay and Colonsay but the stratigraphical gap is thought to be no more than a kilometre (Bentley, 1988). Stewart's maps and sections indicate considerable vertical and lateral facies variations, a situation complicated by some structural repetition.

The lowest 800 m of the succession, exposed on Islay, consist of meta-arkoses and phyllites, interpreted as representing delta-top sheet sands and interdistributary muds. The upper part of the succession on Islay and the lower part of the Colonsay succession are quartz-rich metagreywacke sandstones and phyllites suggesting deeper-water, delta-slope turbidite accumulation. The upper part of the succession on Colonsay shows a change back

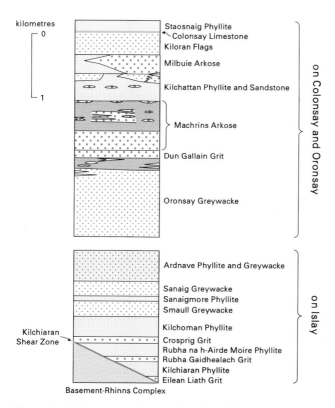

Figure 6   Lithostratigraphical units of the Colonsay Group (modified after Stewart and Hackman, 1973, and Bentley, 1988).

towards shallow-water, mainly siliciclastic sedimentation but with several calcareous developments, notably the 1 to 5 m-thick dolomitic Colonsay Limestone and some calcareous phyllites.

The Colonsay Group sediments were derived mainly from the south, from an area containing deformed high-grade gneisses with sedimentary cover. Saha (1985) pointed out that the source area could not have been remote— the feldspar clasts are angular and fresh—and may have exposed granulite facies rocks which produced the blue quartz clasts found in the Colonsay Group rocks. Their provenance is uncertain since the basement rocks of the Rhinns of Islay are not of granulite facies and lack blue quartz.

The contact of the Colonsay Group with the underlying gneisses on Islay is highly deformed. The units close to the contact are generally coarser grained than elsewhere with local developments of conglomerate. Consequently Wilkinson (1907) and Bentley (1988) interpreted it as a sheared unconformity. However, Stewart and Hackman (1973) recognised that the five lowest units in the Colonsay Group are truncated against the basement without any signs of a basal facies. They also did not identify Wilkinson's basal conglomerate. They therefore interpreted the highly deformed contact to be a tectonic break which they referred to as the Bruichladdich Slide. More recent work (Muir et al., in press) has confirmed this view and has recognised Colonsay Group–basement tectonic interleaving close to the slide, now renamed the Kilchiaran Shear Zone. On North Colonsay, the contact is also highly sheared with a conglomeratic cover sequence containing rounded quartzite clasts and only 'some fragments reminiscent of the local basement' (Bentley, 1988). On this basis it is interpreted as a sheared unconformity by Bentley. However, the cover rocks belong to the Kilchattan Formation (Figure 6), about 4 km above the base of the succession on Islay, thereby requiring a very uneven basement topography.

Stewart (1975) noted that the Colonsay Group cannot be shown to correlate with the Torridonian 'in any sense', although this had been its traditional correlative since the original Geological Survey work, when the basement was thought to be Lewisian. The presence of limestone and dark phyllite in the upper part of the Colonsay Group suggested to Litherland (*in* Stewart and Hackman, 1973) that there could be a correlation with the Appin Group Dalradian, a correlation supported by Rock (1985) who suggested that the chemistry of the Colonsay Limestone was similar to that of the Ballachulish Limestone. Such a correlation would imply that the lower parts of the Colonsay Group were lateral equivalents of the Lochaber Subgroup and the Grampian Group.

The main objection to a correlation with the Dalradian has been on structural grounds, the first phase of deformation of the Colonsay Group apparently being unrepresented in the nearby undoubted Dalradian rocks. Muir et al. (1992) concluded that the stratigraphical affinities of the Colonsay Group remain unclear.

STRUCTURE

Fitches and Maltman (1984) considered Islay and Colonsay to lie within the zone of the Caledonian Front, comparable to the zone of thrusts in the North-west Highlands but with more ductile deformation.

Wright (1908) noted two main periods of movements on Colonsay; subsequent workers have subdivided them. The early events (D_1 and D_2 of Fitches and Maltman, 1984) result from heterogeneous, subhorizontal NNE-directed shear, which produced a grain alignment parallel to bedding and the development of small-scale assymmetrical, and typically recumbent, folds; no large-scale early folds are known. The later events (D_3 and D_4) are typified by upright folds accompanied by spaced, planar fabrics. The D_3 folds are on scales of up to several hundred metres wavelength with NE-trending axial planes which are upright or dip steeply south-east. They are common on Colonsay and control the main outcrop pattern. D_4 is sporadically developed and is represented by chevron folds, kink bands and crenulation cleavages and causes regional swings of the outcrops in the north of Colonsay.

INTRUSIONS

Apart from the late Caledonian lamprophyre and felsite dykes, and Palaeogene basalt dykes, there are intruded into the Colonsay Group of Colonsay a number of small plutonic masses (Figure 5), the main ones of which were described by Craig et al. (1911) as the Kiloran Bay syenite, the Scalasaig diorite and the Balnahard kentallenite (olivine-augite-syenogabbro). Each is associated with a more basic margin (respectively of hornblendite, augite-lamprophyre and augite-diorite) and, in the case of the first two, marginal explosion breccias. Clasts in these breccias contain folds and fabrics attributable to the early deformation of the Colonsay Group, while the intrusions themselves are, in part, foliated by the late deformation, indicating intrusion taking place between the early and late deformation events. This age of intrusion is taken to be about 620 Ma, a result obtained from $^{40}Ar/^{39}Ar$ step-heating of hornblendes from the Kiloran Bay intrusion (Bentley, 1988).

BOWMORE SANDSTONE

The Bowmore Sandstone is separated from the Colonsay Group of western Islay by the Loch Gruinart Fault and from the Dalradian succession of eastern Islay by the Loch Skerrols Thrust (Figure 5; Green, 1924; Amos, 1960; Stewart, 1969; Fitches and Maltman, 1984). The rocks are tightly folded but the locally developed tectonic fabrics are weak and metamorphism is slight. Bedding is generally indistinct and younging indicators are rare. Consequently the stratigraphical age of the Bowmore Sandstone is uncertain and it has been correlated variously with the Moine, the Torridonian and the Dalradian (Fitches and Maltman, 1974, table 1). Those workers who have regarded the Loch Skerrols Thrust as a major structure, possibly equivalent to the Moine Thrust, have correlated the Bowmore Sandstone mainly with the Torridonian, by analogy with the structure of the North-west Highlands (e.g. Johnstone, 1966; Stewart, 1975). More recent work has led to the suggestion that the Loch Skerrols Thrust is a structure of local importance only and hence the Bowmore Sandstone is more likely part of the Dalradian succession. Fitches and Maltman (1984) argue that it is the lateral equivalent of the Crinan Grit on the lower, inverted limb of the Islay

Anticline; it could also be argued that the lithologies are consistent with the Grampian Group.

Lithologies are predominantly grey-brown, feldspathic sandstones which have been divided into two formations, each exceeding 2 km in thickness. The lower, *Laggan Formation*, consists of fine- to medium-grained sandstones with silty and shaly partings, but the upper, *Blackrock Formation*, is mainly coarse-grained sandstones, with pebbly bands. A provenance study of clasts in the Bowmore Sandstone and Colonsay Group has revealed quartz-granulites, acid gneisses, pegmatites and vein quartz of Lewisian type (Saha, 1985). Distinctive blue quartz, characteristic of early Lewisian (Scourian) granulites, is present in both the Bowmore Sandstone and the Colonsay Group and also in Dalradian rocks of north-east Islay. The Blackrock Formation contains in addition pebbles of chert, jasper and ferruginous sandstone, indicative of non-metamorphic or low-grade supracrustal rocks (e.g. Torridonian) in the source area.

4   Grampian Caledonides—introduction

The last two editions of this regional guide emphasised that, despite a long and distinguished history of research, aspects of the stratigraphy and structure of the Grampian sector of the Caledonides remained controversial. To a lesser extent that remains the position today. Continuous geological work since the publication in 1966 of the third edition of this guide has certainly led to an improved understanding of Grampian geology. The general outlines of the stratigraphy and structure are now well established. However, the precise status of a number of stratigraphical units, and the relationships between some of the major divisions still remain conjectural. Also there is no overall consensus about the detailed structural evolution of the Grampian Caledonides. We do now have a reasonably consistent correlation of local successions of metamorphic rocks throughout the Grampian Highlands, as shown in Figures 9 and 10. The distribution of the major lithostratigraphical units across the Grampian Highlands is illustrated in Figure 3 and the generalised overall structure in Figure 19.

STRATIGRAPHY

Much of the Grampian Highlands is underlain by a wide variety of metasedimentary and metavolcanic rocks which are now mostly assigned to the Dalradian Supergroup. However, the basal Grampian Group was previously considered to be part of an older succession which was equated with the Moine north-west of the Great Glen. Migmatites, which are spatially associated with the Grampian Group metasedimentary rocks in the northern part of the Grampian Highlands, were also correlated originally with the Moine (Hinxman and Anderson, 1915; Horne, 1923). Subsequent detailed studies of the migmatites, which referred to them as the 'Central Highland Division', retained their correlation with the Moine. However, this correlation has been questioned and their stratigraphical position is not fully resolved, as the following summary of the history of stratigraphical studies will show.

It was in 1891 that Sir Archibald Geikie first introduced the name 'Dalradian' for the varied group of metamorphic rocks which lies to the east of the Great Glen and which was thought to be younger than the Lewisian Gneiss of the North-west Highlands. The 1892 and 1910 editions of Bartholomew's 10-miles-to-one-inch Geological Map of Scotland, compiled under Geikie's direction, extended the name 'Dalradian' to include all the metamorphic strata east of the Moine Thrust. However, Geikie made it clear in his explanatory notes to the map that the 'Moine Schists' of the Northern Highlands were different in character from the 'Dalradian' rocks south-east of the Great Glen. As research proceeded, quartzofeldspathic rocks of 'Moine Schist' facies were mapped and described south-east of the Great Glen where they were variously referred to as, from north-east to south-west, 'Granulitic Schists of the Central Highlands' (Hinxman and Anderson, 1915), 'Struan Flags' (Barrow, 1904) and 'Eilde Flags' (Bailey, 1910). This left the Dalradian *sensu stricto* as an assemblage of rocks lying mainly in the south-western and south-eastern parts of the Grampian Highlands, and characterised by the presence of black schists, quartzites and limestones with, in its upper part, metagreywackes.

Local successions were established in the Dalradian by detailed mapping throughout the first part of this century. This work was reviewed by Anderson (1948) who noted that the base of the 'Dalradian Assemblage' had been placed at different stratigraphical levels in different parts of the Grampian Highlands. He proposed that all rocks stratigraphically above, and including, a lowermost limestone should be regarded as Dalradian. All underlying strata should be included in a 'Moinian Metamorphic Assemblage'. This comprised a 'Central Highland Psammitic Group' made up of the quartzofeldspathic flags and gneisses regarded unanimously in the various local stratigraphical successions as Moinian, and an overlying 'Pelitic and Quartzitic Transition Group'. The latter included various quartzite/pelitic schist sequences recognised across the Grampian Highlands from Islay to the Banffshire coast and regarded by many of Anderson's contemporaries as Dalradian. In fact, the third edition of this guide referred to the 'Pelitic and Quartzitic Transition Group' as the lowest unit in the 'Dalradian Assemblage', which was subdivided into nine 'groups', the lower six groups forming the Lower Dalradian, the remaining three groups the Upper Dalradian. The various local stratigraphies established by detailed mapping during the first half of this century were all incorporated within this internal stratigraphical framework.

Major revisions, since 1966, of the internal stratigraphy of the Dalradian *sensu stricto*, have amended the internal correlations of the different local stratigraphical sequences. The Dalradian was accorded Supergroup status and subdivided into three groups (Harris and Pitcher, 1975). The correlation of the underlying quartzofeldspathic rocks of the Central Highland Psammitic Group with the Moine was retained, although it was realised that newly acquired isotopic ages of about 740 Ma from the Moine north-west of the Great Glen (van Breemen et al., 1974; Brook et al., 1976) brought this correlation into question. The isotopic ages, mostly from pegmatites, provided a minimum age for the original sedimentation and early deformation. In contrast, the quartzofeldspathic rocks south-east of the Great Glen were regarded as younger. This was based on the fact that they were known to grade locally upwards into the Dalradian sequence and to share, apparently, a common tectonothermal history, which of necessity was younger than the youngest Dalradian rocks, regarded as of uppermost Precambrian or lower Palaeozoic age. To overcome this age problem it was proposed that the Moine could be divided into an 'Older Moine' and a 'Younger Moine' with the possibility of an undetected unconformity separating the two lithologically similar divisions (Johnstone, 1975). The 'Older Moine' referred to rocks north-west of the Great Glen and cut by the pegmatites dated at about 740 Ma, the 'Younger Moine' referring to the quartzofeldspathic rocks which immediately underlie the Dalradian south-east of the Great Glen.

Detailed mapping south of Inverness over the last 20 years initially resulted in a subdivision of the Moine of the Grampian Highlands into an older migmatitic basement referred to as the 'Central Highland Division' and an overlying sequence dominated by unmigmatised quartzofeldspathic flags termed the Grampian Group or 'Grampian Division' (Piasecki and Van Breemen, 1979a; 1979b; Piasecki, 1980). Deformed pegmatites from shears which locally define the contact of the 'Grampian Division' with the migmatitic rocks were dated at about 750 Ma by numerous Rb-Sr isotopic analyses of muscovite, leading to the proposal that the 'Central Highland Division' is a crystalline basement, equivalent to the 'Older Moine' exposed north-west of the Great Glen. Such an interpretation has been challenged (see Chapter 5; Fettes et al., 1981; Harris et al., 1983; Lindsay et al., 1989). Instead it has been proposed that the 'Central Highland Division' comprises the migmatitised lower part of the Grampian Group, which essentially reverts to the conclusion of the original geological surveys (Hinxman and Anderson, 1915; Horne, 1923). To emphasise that the migmatites are a lithodemic unit the informal lithostratigraphical term 'Division' is abandoned and the migmatites in this guide are referred to as the 'Central Highland Migmatite Complex'.

The gradational contact seen locally between the Grampian Group metasedimentary rocks and the basal rocks of the Dalradian Supergroup (as defined by Harris and Pitcher, 1975), together with their common tectonothermal history formed the basis for the subsequent incorporation of the Grampian Division as a formal group in an extended Dalradian Supergroup (Harris et al., 1978). Such a proposal is not accepted unanimously and the separation of the Grampian Group from the Dalradian Supergroup, largely on the basis of perceived lithological contrasts, does attract significant support (Thomas, 1980; Anderton, 1985). The similarities between the Grampian Group, the Central Highland Migmatite Complex and the Moine north-west of the Great Glen remain, although no detailed correlations have been offered and the Grampian Group rocks are no longer referred to as 'Younger Moine'.

The formal hierarchical lithostratigraphy of the Dalradian Supergroup is shown in Figures 9 and 10. Early impetus for a unified Dalradian lithostratigraphy was provided by the recognition of similar sedimentary associations extending right across the Grampian Highlands (Sutton and Watson, 1955; Knill, 1959; Roberts, 1966; Rast and Litherland, 1970). A remarkable agreement of stratigraphical detail is, in fact, preserved in parts of the succession along the whole strike length of over 700 km of the British Caledonides from Western Ireland to the Banffshire coast. Application of modern sedimentological concepts has shown that the initial, predominantly sedimentary pile comprised a complex three-dimensional juxtaposing of different lithofacies and lacked a simple 'layer-cake' structure (Litherland, 1980; Anderton, 1985; Glover and Winchester, 1989). Syndepositional (possibly listric) faulting as well as later polyphase folding and ductile shearing contribute to the overall complexities (Anderton, 1988).

The original gross lenticular nature of the sequence is attributed to the continually changing internal morphology of the depositional sedimentary basins. The polyphase deformation has also both thinned and thickened the original sedimentary-volcanic pile (Borradaile and Johnson, 1973), which now youngs overall from the north-west to the south-east and south-west.

There is also a general temporal change from stable estuarine, coastal or intertidal environments during Grampian and Appin group times to unstable turbidite environments with accompanying volcanicity at the time of deposition of the Argyll and Southern Highland groups (Fettes and Harris, 1986). After early sedimentation on a single shelf, increasing instability, attributed to progressive lithospheric stretching, produced a series of fault-bounded basins. Submarine fan deposits recognised in the Southern Highland Group draped over the faults controlling basin development on a subsiding continental shelf (Anderton, 1985).

Episodes of igneous activity generally took place at the time of the later periods of basin deepening; there was little contemporaneous igneous activity associated with the early sedimentation. The thickest volcanic sequences overlie areas of supposed greatest crustal attenuation (Anderton, 1985). As the chemical composition of the extruded lavas is similar to those formed at extensional plate margins it has been suggested that the attenuation led to actual crustal rupturing with small basins possibly floored by oceanic crust (Graham and Bradbury, 1981).

Several occurrences of syngenetic mineralisation in the Argyll Group have been discovered as a result of detailed geochemical, geophysical and geological mapping and increased mineral exploration activity in the past 20 years. The most important of these are the Aberfeldy baryte deposits, situated near the top of the Ben Eagach Schist in the Argyll Group. The synsedimentation concentration of metals, which extends intermittently over a strike length of at least 90 km, from Loch Lyon to Loch Kander, is thought to have occurred in brine pools at the edge of small fault-bounded basins (Coats et al., 1984).

STRUCTURE AND METAMORPHISM

Current structural interpretations of the Grampian Caledonides are still based upon those proposed by E B Bailey, which drew on the results of the primary mapping by the Geological Survey. In a series of papers, from 1910 to 1938, Bailey demonstrated that the rocks of the South-west and Southern Highlands are disposed in large recumbent folds. The lower limbs of many of these folds are partly replaced by low-angled, extensional faults, termed 'slides', with postulated movements of several kilometres. In 1922 Bailey produced a comprehensive synthesis in which major slides were perceived as fundamental tectonic dislocations separating nappe complexes, each complex having its own stratigraphical succession and structural style. Initially three such nappe complexes were recognised and named, in ascending structural order, the *Ballappel Foundation* (from the type areas of Ballachulish, Appin and Loch Eilde), the *Iltay Nappe* (from Islay and Loch Tay), and the *Loch Awe Nappe*, but these were subsequently reduced to two when Bailey (*in* Allison, 1941) accepted the stratigraphical correlations of other investigators which removed the need to invoke a separate Loch Awe Nappe.

A better understanding of the stratigraphy, following the recognition of sedimentary structures in quartzitic rocks as way-up indicators, led to modifications of the early structural models. Bailey himself used such observations in the Loch Leven area to completely reverse the order of superposition used in his 1922 paper and revised his structural interpretation in a further major review paper (Bailey, 1934). Further developments and modifications resulted from the recognition of graded units and other internal structures in turbiditic rocks, first applied by Shackleton (1958) to the structure of the Southern Highlands and the Highland Border. The concept of 'facing' was introduced as a means to describe the structural 'way-up' of strata (Figure 7). Shackleton (1958) defined 'facing' geometrically as 'the direction normal to the fold axis, along the axial plane, and towards the younger beds'. Thus a synclinal synform is described as 'upward facing', whereas an anticlinal (i.e. inverted) synform is 'downward facing'. Asymmetrical and recumbent folds have a sideways component of facing which is an important descriptive parameter, and which is commonly used to infer the direction of tectonic transport. H H Read, who had commenced work in the north-east Grampians as a member of the Geological Survey, initiated research which resulted in several detailed studies of the Banff Nappe and related structures.

Bailey's model remained the basis of structural descriptions of the Grampian Highlands until the 1960s and was used in the last edition of this regional guide, in which descriptions of the Ballappel Foundation and Iltay Nappe were extended north-eastwards as the 'Northern Grampians Nappe Complex' and 'Southern Grampians Nappe Complex' respectively (Johnstone, 1966).

Continuing detailed stratigraphical and structural studies have resulted in further refinements to the Bailey model. Several alternative structural models have emerged, although many of the major fold structures originally named by Bailey are still recognised. However, detailed interpretations of the relative

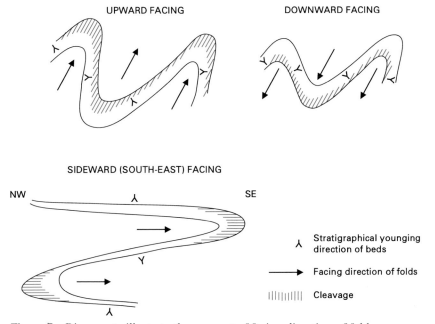

Figure 7 Diagram to illustrate the concept of facing direction of folds as defined by Shackleton (1958)

ages and original geometry of differing sets of folds and their overall tectonic significance remain problematical. Many of Bailey's slides are still recognised as important and complex zones of low-angled tectonic dislocation and stratigraphical attenuation. However, it has been shown that, on a regional scale, both the structure and the stratigraphy of the Grampian Highlands are more integrated and continuous than inferred by Bailey, with elements of the succession being traced vertically through the nappe pile as well as laterally along the full strike length. Hence most of the Dalradian must be regarded as autochthonous or parautochthonous and the slides may have lost much of their original significance as major tectonic and stratigraphical boundaries. In most current models, where tectonostratigraphical boundaries are recognised they are commonly defined by cross-strike lineaments. Such lineaments are taken to represent fundamental, deep-seated crustal fractures which have influenced sedimentation, igneous activity and structural development over a long period of time. They can have a marked influence on the outcrop pattern and are recognised by various combinations of features such as abrupt lateral facies changes in the metasedimentary sequence, sites of igneous activity, geochemical and geophysical changes and zones of structural discontinuity.

The structure of the Grampian Caledonides is discussed in detail in Chapter 6. However, in order to describe the stratigraphy of the area in Chapter 5 it is necessary to have some knowledge of the major fold structures and tectonic dislocations that influence the disposition of the lithostratigraphical units. Most were established by E B Bailey and are shown on Figure 19. They are described below without recourse to complex and controversial discussions of deformation phases, relationships with adjacent or other structures, metamorphic textures or related fabrics.

The overall structure of the Grampian Highlands is most easily described and understood with reference to structures in the South-west Highlands (Roberts and Treagus, 1977). Here the major folds are seen to diverge on either side of a central *Loch Awe Syncline* which is interpreted as an early primary structure. Apparently geometrically similar structures farther to the north-east include the *Ben Lawers Synform* and *Sron Mhor Syncline* but these are actually due to later refolding events. To the south-east of the Loch Awe Syncline, the *Ardrishaig Anticline* is interpreted as the core of a large SE-facing nappe, the *Tay Nappe*, which dominates the overall structure of the Southern Highlands. The Tay Nappe is flat-lying, although in the south-west it is folded across a broad arch known as the *Cowal Antiform*. The erosion level is such that most of the outcrop constitutes part of the inverted limb of the nappe, so that stratigraphical sequences are inverted. Structures underlying this inverted limb are seen only in the area of the Angus glens, where a largely right-way-up sequence has been interpreted as a separate *Tarfside Nappe* (Harte, 1979). Close to the Highland Boundary Fault Zone, the Tay Nappe is bent downwards to form the *Highland Border Steep Belt*. The hinge zone of the nappe thus becomes downward facing as a synformal anticline, recognised in the Southern Highlands as the *Aberfoyle Anticline* (Shackleton, 1958).

To the north-west of the line of the Loch Awe Syncline is a pile of apparently NW-facing folds, including the *Islay Anticline* and the folds of the original 'Ballappel Foundation' such as the *Beinn Donn Syncline, Ballachulish Syncline* and *Appin Syncline* (Treagus, 1974; Roberts, 1976). Several major slides are recognised within this nappe pile. The *Boundary Slide* transgresses several fold limbs in the upper part of the overall pile resulting in the attenuation or removal of large parts of the Dalradian succession. For most of its length it is coincident with the boundary between the Grampian Group and Appin Group. Towards the base of the pile in Lochaber, the *Fort William Slide* locally forms the boundary between the Grampian Group and lower parts of the Appin Group. These two slides were the major tectonostratigraphical boundaries in the early stratigraphical interpretations of Bailey.

Because of the mushroom-like structure resulting from the divergent facing directions of the major folds, many of the early models suggested that a fundamental 'root zone' lay beneath the Loch Awe Syncline (Rast, 1963). This zone was then projected north-eastwards beneath the Sron Mhor Syncline and the associated *Tummel Steep Belt* (Sturt, 1961; Harris, 1963). However, it has subsequently been shown that these folds, and the steep belt, are later structures which postdate, and therefore fold, the early major recumbent nappes (Roberts and Treagus, 1979; Bradbury et al., 1979). Folds immediately to the north-west of the Loch Awe Syncline are now known to face downwards and hence south-eastwards, including those of the Glen Orchy area (Thomas and Treagus, 1968) and the *Atholl Nappe* (Thomas, 1979; 1980). Thus, if a root zone does exist, it must lie farther to the north-west, possibly in the region of a 4 km-wide zone of upright, isoclinal folds known as the *Ossian–Geal Charn Steep Belt* (Thomas, 1979; 1980). Other workers, whilst recognising the existence of the steep belts, reject the idea entirely of any connection with a root zone.

Structures in the northern part of the Grampian Highlands are still being elucidated, but it has been suggested that the Grampian Group and the Central Highland Migmatite Complex contain comparable structures and share a common history of structural development with those of the overlying Atholl Nappe (Lindsay et al., 1989). The status of the *Grampian Slide* which separates the migmatitic from the non-migmatitic units is a matter of current debate and is discussed in Chapters 5 and 6. Major folds which influence the outcrop pattern considerably are recognised in the west of this area, above the Fort William Slide. They include the *Stob Ban Synform* and the *Appin Syncline* which are responsible for large infolds of Appin Group rocks within the Grampian Group.

In the north-east Grampians, recumbent folds which may be correlated with those of the Tay Nappe have been identified in sections around Collieston on the east coast. However, the area is dominated by broad NNE-trending late folds, principally the *Turriff Syncline* and *Buchan Anticline*. Within these structures the upper parts of the Dalradian succession occur in right-way-up sequences. Early interpretations classed this succession as a separate 'Banff division' in an allochthonous 'Banff Nappe', which was separated by a slide from an underlying Dalradian sequence continuous with that of the Central Highlands (Read, 1955; Read and Farquhar, 1956). Although other authors have also suggested that this area is allochthonous (Sturt et al., 1977; Ramsay and Sturt, 1979), current interpretations suggest that the succession is essentially autochthonous, passing downwards from the Southern Highland Group into an Argyll Group succession which correlates with that farther to the south-west (Harris and Pitcher, 1975; Ashworth, 1975; Harte, 1979; Treagus and Roberts, 1981; Ashcroft et al., 1984). If this is so, then it is not necessary to invoke the presence of a Banff Nappe and underlying slide. However, the area is bounded to the west and south by major shear belts and it is distinguished from the remainder of the Scottish Dalradian by different stratigraphical, metamorphic, igneous and geophysical features. The western boundary of this *Buchan Block* is marked by the *Portsoy–Duchray Hill Lineament*, a major tectonic and stratigraphical boundary which can be traced from the north coast to the Glen Shee area (Fettes et al., 1986). To the west of this lineament Appin and Argyll group rocks are involved in a series of NW-facing folds which can be traced down sequence into the underlying Grampian Group.

The main outcrop pattern of the Dalradian as depicted in Figure 3 is largely governed by the major folds described above. During the later stages of orogenesis, many of these structures were further modified by block uplift and faulting, which occurred at different rates and at different times throughout the Dalradian outcrop (Harte et al., 1984; Dempster, 1985). The differential uplift resulted in an outcrop pattern that reveals wide variations in levels of exposure. Later, post-orogenic, brittle fault movements have displaced boundaries, by several kilometres in some cases (Chapter 16). Between the major boundary faults of the Great Glen and the Highland Boundary lie a whole series of parallel, NE-trending faults showing sinistral displacement. Of these, the *Glen Markie, Ericht–Laidon, Tyndrum, Bridge of Balgie* and *Loch Tay faults* have particularly significant displacements (Figure 42). In some areas, appreciable displacements, largely of a vertical nature, also occur on NW-trending faults such as the *Rothes* and *Pass of Brander faults*, some of which may reflect fundamental deep level tectonostratigraphical lineaments (Chapter 6).

The Dalradian rocks occur at the surface in a variety of metamorphic states, illustrated best by the pelitic rocks which range from slate and phyllite, especially in the south and east, to coarse-grained schists, gneisses and migmatites in the north. In detail, as outlined in Chapter 7, the metamorphic development in space and time was complex, but the overall pattern exposed can be described in terms of two main components:

—an overall thermal anticline plunging towards the south-west (Kennedy, 1948), resulting in greenschist and epidote-amphibolite facies rocks in the South-west Highlands, along the Highland Border and in Buchan; the rocks there are generally at the higher structural and stratigraphical levels. Amphibolite facies rocks in the north are mainly of the lower stratigraphical units, the Grampian Group and the Central Highland Migmatite Complex.

—a superimposed area of high-grade rocks in the eastern Grampian Highlands, where there are also numerous layered basic/ultrabasic intrusive complexes.

This overall rather simple pattern obscures the fact that the pressure conditions under which the metamorphism occurred varied from relatively high pressure in the south-west, south-east and north (Barrovian metamorphism) to intermediate/low pressure in the north-east (Buchan metamorphism). It has also been shown, from mineral transformations, that in some places there were pressure increases during the metamorphism, possibly as a result of tectonic thickening.

Over most of the Grampian Highlands the peak metamorphic mineral assemblages and textures, developed during the nappe folding, are largely preserved; there is, however, widespread, but mostly minor, retrogression of the rocks associated with the final phases of deformation and uplift.

5 Lithostratigraphy of the Grampian Caledonides

All the metamorphic rocks of the Grampian Highlands are here assigned to the Dalradian Supergroup although locally there are successions of uncertain stratigraphical position and the stratigraphical affinities of the oldest subdivision, the Central Highland Migmatite Complex, is currently a matter of debate. The succeeding Grampian, Appin, Argyll and Southern Highland groups form, in places, a continuous stratigraphical succession. The Appin and Argyll groups are divided into a number of subgroups (Figure 10).

Our understanding of the stratigraphy of these rocks is still far from complete and few units at formation and member level have been formally defined. In this handbook we use formal names where these have been established (e.g. Appin Phyllite and Limestone Formation) but of necessity many existing informal names of no defined status are retained (e.g. Islay Limestone). Some of the informal names do not accord with accepted stratigraphical practice but continue to be used because the name is well established and formal alternatives have not yet been proposed. Old terms which are not in accord with modern, more specific usage are placed in parentheses and not capitalised (e.g. Portsoy 'group'). The term 'division' is used here for major coherent units of mixed lithologies which are at present not formally defined (e.g. the Stuartfield 'division' of the north-east Grampians).

The rocks of the Dalradian Supergroup are mainly metamorphosed clastic sediments (sandstones, siltstones, mudstones) and limestones and the names used to describe individual units are dependent upon metamorphic grade and the ease with which primary sedimentary features can be identified. Thus in the South-west Highlands, where metamorphic grade is low, many authors have used sedimentary rock terms (e.g. sandstone, siltstone). Where the rocks are more metamorphosed, the terms pelite, semipelite, psammite and (meta) quartzite are used to represent original compositions ranging from argillaceous to arenaceous respectively. To these may be added the textural terms slate, phyllite, schist and gneiss. Carbonate rocks are usually referred to as 'limestones' or 'dolostones' where they are relatively pure and consist essentially of recrystallised carbonate minerals, regardless of metamorphic grade. The term 'marble' is rarely found in modern descriptions of Dalradian rocks. 'Calc-silicate rocks' represent originally impure calcareous (or magnesian) carbonate rocks and contain a high proportion of calcium-magnesium silicate minerals such as tremolitic amphibole, grossular garnet, epidote, zoisite and idocrase. Such rocks grade into (metasedimentary) para-amphibolites. Metamorphosed igneous rocks are classified, whenever possible, according to their nonmetamorphosed equivalents (e.g. metabasalt, metadolerite, metagabbro) or given a general compositional term such as 'metabasite'. Where the non-metamorphosed equivalent cannot be identified general descriptive terms such as 'hornblende-schist' or (igneous) ortho-amphibolite have been retained. The term 'epidiorite', formerly used for various metabasic rocks and hornblende-schists, has been abandoned.

The intensity of metamorphism and deformation varies considerably through the Grampian Highlands (Chapter 7). In the South-west Highlands, and on Islay and Jura, a generally low metamorphic grade has enabled detailed studies of the sedimentology to be made which have revealed much information concerning palaeogeography, the origin of the sediments and lateral facies changes (Knill, 1963; Roberts, 1966; Hickman, 1975; Anderton, 1976; 1979; Litherland, 1980). Petrological and geochemical studies of penecontemporaneous igneous rocks have also been interpreted by comparison with nonmetamorphosed equivalents (Borradaile, 1973; Graham, 1976). As a consequence of this detailed knowledge, many reviews of Dalradian evolution have been heavily biased towards the interpretations from the areas of low-grade metamorphism (e.g. Knill, 1963; Harris et al., 1978; Johnson, 1983; Anderton, 1982; 1985; 1988). In the areas of intermediate- and high-grade metamorphism (Figure 25) detailed interpretations of the original nature of the rocks are more difficult, although where sedimentary structures are preserved some sedimentological studies have been made (Winchester and Glover, 1988; Glover and Winchester, 1989).

Accounts of the stratigraphical successions are now published for the South-west Highlands, the Southern Highlands and parts of the Central Highlands. Most of the North-east Highlands and substantial parts of the Monadhliath Mountains of the Central Highlands have only recently been, or are still being, mapped. Much of the following account for these areas is based upon this, as yet unpublished, mapping.

Correlation between local successions is complicated by lateral facies changes, diachronous boundaries, local unconformities, non-sequences, tectonic discontinuities and changes in metamorphic grade. However, certain key units of distinctive lithology have been traced throughout the Grampian Highlands and, in some cases, through north-western Ireland, for distances of up to 700 km. Such key units may have different names in different areas. Intervening beds are generally less traceable but they may make up distinctive lithostratigraphical sequences traceable over various distances. Figure 10 presents a correlation table for the Dalradian Supergroup which reflects the general consensus among current workers. This builds on the original subdivision of the Dalradian by Harris and Pitcher (1975). The Dalradian Supergroup has a total aggregate thickness of more than 25 km. Such a considerable thickness is not developed in any one continuous section or in any one area because deposition occurred in basins of different and changing morphology. The sedimentary succession at various places has been modified by tectonic thickening and thinning during the orogenic deformation which followed the original deposition (Borradaile and Johnson, 1973).

CENTRAL HIGHLAND MIGMATITE COMPLEX

Large areas of migmatites occur in the northern part of the Grampian Highlands (Figure 8). Those east and south of Inverness were generally regarded as stratigraphically equivalent to the surrounding psammitic rocks,

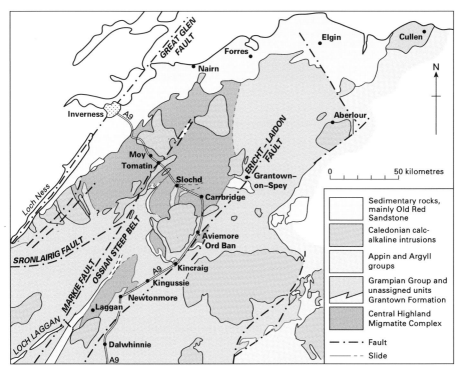

Figure 8   Distribution of the Central Highland Migmatite Complex (modified after Piasecki and Temperley, 1988).

and all were equated with the younger part of the Moine succession north-west of the Great Glen Fault. Detailed mapping, with concomitant isotopic age determinations, of the migmatites in the 1970s led to the radical re-interpretation of their stratigraphical and structural significance (Piasecki and van Breemen, 1970a; 1970b; Piasecki and Temperley, 1988). The migmatites were recognised as a distinct tectonostratigraphical unit referred to as the 'Central Highland Division', and regarded as an older basement to the adjacent psammites. Recently this interpretation has been challenged and the traditional view that the 'Central Highland Division' comprises migmatised versions of the adjacent rocks has been restated (Lindsay et al., 1989); here it is referred to as the Central Highland Migmatite Complex.

The main outcrop of the Central Highland Migmatite Complex covers about 100 km^2 to the south-east of Inverness with isolated outcrops occurring at Ord Ban, Kincraig and Laggan. The base of the migmatites is not exposed but the highest parts are infolded with Grampian Group metasedimentary rocks. This upper contact is defined as being at the limit of migmatisation but in many places the contact is at a tectonic break, the Grampian Slide Zone (see next section and Chapter 6). Within the migmatite complex are tectonic slices of non-migmatitic, recognisable Grampian Group lithologies, and podded metagabbroic sheets (Highton, 1992). Pegmatites and quartz veins are particularly well developed in shear zones such as the Grampian Slide and are of great significance in the dating of tectonothermal events (Chapter 6).

The migmatitic complex consists of coarse-grained psammites, quartzites and pelitic and semipelitic gneisses at upper amphibolite-facies metamorphic grade. The gneisses are generally migmatitic, with granitic (quartz-alkali feldspar) leucosomes predominant in the psammitic rocks and trondhjemitic (quartz-plagioclase) leucosomes in the semipelites (Plate 2). Metasedimentary calc-silicate bands and pods are usually confined to notably striped psammitic units.

Four main lithofacies were recognised by Piasecki and Temperley (1988):

(i)   Coarse quartzofeldspathic gneisses, with alternating feldspathic, siliceous and biotitic bands, commonly accentuated by parallel quartzofeldspathic veins.

(ii)   Coarsely foliated, massive to regularly layered semipelitic gneisses; usually migmatitic and commonly with quartzofeldspathic or siliceous ribs, grading into other gneissose lithologies.

(iii) Psammitic, often migmatitic, gneisses, in which quartz-rich layers are interbanded with layers containing more biotite and/or feldspar. In rare non-migmatitic areas of low strain, rhythmic banding is present, with the preservation of thin, graded units.

(iv)   Very coarse-grained, usually massive, feldspathic quartzites, typically conspicuously migmatitic. With the development of biotitic layers the quartzites grade into psammitic gneisses.

Pods and lenticular bodies of gneissose amphibolite (metagabbro), with or without garnet, are common in the semipelitic and quartzofeldspathic gneisses.

Plate 2   Migmatitic semipelite showing both fine- and coarse-grained contorted leucosome segregations. Central Highland Migmatite Complex, Creag Bhuide, Strathnairn south of Inverness (D 4349).

No attempt has yet been made to establish an overall internal stratigraphical framework within the migmatite complex. The intense high-grade tectothermal deformation, lack of marker horizons, rarity of preserved sedimentary structures and variable degrees of migmatisation are such that the original stratigraphy within the complex is very difficult to identify. Some local structural–lithological successions have been recognised and, in Glen Banchor, Piasecki and Temperley (1988) produced the following tripartite subdivision of a well-exposed 1000 m-thick section.

(iii) *Upper Psammitic group.* Psammites containing discontinuous quartzites pass structurally upwards into micaceous psammites lacking interbedded quartzites. These psammites are overlain by a semipelite unit which contains numerous pods of garnet-amphibolite. The structural top of the group is defined by a psammitic unit with interbanded thin quartzites and pods of garnet-amphibolite.

(ii) *Middle Pelitic group* Essentially coarse-grained, biotite-gneisses with associated fine-grained schists rich in muscovite and almandine garnet. The gneisses occur in lensoid areas of low ductile strain, and the schists are interpreted as their reworked equivalents. The upper part of the group is highly sheared and contains at least three zones of very high strain.

(i) *Lower Siliceous group* Thick units of psammitic gneisses, massive feldspathic quartzites and only minor semipelitic units.

Throughout the Central Highland Migmatite Complex sedimentary structures have been largely obliterated and replaced by ductile tectonic fabrics and metamorphic textures. An early widespread gneissification produced migmatites with regularly spaced and sharply defined layers. The layering, which ranges from several millimetres to centimetres in thickness, is defined by varying proportions of feldspar, quartz and biotite produced by quartz–plagioclase (trondhjemitic) segregation. The gneissose banding appears to be broadly concordant with the original bedding. A second migmatisation, also with trondhjemite segregations, accompanied a later localised development of ductile shearing and folding. Minor folds of the early migmatitic banding, and ductile shear zones are the main structures seen at outcrop. The folds include early recumbent isoclines and later more open asymmetric upright folds. The shears, including the Grampian Slide, are developed preferentially along lithological boundaries or within the least competent lithological horizons. Within the shear zones a mylonitic fabric largely obliterates the gneissose banding and is planar and penetrative in the psammitic lithologies but is commonly lenticular or ribbon-like in the more pelitic rocks. Numerous *en échelon* lenticular and laterally persistent quartz veins, and boudinaged pegmatites, are developed subparallel to the mylonitic foliation.

GRAMPIAN GROUP

The Grampian Group crops out over an area of approximately 4250 km^2 in a broad NE-trending zone extending from Glen Orchy to near Elgin, with an isolated outcrop on the Moray coast around Cullen (Figure 8, Plate 3). The

Plate 3   Grampian Group country.
In the distance are the rounded peat-covered Drumochter Hills, formed of Grampian
Group psammites. General Wade's Military Bridge spans the River Truim with its
alluvial plain and glacial deposits; the A9 road is on the left (D 4967).

group consists mainly of psammites and semipelites at amphibolite-facies
metamorphic grade.

The nature of the contact between the migmatitic, gneissose lithologies of
the Central Highland Migmatite Complex and the essentially non-gneissose
lithologies of the Grampian Group has become the subject of differing interpretations. The two groups of rocks are separated by a complex zone of high
strain and ductile shears of regional extent, the Grampian Slide Zone
(Piasecki and Temperley, 1988). The Slide is well exposed at Lochindorb and
at Kincraig House there are thin zones of mylonite and ultramylonite. The
zones of high strain are characterised by the presence of concordant, highly
deformed, apparently syntectonic pegmatite and quartz-muscovite veins
which have yielded Rb/Sr mineral ages of 750 Ma (Piasecki and van
Breemen, 1983).

Piasecki (1980) recognised early structures only in the migmatitic rocks
which led him to interpret the relationship with the Grampian Group as one
of basement and cover. In the road cutting at Slochd Summit, the contact
between migmatitic and non-migmatitic rocks is well exposed and has been
interpreted as a locally preserved unconformable relationship (Piasecki and
Temperley, 1988).

An alternative interpretation of the contact has been made by Lindsay et al.
(1989) who did not recognise the presence of early structures confined to the
migmatites, instead describing an apparent continuity of stratigraphy and
early tectonic structures from the Grampian Group of the Atholl Nappe
through the Central Highlands area to the Great Glen. The contrasts
between the Grampian Group and the migmatites was attributed to a gradual

increase in metamorphic grade from south-west to north-east, with the contact being interpreted in terms of a sedimentary passage locally modified by zones of higher tectonic strain and selective migmatisation, i.e. downgrading the role and significance of the Grampian Slide.

The upper contact of the Grampian Group, along the southern and eastern margins of its outcrop, is with the Appin Group. In part, the contact is defined tectonically by the Boundary Slide (Chapter 4). In the north-east, however, there is a gradational passage upwards from the Grampian Group into the Appin Group, so that the boundary is arbitrarily defined. A conformable upper contact is also recognised in the south-west where it is traditionally placed at the base of an impersistent quartzite (Eilde Quartzite). However, in this area, different lithologies in both the Grampian and overlying Appin groups are in contact and it is possible that local unconformities exist (Winchester and Glover, 1988; Glover, 1993). The western limit of the Grampian Group is defined by the Great Glen Fault Zone, although the stratigraphical status of some clastic metasedimentary rocks south-east of the fault zone is ambiguous. There are also some brecciated high-grade psammitic rocks within the fault zone near Fort Augustus which may be Moinian.

The Grampian Group is composed mainly of micaceous to quartzose psammites and semipelitic schists. The psammites occur as massive beds of varying thicknesses, with graded bedding developed locally. The bedding is generally accentuated by bedding-parallel mica foliation planes. The semipelites are internally more homogenous, but variably migmatitic with quartzofeldspathic segregations, and appear to be more continuous laterally than the psammites. Quartzites are widespread and, notably in the north-east, form a persistent lithology towards the top of the group. White (clinozoisite-bearing) and green (actinolite-bearing) calc-silicate lenses, mostly concordant to bedding, occur in both psammitic and semipelitic lithologies.

The association of impure diopsidic limestones, pelitic schists (locally with kyanite, tourmaline and garnet) and concordant foliated tholeiitic amphibolites forms a distinctive assemblage (the *Grantown Formation*) overlying the migmatites in the Strathspey and Strathdearn areas. This formation has now been traced, in whole or in part, from Grantown south-westwards, possibly to Kinlochlaggan (Figure 8).

Lithofacies identified in the psammites and semipelites include various types of turbidites, rhythmically interbedded sequences and laminated, lenticular, channel-fill and wave-rippled units, generally within micaceous psammites (Winchester and Glover, 1988). These units can be parallel sided, wedge shaped or lenticular with conformable or erosional bases. Preserved sedimentary structures are abundant locally, despite widespread ductile deformation and greenschist to amphibolite-facies metamorphism; they commonly include grading, trough and tabular cross-bedding and, less abundantly, load and water escape structures, scouring, mudflake-breccias and various types of internal lamination. Bottom structures are either absent or have been removed by bedding-parallel slip.

The semipelites and psammites are composed essentially of quartz, plagioclase and mica, biotite being more common than muscovite, with variable amounts of mainly detrital K-feldspar. Apatite-bearing heavy-mineral bands with various opaque phases are recognised in the psammites. The semipelites, and less commonly the psammites, are locally garnetiferous. Several hundred whole-rock analyses of the psammitic and semipelitic rocks are published (Lambert et al., 1982; Haselock, 1984; Winchester and Glover, 1988). There are slight chemical differences between semipelites at different stratigraphical levels, although as a group they are chemically distinguishable from those in the overlying Appin Group. All the psammites have broadly similar chemical compositions.

Stratigraphical successions have been established in a number of separate parts of the Grampian Group outcrop but at present no complete picture of the sedimentation and stratigraphy has emerged. Correlation

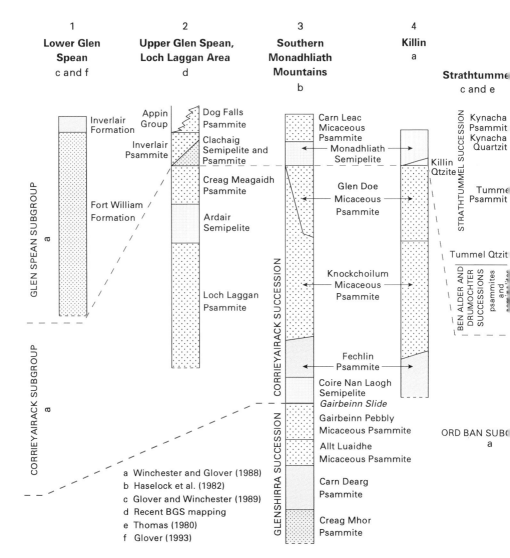

Figure 9   Lithostratigraphical units in the Grampian Group.

between the established local successions is incomplete owing to the possible effects of lateral changes of lithofacies, of the high-strain zones and thrusts, and of penecontemporaneous faults which exerted an influence on sedimentation.

The local stratigraphies generally have been divided into formations which may be amalgamated into informal successions (Figure 9) such as the Ben Alder, Drumochter and Strathtummel successions for the area north of Schiehallion (Thomas, 1980). Further work is needed before a more formal stratigraphical framework for the whole of the Grampian Group is possible, although Winchester and Glover (1988) have suggested

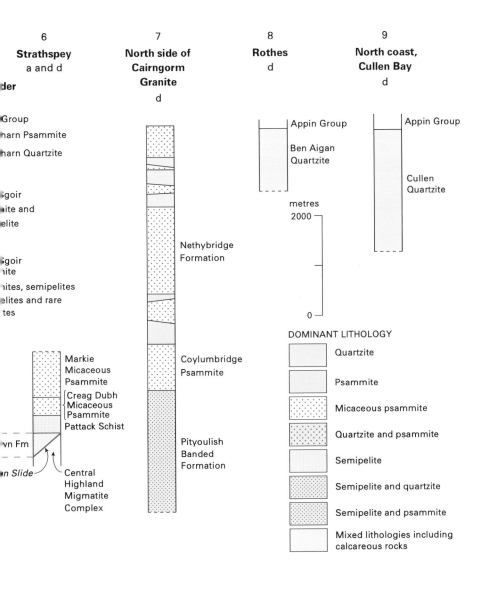

a regional tripartite subdivision, into the Ord Ban, Corrieyairack and Glen Spean subgroups.

Ord Ban Subgroup

This subgroup consists of the distinctive assemblage of limestone, pelitic schist (partly kyanite-bearing) and amphibolite immediately overlying the Central Highland Migmatite Complex in Strathspey; it was originally called the Grantown Series by Hinxman and Anderson (1915) and, in recognition of this, it has been renamed the *Grantown Formation*. It is overlain by rhythmites and thick psammites but the relationship of these units to the Corrieyairack Subgroup farther west is uncertain at present.

Corrieyairack Subgroup

This subgroup is the major component of the Grampian Group of the southern Monadhliath Mountains where it is about 4500 m thick and consists of a thick succession of psammitic rocks overlying a basal semipelite (Haselock et al., 1982; Okonkwo, 1988) (Figure 9, column 3). Here it is separated from the underlying Glenshirra succession, which has uncertain stratigraphical status (see next section). Southwards, towards Glen Spean, quartzite forms several distinctive units of varying thickness.

Glen Spean Subgroup

This subgroup consists of a mixed sequence of semipelite and psammite with quartzite. At Spean Bridge the subgroup is approximately 4000 m thick (Glover and Winchester, 1991) but farther south, in the River Leven, and on the Black Mount, the exposed psammites and semipelites of the subgroup are only about 100 m in thickness. The rocks of this subgroup are thought to reflect a change of depositional environment from the deep water turbidites of the Corrieyairack Subgroup to shallow marine shelf sedimentation (Glover, 1993). Similar lithologies farther north-east, exposed in the mountainous tract bisected by the A9 road (which provides excellent exposures in cuttings), have been described by Thomas (1980; 1988). Here a *Drumochter succession* of monotonously flaggy psammites and semipelites is overlain by the predominantly psammitic *Strathtummel succession* (Figure 9, column 5); they may be lateral equivalents of the Glen Spean Subgroup (Glover and Winchester, 1989). Shallow water sedimentary structures are well preserved in the upper psammites which are up to 3000 m thick.

In Strathspey the relations between the Grantown Formation of the Ord Ban Subgroup and the overlying succession are uncertain and the contact is at least in part tectonic. The overlying succession consists for the most part of micaceous psammites with some white quartzites but there is one distinctive unit consisting of an alternation of quartzites and semipelites. Exposure in the area is generally poor and the structure is consequently not fully understood but the succession appears to be several kilometres in thickness (Figure 9, column 6).

Quartzite becomes the dominant component of the upper Grampian Group succession at the north-east limit of its outcrop, around Rothes and on the Moray Firth coast at Cullen. Here, quartzites up to 2500 m thick, directly underly Appin Group lithologies (Figure 9, columns 8 and 9). The *Cullen Quartzite Formation* crops out along 12 km of the Moray Firth coast, between Buckpool and Logie Head. It is divided into the *Findochty Quartzite Member*, consisting of hummocky-bedded and thickly bedded quartzites with garnet-mica-schist interbeds, and an overlying *Logie Head Quartzite Member*, comprising planar-bedded, flaggy quartzites, also with finely interbanded garnet-mica-schists. Shallow water sedimentary structures are preserved in the various quartzites. Inland, the *Ben Aigan Quartzite Member* is equivalent to the Logie Head Member.

Metasedimentary rocks of uncertain stratigraphical affinity

Structurally underlying the Corrieyairack Subgroup in the area south-east of Loch Lochy and Loch Ness is a succession of arkosic psammites with meta-conglomerates and pebbly psammites that is more than 2000 m thick north-east of Fort Augustus (Parson, 1982). Near Loch Lochy thin lenses of dolomitic marble, black schist, quartzite and rare metabasite also occur. These lithologies have been grouped together with the psammites as the *Glen Buck Pebbly Psammite Formation*, which is separated from the overlying Corrieyairack Subgroup by the Eilrig Shear Zone, a zone of mylonites up to 1 km thick locally (Phillips et al., 1993). The stratigraphical affinity of a fault-bounded outcrop of gneissose micaceous psammites with minor siliceous marble at Gleann Liath, near Foyers (Mould, 1946) remains problematical.

Farther to the south-east the *Glenshirra succession* is separated from the overlying Corrieyairack Subgroup by a slide, or high-strain zone, the Gairbeinn Slide (Haselock et al., 1982) (Figure 9, column 3). The Glenshirra succession is subdivided into four formations in its type area between the Allt Crom Granite and the Corrieyairack Igneous Complex and is at least 2500 m thick. Psammites with interbanded semipelites are the main rock types in the succession which includes an upper unit of pebbly psammites. Rapid lateral facies changes occur and Haselock (1984) described geochemical and sedi-mentological differences between the Glenshirra succession and the overlying Corrieyairack Subgroup. Both the Glenshirra succession and the Glen Buck Pebbly Psammite Formation are separated from the Corrieyairack Subgroup by tectonic discontinuities and their stratigraphical relationships with the overlying Grampian Group are consequently uncertain.

On Islay, it is possible that the Bowmore Sandstone may be assigned to the Grampian Group. This problematical unit is discussed in Chapter 3.

APPIN GROUP

The Appin Group consists for the most part of a sequence of shelf sediments comprising pelites, semipelites, quartzites, calc-silicate rocks, limestones and dolostones, usually with rapid alternations of facies (Wright, 1988). Local suc-cessions are easily established and the group has been divided into three subgroups. Lateral facies changes are well documented in several areas, but certain key beds can be traced over large distances and there is an overall general consistency of facies from Connemara in western Ireland to the Moray

Firth coast. Correlations between local successions have thus been made with reasonable confidence throughout the Grampian Highlands (Figure 10), aided in some areas by detailed studies of the whole-rock geochemistry of a variety of lithologies. Such studies have been more successful in the Appin Group than in other parts of the Dalradian succession (Lambert et al., 1981; 1982; Hickman and Wright, 1983; Rock et al., 1986). Of particular use are the geochemical studies of carbonate units, some of which retain distinctive geochemical characteristics over considerable distances (Rock, 1986; Thomas, 1989).

Rocks of the Appin Group crop out over some 2100 km^2 in a relatively narrow outcrop extending across the Grampian Highlands (Figure 3). Thick developments occur in Lochaber and around Appin, which are type areas for the two lowest subgroups, the Lochaber and Ballachulish subgroups (Figure 11). In the south-west a complete sequence, which continues up into the overlying Argyll Group, is recognised in the core of the Islay Anticline. South-eastwards from Appin, rapid facies changes and considerable attenuation occur (Figure 12). Higher parts of the group were either not deposited, or are cut out by unconformities, or have been excised by tectonic dislocation in the Boundary Slide Zone. As a result, only a condensed and possibly incomplete sequence of Lochaber and Ballachulish subgroup rocks is present from Glen Orchy to Glen Lyon. A more complete although still condensed sequence, which passes up conformably into the Argyll Group, reappears to the north of Schiehallion and expands rapidly eastwards to Blair Atholl, the type area of the highest, Blair Atholl Subgroup. The complete sequence is then traceable north-eastwards to Braemar. To the north of the Cairngorm and Glengairn granites a similar succession has been traced northwards to link with the well-known Appin Group succession on the Moray Firth coast.

Lochaber Subgroup

In many areas a sequence of semipelitic and pelitic schists with lenticular interbedded quartzites at the base of the Appin Group conformably succeeds the underlying more psammitic rocks of the Grampian Group. This sequence, the Lochaber Subgroup, represents a depositional facies of alternating clean sand, silt and mud deposition in which individual units are traceable over strike lengths of tens of kilometres. However, individual quartzites do taper out laterally and the facies they represent may be diachronous. In the type area of Lochaber there are three major quartzite units, whereas elsewhere typically only one quartzite horizon is developed. Concomitant with the lateral thickness changes there are also observable detailed facies variations within each of the three quartzites of the type area. Consequently isolated quartzite units cannot readily be correlated with any of the three type-area quartzites.

The difficulties of correlation posed by sedimentary facies changes are compounded by the north-eastwards increase in regional metamorphic grade and by superimposed folding and shearing. Within the greenschist facies, isograds cut across the regional strike and the development of garnet in semipelitic schists increases towards the north-east. The polyphase deformation has caused repeated and locally overturned statigraphical sequences resulting in complex outcrop patterns (as shown in fig. 1 of Hickman and Wright, 1983).

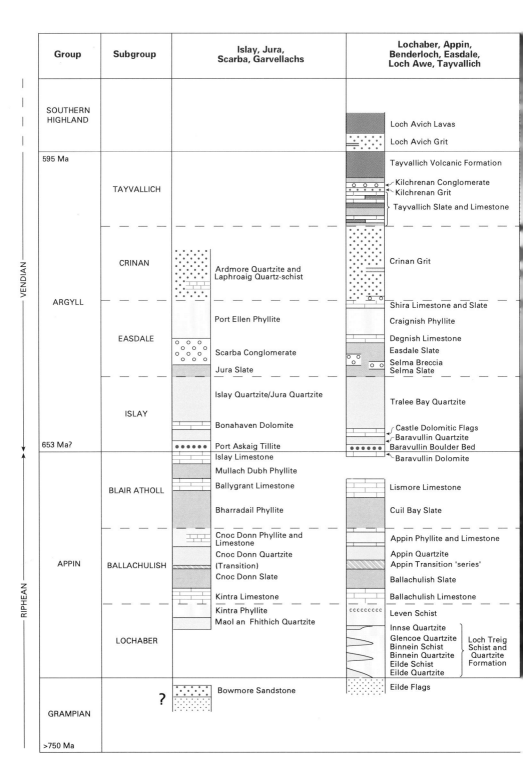

Figure 10 Composite lithostratigraphical sections (not to scale) of the Appin, Argyll and Southern Highlands groups.

The red line encloses successions in the North-east Highlands which are separated from the more coherent successions by major tectonic discontinuities (e.g. the Portsoy Lineament).

The main lithologies repeated in each formation within the Lochaber Subgroup retain common characteristics. The major quartzites are generally white and all are well bedded. Internal shallow-water sedimentary structures, such as cross-bedding, grading, ripple-marks, slump and de-watering structures, are well preserved. Variable amounts of K-feldspar and plagioclase are present, and both biotite and muscovite are commonly concentrated along bedding-parallel foliation planes. The quartzites become finer grained and less feldspathic towards the north-east, a mineralogical change that is reflected in their whole-rock chemistry (Hickman and Wright, 1983). Intraformational contacts between quartzites and schists are commonly transitional over several metres with fine-scale interleaving of the two lithologies.

The schist formations generally consist of a groundmass of muscovite, biotite, quartz, plagioclase and K-feldspar with scattered porphyroblasts of biotite and, less commonly, garnet. All are less feldspathic than the underlying Grampian Group semipelites, a feature that is reflected by marked differences in whole rock chemistry between Lochaber Subgroup and older schists (Lambert et al., 1982; Winchester and Glover, 1988).

The major development of the Lochaber Subgroup lies between Glen Roy and Port Appin and includes the type area in the mountainous ground between Glen Nevis and Loch Leven (Figure 11). Here the contact between the Grampian and Appin groups is usually conformable, although lateral facies changes in both groups means that different lithologies are adjacent along the contact. In the north-western part of the type area an attenuated Lochaber Subgroup sequence is sandwiched in the Fort William slide zone between the Grampian Group and the Ballachulish Subgroup. The Grampian Group/Lochaber Subgroup contact, previously regarded as tectonic, has been re-interpreted by Glover (1993) as a localised unconformity. The upper contact of the Lochaber Subgroup with the Ballachulish Subgroup is conformable; in the Glen Roy area the two subgroups appear to interfinger.

In its type area the Lochaber Subgroup is divided into six units, with a maximum aggregate thickness of 4200 m (after Carruthers in Hinxman et al., 1923; Bailey, 1934). The succession has been partly formalised by Hickman (1975) who defined type sections for some of the units within a continuous section along the River Leven and Loch Leven around Kinlochleven. The basal *Eilde Quartzite* (600 m), best exposed around Loch Eilde Mor, consists essentially of flaggy, pink to white, feldspathic quartzites with schist and psammite intercalations. A pebble bed has been identified at about 200 m above the base. The *Eilde Schist* (400 m) comprises interbedded pelitic and semipelitic schists which become increasingly pelitic north-eastwards. The overlying *Binnein Quartzite* (400 m), which is best exposed on the hills around Kinlochleven, comprises relatively distinctive, fine-grained well-bedded orthoquartzites. These are overlain by semipelitic schists with quartzite intercalations forming the *Binnein Schist* (400 m) in which the uppermost schists are locally graphitic with calcareous seams. In its type area on the southern slopes of Mam na Gualainn, the *Glencoe Quartzite* (400 m) is lithologically very similar to the type Eilde Quartzite. In parts it is also coarse grained and arkosic with discrete pebble beds. North-eastwards the upper quartzite beds become less feldspathic and resemble the Binnein Quartzite. The *Leven Schist,* around its type area near North Ballachulish, comprises a 2000 m-thick

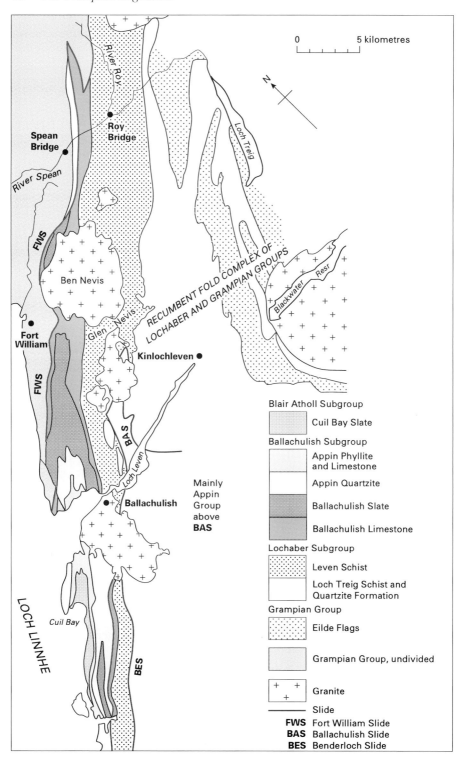

0   5 kilometres

River Roy

Loch Treig

Roy Bridge

Spean Bridge

River Spean

FWS

Ben Nevis

Glen Nevis

RECUMBENT FOLD COMPLEX OF LOCHABER AND GRAMPIAN GROUPS

Blackwater Resr

Fort William

FWS

Kinlochleven

BAS

Loch Leven

Mainly Appin Group above **BAS**

Ballachulish

Cuil Bay

LOCH LINNHE

BES

Blair Atholl Subgroup

Cuil Bay Slate

Ballachulish Subgroup

Appin Phyllite and Limestone

Appin Quartzite

Ballachulish Slate

Ballachulish Limestone

Lochaber Subgroup

Leven Schist

Loch Treig Schist and Quartzite Formation

Grampian Group

Eilde Flags

Grampian Group, undivided

Granite

Slide

FWS Fort William Slide
BAS Ballachulish Slide
BES Benderloch Slide

homogeneous sequence of greenish grey phyllites or schists (Plate 4) with thin quartzites (including the *Innse Quartzite*) confined to the lower transition zone with the underlying Glencoe Quartzite. However, north-east of the type area, a tripartite division is recognised in which basal dark phyllites or schists are overlain by striped siltstones with thin carbonates which in turn are overlain by pale greyish green phyllites or schists (Hickman, 1975; Litherland, 1980). The uppermost schists become increasingly calcareous and contain thin limestone bands in the Glen Spean area.

The lateral and vertical facies variations in all six units, coupled with different interpretations of local and regional structures, have caused much debate about the stratigraphical framework of the subgroup in the Lochaber area, so far without complete agreement being reached (see Bailey and Maufe, 1960; Hickman, 1975; Treagus and King, 1978; Litherland, 1980; Anderton, 1985, 1988 and references therein). Consequently the policy of giving structurally isolated schists and quartzites local informal names remains valid at present.

In the northern part of the Lochaber area, near Loch Treig and Glen Spean, a local stratigraphy has been established, made up of the Loch Treig Schist and Quartzite Formation and the overlying Leven Schist Formation (Key et al., in press). In this succession the three main quartzites of Lochaber are thin and are classed as members within the *Loch Treig Schist and Quartzite Formation*. North of Glen Spean the three quartzites die out. The overlying *Leven Schist Formation* is some 2200 m thick in the Glen Spean– Loch Treig area.

Between Ben Alder and Kingussie a series of tight, upright folds occurs at the north-western edge of the Ossian–Geal Charn Steep Belt and these include lithologies similar to those of the Leven Schist (Anderson, 1956). An informal local succession in the Kinlochlaggan Syncline has been assigned largely to the Lochaber Subgroup (Treagus, 1969; 1981). However, the rocks are at high metamorphic grade and are strongly deformed, and there is the possibility of an alternative correlation with the Grampian Group, in particular the limestone-amphibolite parts of the sequence being correlated with the Grantown Formation. The *Kinlochlaggan Boulder Bed* consists of about 40 m of bedded psammites and feldspathic quartzites, within which are two 7 m-thick beds of massive, unbedded psammite containing clasts up to 33 cm in maximum diameter. The clasts are sparsely distributed, unsorted and consist of a variety of intra- and extrabasinal rock types including granite. In common with the similar, more widespread boulder beds at the base of the Argyll Group, these have been interpreted as tillites, deposited from ice sheets (Treagus, 1981).

On Islay a regional plunge reversal exposes the upper part of the Lochaber Subgroup in the core of the Islay Anticline (Rast and Litherland, 1970; Wright, 1988). Here the subgroup is divided into two units. The lower, *Maol an Fhithich Quartzite*, consists of massive, coarse-grained, cross-bedded quartzites with phyllites and pebble beds containing extrabasinal granite clasts. The overlying *Kintra Phyllite* or *Glenegedale Slate Formation* is composed

Figure 11 Grampian and Appin groups in the Lochaber and Appin area (modified from Hickman, 1975 and incorporating results of BGS mapping to the north).

Plate 4 Leven
Schist, Lochaber
Subgroup, Allt
Ionndrainn, Glen
Roy district.
Asymmetrical folds
defined by an early
foliation and lenses
of dolomitic
limestone (D 4175).

of striped greenish, semipelitic phyllites or slates which become more calcareous upwards.

In the Southern Highlands, between Glen Orchy and Braemar, a thin development of the Lochaber Subgroup has been recognised, rarely exceeding a few hundred metres in thickness, in which few formations can be correlated with the type area (Roberts and Treagus, 1979; Treagus and King, 1978; Treagus, 1987; Upton, 1986). Throughout much of this area the junction with the underlying Grampian Group lies within a zone of high strain, some 500 m to 2000 m in thickness (formerly the Iltay Boundary Slide Zone—Chapter 4.) In this zone several formations are strongly attenuated or even excised locally along slides. However, in some areas, there is a continuous overall stratigraphical transition from the Grampian Group into the Lochaber Subgroup. The reduced thickness of the latter is not solely attributed to deformation in the slide zone; there would seem to be considerable sedimentological thinning in this central area of the Southern Highlands as there is in many of the succeeding Appin Group formations. In the Schiehallion area the *Beoil Quartzite* is overlain by highly tectonised pelites of the *Beoil Schist* and by calcareous pelites of Leven Schist type, the *Meall Dubh Striped Pelite* (Treagus and King, 1978; Treagus, 1987). Leven Schist-type lithologies follow on from the Grampian Group Struan Flags in the '*Banvie Burn Series*' of Glen Tilt (Bailey, 1925); and south-west of Braemar highly strained mica-schists at the base of the Lochaber Subgroup are succeeded by striped psammites, silty mica-pelites and pelites with calc-silicate lenses (Upton, 1986).

South and west of Tomintoul a poorly defined association of micaceous psammites with thin lenticular quartzite units overlies the Grampian Group conformably. Farther north, between Bridge of Brown and Glenlivet, Grampian Group psammites grade upwards into slaty semipelites and calcareous semipelites. These beds probably represent a thin development of the Lochaber Subgroup.

The Lochaber Subgroup thickens markedly farther north around Dufftown and between Keith and the north coast around Cullen (Peacock et al., 1968; Read, 1923; 1936). Here a thick sequence of flaggy, micaceous psammites and semipelites, the *Findlater Flag Formation*, rests conformably upon massive psammites and quartzites of the Grampian Group. At least two prominent flaggy quartzites occur within the Findlater Flags and bands of garnet-biotite-schist and gneiss are recorded. Calc-silicate lenses mark a transition into more persistent banded grey, cream and pale green calcareous psammites and semipelites locally termed the *Pitlurg Calcareous Flag Formation* and *Cairnfield Calcareous Flag Formation*. These are characterised locally by tremolitic amphibole occurring either as pervasive aggregates or as monomineralic bands (Stephenson, 1993). Limestone bands occur locally in the upper parts of the formations. These herald the more extensive limestone development in the Ballachulish Subgroup as in the type area of Lochaber.

Ballachulish Subgroup

This subgroup, more than any other in the Dalradian, exhibits a remarkable continuity of lithological type; key elements of the classic sequence in the Lochaber–Appin area can be traced from north-west County Mayo to the Moray Firth coast.

Four formations are recognised in the Lochaber–Appin area (Figures 10 and 11), each of which exhibits a transitional passage into the overlying formation (Bailey, 1960; Hickman, 1975; Litherland, 1980). The lowest, the *Ballachulish Limestone Formation*, comprises grey-green calcareous phyllites, cream and grey dolostones, dark bluish grey limestones and intercalations of slaty pelite. The phyllites resemble those of the upper part of the Leven Schist and are commonly amphibolitic. In its type area, around Ballachulish and Onich, the formation is about 250 m thick. In the north of Lochaber the formation can be traced around fold closures in the area of Glen Roy, but is attenuated and terminates along the Fort William Slide (Figure 11). Farther east, in the core of the Kinlochlaggan Syncline, the *Kinlochlaggan Limestone* has been equated with the Ballachulish Limestone, but this correlation is not yet confirmed. Towards the south, in Appin, the formation is more psammitic. The *Ballachulish Slate Formation* has a similar areal distribution to the preceding formation. In the type area at Ballachulish, a transitional zone of intercalated slate and carbonate is followed by up to 400 m of black slates and graphitic phyllites. These bear large cubes of pyrite and were quarried extensively for roofing slates which may be seen throughout Britain. In the top 100 m, graded psammite and quartzite intercalations on scales from a few millimetres to about a metre become numerous, forming a distinctive *Appin Transition 'series'*.

The *Appin Quartzite Formation* is a massive to blocky, well-bedded, locally feldspathic quartzite which is about 300 m thick in its type area, but it thins considerably to the north-east like the quartzites of the Lochaber Subgroup. Grain size and feldspar content increase both upwards and laterally to the south-west, where the quartzite becomes markedly pebbly. Sedimentary structures such as cross-bedding, ripple marks and graded bedding are ubiquitous. The overlying *Appin Phyllite and Limestone Formation* consists of an alternating sequence of carbonate rocks and phyllites with flaggy psammites and thin quartzites. It attains a total thickness of up to 400 m in the type area. Carbonates, often dolomitic, are more prevalent in the lower part and include the pure, white Onich Limestone. The distinctive 'tiger-rock' of Bailey (1960), which consists of regularly spaced 5 to 10 cm layers of deep yellow-weathering dolostone and dark phyllite, occurs at several stratigraphical levels. Marked facies changes occur south-westwards towards the Firth of Lorn where the phyllites grade into more psammitic units and pass locally into calcareous quartzites.

South-west of the type area, representatives of the Appin Quartzite and the Appin Phyllite and Limestone Formation crop out on small islands in the Firth of Lorn, east of Lismore. Farther to the south-west, after a gap of 75 km, the subgroup crops out in the core of the Islay Anticline. Here the *Kintra Limestone, Cnoc Donn Slate, Cnoc Donn Quartzite* and the *Cnoc Donn Phyllite and Limestone* can be matched confidently with the formations of the Appin district, with little variation in facies (Rast and Litherland, 1970).

To the south-east of Appin the type succession thins rapidly across strike and changes in character (Figure 12). Hence, in Glen Creran the Ballachulish Limestone Formation become limestone-free, the limestones of the Appin Phyllite and Limestone Formation are reduced to small lenses, the accompanying slates and phyllites pass into flaggy semipelites, and the Appin Quartzite dies out (Litherland, 1980). The lower part of the sequence may be cut out by an unconformity and local non-sequences within the Appin Phyllite and Limestone Formation suggest some syndepositional tectonic control.

Farther to the south-east, beyond the Etive Granite Complex, the Ballachulish Subgroup is generally absent, apart from a few small outliers of calcareous schists above the Lochaber Subgroup in the Loch Dochard, Glen Orchy and Ben Dorain areas. These are probably Ballachulish Limestone equivalents (Bailey and Macgregor, 1912; Thomas and Treagus, 1968; Roberts and Treagus, 1979). The outliers underlie the main Boundary Slide, which in this area is thought to have excised the remainder of an Appin Group succession already attenuated by sedimentological factors as in the Glen Creran area.

The attenuated Ballachulish Subgroup reappears eastwards as a condensed sequence totalling only 100 m at Errochty Water, but increasing

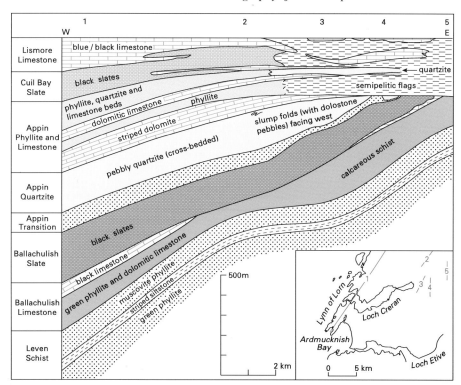

Figure 12 Appin Group facies changes in the Loch Creran area, based on correlation of strike sections and estimation of pre-tectonic thicknesses (from Litherland, 1980).

to 700 m in Strath Fionan, north of Schiehallion (Treagus and King, 1978). The subgroup continues to the Loch Tay Fault at Foss where it is displaced north-eastwards to the Blair Atholl area (Smith and Harris, 1976). It may then be traced in continuous outcrops north-eastwards to Braemar (Upton, 1986). Throughout this continuous strike length of some 65 km the succession can be matched almost bed for bed with that in the type areas of Lochaber and Appin (see Figure 10 for local names). Such distinctive lithologies as the graphitic schists, the Appin Transition 'series' at the base of the main quartzite and the crystalline white limestones and 'tiger rock' in the topmost limestone and phyllite formation are found throughout this section.

The subgroup is well developed to the north of the Cairngorm and Glengairn granites and can be traced as far as the Keith area. The *Mortlach Graphitic Schist* is several hundred metres thick in Glenlivet but thins locally to 5 or 6 m. A dark limestone, the *Dufftown Limestone*, is commonly present at or near the base of the schist, and other limestones occur locally in the lower part. Minor quartzites also occur locally. At the base of the *Corryhabbie Quartzite* there is a transitional unit of interbedded pelite and psammite which is thick in the south but is reduced to a few metres farther north, where thickly bedded psammites pass up into the main clean, cross-bedded quartzite, typical of the subgroup. More psammites are followed by the *Ailnack Phyllite and Limestone Formation*, consisting of phyllitic semipelites, with several thin white limestones, calc-silicate bands and one more-persistent banded limestone member.

North of Keith marked facies changes, probably associated with NW-trending growth faults (Chapter 16), and increased structural complexity make individual units difficult to trace, so that formations become ill defined. The graphitic character of the lower part of the subgroup is locally much reduced and thick, persistent limestones are absent. A condensed sequence of limestone and graphite-schist is seen around Deskford, but in boreholes on the coast at Sandend Bay the graphitic schist is over 300 m thick and is overlain directly by phyllites and limestones with no intervening quartzite.

Blair Atholl Subgroup

This subgroup maintains a generally constant lithology of dark pelites and limestones from Connemara to the Moray Firth, although local successions differ in detail, making bed-for-bed matching difficult. In some areas, notably Islay and the Central Highlands, the upper part is less pelitic with thinly banded phyllites, psammites, limestones and dolostones comprising a distinctive 'pale group'.

In the Appin area, the Appin Phyllite and Limestone Formation of the Ballachulish Subgroup is overlain by the distinctive *Cuil Bay Slate Formation* (Figure 11). This comprises some 300 m of dominantly dark grey, pyritiferous pelitic and semipelitic slates with minor dark grey limestone beds and some more-psammitic bands (Hickman, 1975). The slates pass upwards through a finely banded passage series into the *Lismore Limestone Formation*, a 1 km-thick sequence of banded, blue-grey, flaggy limestones with thin black slate members, which forms the island of Lismore. It is also seen as an inlier in the core of the Loch Don Anticline in south-eastern Mull. Hickman (1975) has divided the formation into 15 members and recognises several limestone–slate cycles, in each of which a thick limestone passes through a transitional, banded, argillaceous limestone into a slate. Both the limestones and the slates are pyritiferous and graphitic. Slump folds and syndepositional breccias indicate periods of sediment instability.

Owing to the south-westerly plunge of the major folds, any higher beds of the subgroup which may have been deposited in the Appin–Lismore area lie beneath the Firth of Lorn, but an extended sequence has been recognised in the Islay Anticline (Rast and Litherland, 1970). Here, dark graphitic slates and phyllites (*Bharradail Phyllite Formation*), followed by the bluish grey *Ballygrant Limestone* are equated with the Cuil Bay Slate and Lismore Limestone respectively (Figure 10). They are overlain by more dark grey phyllites with graded quartzose or calcareous bands, the *Mullach Dubh Phyllite*, and a distinctive banded, partly oolitic and stromatolitic, thin-bedded limestone, the *Islay Limestone* (Spencer, 1971).

To the south-east of the Appin area, in Glen Creran, the Blair Atholl Subgroup thins and the slates pass into semipelitic flags, although the limestones persist (Figure 10; Litherland, 1980). The facies changes are similar to some of those in the underlying Ballachulish Subgroup and the Blair Atholl

Subgroup is also absent eastwards from Glen Creran due to a presumed combination of sedimentary thinning and movement on the Boundary Slide.

In the Schiehallion area, rocks of the Blair Atholl Subgroup reappear east of Loch Errochty in stratigraphical continuity with the Ballachulish Subgroup and a complete sequence, 250 to 350 m thick, is present between here and the Loch Tay Fault at Foss (Figure 10) (Treagus and King, 1978). A lower sequence consisting of three alternations of dark pelites and dark grey limestones is equated with the Cuil Bay Slate and Lismore Limestone. The pelites are commonly graphitic and kyanite-bearing. The overlying non-graphitic pale 'group' is generally composed of psammitic ribs in a graded pelitic or semipelitic matrix. Local bands of pure quartzite contain ripple cross-laminations, channels and sedimentary dykelets. At the top of the subgroup is a pale cream-weathering dolomitic marble containing tremolite and phlogopite, equated with the Islay Limestone.

Across the Loch Tay Fault, in the area around Blair Atholl, a similar continuous succession of Ballachulish and Blair Atholl subgroups has been demonstrated (Smith and Harris, 1976). This constituted the original type succession for the Blair Atholl 'series' (Bailey, 1925; Pantin, 1961). Thick bands of dark graphitic limestone in the lower part of the Blair Atholl Subgroup are a distinctive feature in this area and have been quarried extensively. The main formations can be followed north-eastwards through the Glen Shee area almost to Braemar (Upton, 1986).

North of the Cairngorm and Glengairn granites the Blair Atholl Subgroup consists mainly of mid to dark grey semipelitic schists, locally pelitic, graphitic and calcareous. A prominent thick bluish grey limestone formation, the *Inchrory Limestone*, occurs in its central part and minor limestones occur locally in the lower part. Maximum developments of the subgroup occur in Upper Donside and the Braes of Glenlivet. Farther north, around Edingight, black graphitic pelites with staurolite are interbedded with thin bands of blue-grey limestone. The limestones thicken considerably around Fordyce and constitute the only part of the subgroup exposed on the north coast.

ARGYLL GROUP

The Argyll Group has been divided into four subgroups (Figure 10). In the oldest, Islay Subgroup, conditions of deposition were similar to those of the preceding Appin Group. The deposits accumulated on or close to an extensive continental shelf and can be traced and correlated along the whole Dalradian outcrop. The base of the group as originally defined was marked by a tillite or a sequence of tillites, deposited from grounded ice sheets. This distinctive unit was, and continues to be, regarded as an important chronostratigraphical marker, not only in the British Isles but throughout the Caledonides. However, its use in defining a major *lithostratigraphical* boundary has led to problems where tillites occur in diachronous units (for example within the Ladder Hills Formation). The tillite-bearing sequence is followed by thick shallow-water shelf and deltaic quartzites which mark the end of stable conditions. The succeeding Easdale, Crinan and Tayvallich subgroups are generally characterised by basin deposits, turbidites and unstable basin margin slump deposits, with only one widespread major shallow-water interlude, in the upper Easdale Subgroup. It is rarely possible to trace individual beds in the Easdale and Crinan subgroups for any great distance and correlations are made on the grounds of general similarity of facies (Figure 10). In contrast, for most of its outcrop the Tayvallich Subgroup is dominated by a carbonate unit which forms one of the principal marker bands of the Dalradian succession. Further evidence for tectonic instability in the Argyll Group comes from the widespread syngenetic mineralisation in the Easdale Subgroup and from the penecontemporaneous igneous activity, evidence of which is found throughout the group. The maximum development of volcanic rocks lies at the top of the Tayvallich Subgroup in the South-west Highlands.

Rocks of the Argyll Group crop out over an area of some 5700 km^2 (Figure 3). Extensive outcrops occur on Islay and Jura, and Argyll Group rocks constitute almost the whole of the South-west Highlands to the north-west of Kintyre and Loch Fyne. The type successions for all four subgroups are found in this latter area where outcrops can be described in relation to three major structures, the Islay Anticline, the Loch Awe Syncline and the Ardrishaig Anticline (Figure 19). To the north-east of these structures, between the Etive Granite Complex and the Bridge of Balgie Fault, the Islay Subgroup, like much of the preceding Appin Group, is absent, although younger units are continuous. Farther to the north-east a full succession is present through the Tummel Steep Belt to the Glen Shee area, with the higher units continuing to middle Deeside. In the North-east Highlands, west of the Portsoy Lineament, the lowest units have been traced intermittently to the north coast. East of the Portsoy Lineament, the higher parts of the group are believed to be present in an undivided gneissose sequence forming a horseshoe outcrop pattern in the Turriff Syncline.

Islay Subgroup

This largely psammitic subgroup is dominated in most areas by a thick quartzite formation. The quartzite and the underlying basal tillite-bearing formation are persistent and distinctive, enabling a good correlation of beds from Connemara to the Moray Firth.

In the type area for the subgroup on Islay the basal *Port Askaig Tillite Formation* consists of a sequence of sandstones, siltstones, conglomerates and dolostones in which boulder beds have been recognised (Spencer, 1971; 1981; Plate 5). These beds range from 0.5 m to 65 m in thickness and contain boulders up to 2 m in diameter. One particular bed, the 'Great Breccia' contains large rafts of dolostone up to 320 m long. The lower beds contain clasts of dolostone, probably derived locally from within the formation or from the underlying Blair Atholl Subgroup. However, the higher boulder beds contain clasts of granite and granite-gneiss of extrabasinal origin. This division on the basis of clast content is also recognisable in boulder bed sequences outwith the type area. The boulder beds are generally regarded as tillites but isolated clasts in associated varved siltstones have been interpreted as drop-stones from floating ice. Polygonal sandstone wedges have been interpreted as ice wedges formed during periods of emergence (Eyles and Clark, 1985).

Plate 5   Boulder bed in the Port Askaig Tillite Formation, Argyll Group, Islay; cleaved and containing large clasts mainly of granite (C 1276).

Within the type area the formation varies considerably in thickness and in the number of tillites recognised. Around Port Askaig on Islay and on the Garvellach Islands, the formation is about 750 m thick and the combined sequence from the two areas may be up to 870 m, with up to 47 separate tillites recognised. However, to the south-west, on the Mull of Oa, the sequence is incomplete and less than 40 m thick. The presence of glacial deposits and their relationship to the adjacent dolomitic beds has consider-able implications for the palaeogeography of Argyll Group time and for the age of the group, both of which are discussed in later sections.

In the Port Askaig area of Islay tidal sandstones at the top of the tillite formation pass transitionally into the succeeding *Bonahaven Dolomite Formation* (Klein, 1970) which consists of up to 295 m of sandstones, pelites, dolostones and impure dolomitic rocks. It has been divided into four members by Spencer and Spencer (1972). Member 3 is the principal dolomitic unit, in which algal stromatolites occur at ten horizons in a total thickness of 150 m. The stromatolites have been described by Hackman and Knill (1962) and a sedimentological account of the whole member is given by Fairchild (1980a). More detailed studies of the diagenesis, petrography and geochemistry of the dolostones are described by Fairchild (1980b, 1985).

The succeeding *Islay (or Jura) Quartzite* marks an abrupt change to a thick succession of cross-bedded and pebbly quartzites characteristic of a tidal shelf environment (Anderton, 1976). The quartzites form almost all of the islands of Jura and Scarba (Figure 13) and crop out on both limbs of the Islay Anticline. Even in their thickest development of over 5000 m on Jura there are no marker bands by which they may be subdivided. However, in addition to the coarse sandstones and pebbly conglomerates, in parts there occur laminated or rippled fine sandstones and interbedded mudstones. Sedimentary structures imply dominant palaeocurrent flow directions towards the NNE throughout the formation, with a general change towards the finer-grained facies observed in the same direction. Within the type area the formation thins markedly along strike, to both the south-west and to the north-east (Figure 13; Knill, 1963).

Farther along strike to the north-east in the Benderloch area (Figure 14), the Islay Subgroup is reduced in thickness to between 300 m and 800 m in the Ardmucknish Succession (Litherland, 1980). However, the sequence here is similar to that of the type area in that it includes two dolomitic boulder beds at the base, followed by dolomitic flags and pebbly, cross-bedded quartzites. From Benderloch the subgroup increases in thickness across strike (Figure 15). Around Glen Creran, some 20 km to the east, the facies is markedly different to that of both the type area and Benderloch, the succession consisting predominantly of banded or flaggy quartzites and semipelites with no boulder beds (Litherland, 1980). Correlation is made mainly on the basis of stratigraphical continuity of distinctive formations both above and below. The exact limits of the subgroup are hence difficult to define here, but it is probable that between 2000 and 4500 m are present. Graded turbidites and 'green beds' of volcanic origin, which are

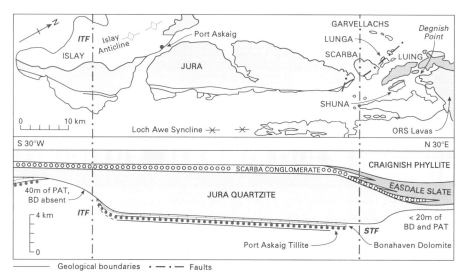

Figure 13   Geological sketch map and section of the Jura area—with along-strike stratigraphical section showing relationships at the end of Craignish Phyllite times (modified from Anderton, 1979; 1985). STF—Scarba Transfer Fault and ITF—Islay Transfer Fault are interpreted growth faults and do not crop out at the surface. Note that the sense of movement on the STF reverses after the deposition of the Jura Quartzite. Hence the thick accumulation of Jura Quartzite lies to the south of the fault, whereas the thick accumulation of Easdale Slate and Craignish Phyllite lies to the north. BD = Bonahaven Dolomite; PAT = Port Askaig Tillite.

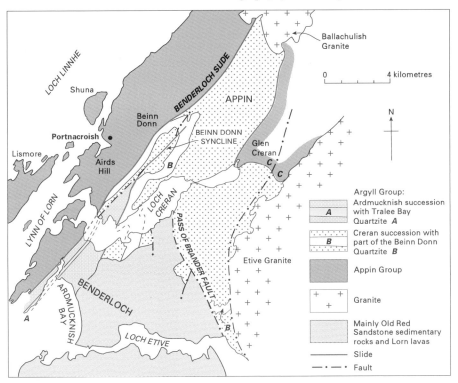

Figure 14   Geological sketch map of Benderloch and south Appin to show the distribution of the equivalent Ardmucknish and Creran successions of the Argyll Group (see Figure 15). For details of the Appin Group see Figure 12. At C–C there is a sedimentary passage between the Appin and Argyll groups. (Simplified from Litherland, 1980.)

recognised in the lower part of the Creran succession, are features which become more common in the subgroup in the Southern and North-east Highlands.

Eastwards from the Etive Granite Complex to the Bridge of Balgie Fault in the Southern Highlands, the Islay Subgroup is not present. Rocks of the overlying Easdale Subgroup rest on a strongly deformed zone representing the Boundary Slide, beneath which are lower Appin Group rocks (Roberts and Treagus, 1979). It is impossible to determine if rocks of the Islay Subgroup were deposited in this area. If there was little or no deposition it is necessary to invoke a fault-bounded structural 'high', separating rapidly subsiding basins on either side, to account for the great thicknesses observed in the adjoining areas. Large parts of the succession have undoubtedly been excised tectonically, but it has also been suggested that contemporaneous erosion and resultant unconformities may account for many gaps in local Islay Subgroup successions (Pantin, 1961; Harris and Pitcher, 1975).

East of the Bridge of Balgie Fault, in the Schiehallion and Pitlochry areas, the Islay Subgroup is well developed, with a succession comparable with that of the type area (Bailey, 1925; Bailey and McCallien, 1937; Pantin, 1961; Harris, 1963; Figure 10). The *Schiehallion Boulder Bed* may be traced throughout most of the area, but is particularly well developed on the northern slopes of Schiehallion itself. The lower part has a matrix of calcareous mica-schists with only carbonate clasts, whereas the upper part is more siliceous with large clasts of granite, syenite and quartzite. The overlying *Schiehallion Quartzite* is typically massive, fine-grained and rarely cross-bedded. In the lower part, conglomeratic bands contain boulders of granite identical to those of the boulder bed; dolomitic beds, consisting of tremolitic limestone and calcareous pelites, are well developed locally.

Farther north-east, towards Braemar, thin developments of pebbly, granitic 'boulder bed' occur locally at the base of the massive *Creag Leacach Quartzite* (Upton, 1986). To the north of the Cairngorm and Glengairn granites the lower part of the Islay Subgroup consists of two interdigitating and diachronous formations. On Donside, semipelites and pelites with thin limestones and dolostones comprise the *Nochty Semipelite and Limestone Formation*. These lithologies pass laterally northwards into striped psammites with graded units, semipelites and minor pelites which together comprise the *Ladder Hills Formation*. This formation is several kilometres thick in its type area, but is absent or only thinly developed elsewhere. Boulder beds, typically underlain by a thin dolostone, occur locally towards the top of the Ladder Hills Formation. The best section of these boulder beds is found in the Muckle Fergie Burn, south of Tomintoul, where a limestone and a dolostone are succeeded by a 10 m-thick boulder bed containing clasts of dolostone in its lower, calcareous part and of granite and quartz-syenite above. Minor basic tuff bands also occur locally and in the Muckle Fergie Burn pillow lavas have been recognised just below the boulder beds. In some areas, for example in upper Donside near Corgarff, boulder beds occur within the lower units of the overlying *Kymah Quartzite* and in the Kymah Burn section thin basic lavas and tuff lenses are found near the base of the quartzite. The quartzite varies considerably in thickness, from only 10 m over fault-controlled structural 'highs' (e.g. Lecht–Cockbridge) to a more typical development of 300 to 500 m in adjacent basins.

The subgroup is probably cut out structurally to the west of Huntly, and to the east of Keith correlations are as yet uncertain. However, north of the River Isla the *Durn Hill Quartzite* is confidently assigned to this subgroup on account of loose blocks of tillite which have been found close to its base in several locations around Fordyce and Edingight (Spencer and Pitcher, 1968).

Easdale Subgroup

The base of the subgroup is marked in most places by rapid change to finer-grained rocks showing features typical of deep-water sedimentation and turbidity currents, with local incursions of coarse-grained mass-flow deposits. Higher parts of the subgroup are dominated by calcareous pelites and semipelites with local limestones, representing shallower-water sedimentation. These general characteristics are preserved throughout the outcrop from Islay to the north coast and individual units can be traced for up to 100 km, but the detailed stratigraphy is less continuous than in preceding subgroups.

In south-east Islay and south Jura the Islay and Jura quartzites are overlain by the *Jura Slate* which constitutes the basal part of the *Scarba Conglomerate Formation* (Figures 10 and 13; Anderton, 1979). Here this formation is about 450 m thick consisting of grey and black slates with thick lenses of quartzite and conglomerate which pass upwards into persistent quartzite and conglomerate units. Sedimentary structures such as graded beds and erosional bases suggest deposition from turbidity currents on a subsiding offshore platform shelf. Farther north, on Scarba and adjoining islands, there is a pronounced increase in thickness and a change of facies. Proximal turbidites similar to those of the southern outcrops are interbedded with thick, coarse debris flows which are considered to have slumped downslope northwards into a fault-bounded basin. The debris flows consist of locally derived blocks, up to 6 m in diameter, consisting of quartzite, limestone and phyllite, all in a gritty matrix. To the north of Scarba the debris flows and associated quartzites become finer grained and thinner as they become interbedded with and pass upwards into the deeper-water *Easdale Slate* (Anderton, 1979; Figure 13). The Easdale Slate consists predominantly of black, carbonaceous, pyritic slates, best developed on Seil and Luing where they have been quarried extensively (Baldwin and Johnson, 1977). Incursions of distal turbidites are seen as thin, poorly graded sandstones and occasional dolomitic carbonate beds (Plate 6).

On the island of Shuna and on Degnish Point, the Easdale Slate passes upwards into the 20 m-thick *Degnish Limestone* (Figure 10), a complex unit composed of interbedded limestones, mudstones and calcareous sandstones which marks a passage into the shallow marine facies of the *Craignish Phyllite* (Knill, 1959; Roberts, 1966; Borradaile, 1973; Anderton, 1975). This phyllite unit, which is correlated with the *Port Ellen Phyllite* of Islay and Jura, consists of abundant alternations of laminated phyllites, quartzites, limestones and pebbly sandstones, with a total thickness of up to 4500 m. Many of the sediments were deposited on tidal flats and in subtidal environments with pseudomorphs after gypsum that indicate periodical emergence. Thick cross-bedded, dark grey limestones, intercalated with dark phyllites at the top of the formation comprise the *Shira Limestone and Slate*, which is a persistent marker unit, up to 300 m thick, in parts of the South-west Highlands.

In the Benderloch area (Figure 14, 15), the top part of the Ardmucknish succession (600 to 1200 m thick) consists predominantly of black slates with a few thin black limestones and pebbly mudstones, the *Selma Slate Formation* (Litherland, 1980). The slates contain several beds of calcareous grit and a sedimentary breccia with a variety of exotic clasts, the *Selma Breccia* (up to 100 m thick). The whole sequence is interpreted as equivalent to the Jura Slate and Scarba Conglomerate to the south-west. At the top of the Ardmucknish succession is a 200 m-thick, graded, pebbly quartzite, the *Culcharan Quartzite*. Correlation with the top part of the Creran succession to the east is difficult. Here black slates and striped siltstones of the *Salachan Formation* (100 m thick) are overlain by the *Beinn Donn Quartzite Formation* (1500 m thick) (Litherland, 1980). South of Loch Creran the quartzites are pebbly and are intercalated with pelites, pebbly mudstones, calcareous grits and thin

Plate 6   Folded black slates and limestones of the Easdale Slate Formation, Argyll Group, Isle of Kerrera, Oban (D 1564).

limestone breccias; Anderton (1985) regards the whole succession as typical of the Easdale Subgroup. A massive, pebbly quartzite with graded bedding and intraformational breccias, within the Beinn Donn Quartzite Formation, is correlated by Litherland (1980) with the Cairn Mairg Quartzite of the Southern Highlands However, the same quartzite can be traced around the Beinn Donn Syncline (Figure 14) where it changes in facies to cross-bedded quartzites and flags, regarded by Litherland (1980; 1987) as equivalent to the Islay and Jura quartzites. Regardless of whether this part of the succession is assigned to the Islay or the Easdale subgroup, it seems likely that the incoming of Easdale-type facies, marked by slide deposits, pebbly quartzites, pebbly limestones and pebbly mudstones, is diachronous, occurring earlier in the east than the west.

On the south-eastern side of the Loch Awe Syncline, only the top part of the South-west Highlands Easdale Subgroup succession is exposed in the core of the Ardrishaig Anticline. Here, the *Ardrishaig Phyllite* is lithologically similar to the Craignish Phyllite and is regarded as stratigraphically equivalent (Roberts, 1966; Borradaile, 1973). Sedimentary structures are not so well preserved as in the Craignish Phyllite owing to a higher metamorphic grade but pseudomorphs after gypsum have been recorded and sulphur isotopes are consistent with a shallow-water, mildly reducing marine environment (Hall et al., 1994). The amount of deformation makes estimates of thickness difficult but over 4000 m of succession are recorded in places. In Knapdale a diachronous, predominantly quartzitic unit, the *Erins Quartzite*, has an apparent thickness of up to 5000 m. The lower part is included in the Easdale

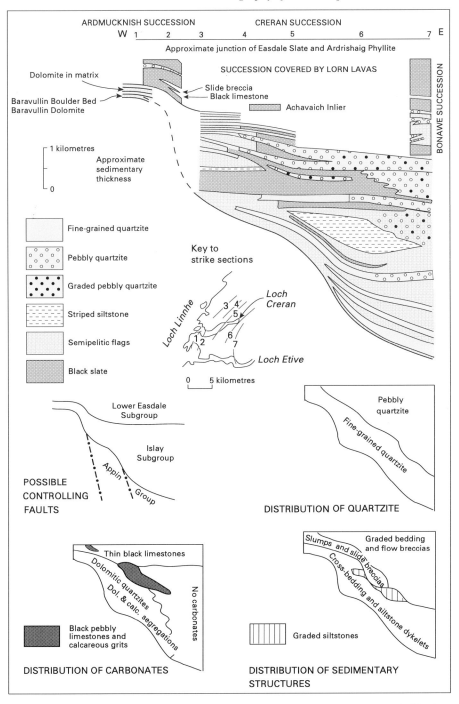

Figure 15   Variation in thickness and lithology over a possible fault-controlled basin margin, lower Argyll Group of Benderloch. (Modified after Litherland, 1980. Several beds are omitted.)

Subgroup, as are two more-pelitic lithologies, the *Stornoway Phyllite* and *Stronachullin Phyllite*, which locally separate the quartzites into lower and upper developments. The Shira Limestone is not present in Knapdale but farther to the north-east it appears on the north-western limb of the Ardrishaig Anticline as a gritty black limestone, more than 100 m thick. The whole subgroup becomes more pelitic towards the north-east around upper Loch Fyne, where the Ardrishaig Phyllite is overlain directly by the *St Catherine's Graphitic Schist* (apparent thickness 200 m), consisting of black, graphitic schists with thinly bedded limestones, and probably equivalent to the Shira Limestone (Roberts, 1966). Minor 'green beds' of tuffaceous or detrital volcanic material occur in the south-eastern parts of the Ardrishaig Phyllite and Erins Quartzite outcrop, and numerous metabasic sills may be near-contemporaneous with sedimentation (Chapter 8).

The Easdale is the earliest subgroup in the Dalradian which can be traced continuously from the South-west Highlands through the Southern Highlands to the Pitlochry area. In the area around Dalmally, Roberts and Treagus (1975) recognise a succession above the Boundary Slide Zone consisting of pebbly quartzites overlain by black slates and Ardrishaig Phyllite. The quartzites can be equated with the Carn Mairg Quartzite of the Southern Highlands and the black slates with both the Easdale Slate to the west and the Ben Eagach Schist to the east, hence providing a crucial link. East of the Tyndrum Fault, published information on the detailed stratigraphy is sparse. By contrast, the presence of extensive stratabound and vein mineralisation within the subgroup between here and Pitlochry has resulted in many publications which include brief descriptions and maps of the general geology (see Chapter 17). More detailed accounts of the stratigraphy have been provided by Nell (1984) and Scott (1987). Farther to the north-east, some stratigraphical details around Killin are given by Johnstone and Smith (1965) and detailed accounts are available for Ben Lawers (Elles, 1926), Lower Glen Lyon/Schiehallion (Bailey and McCallien, 1937; Rast, 1958; Treagus, 1987), south of Loch Tummel (Sturt, 1961) and north of Loch Tummel/Pitlochry (Bailey, 1925; Harris, 1963).

Throughout the areas listed above a common succession may be recognised which, with only local absences, consists of Killiecrankie Schist–Carn Mairg Quartzite–Ben Eagach Schist–Ben Lawers Schist (Figure 10). Contrary to many previous accounts, the base of the subgroup is now drawn below the Killiecrankie Schist, which can be regarded as equivalent to the Easdale Slate and is a comparable deep-water facies. The *Killiecrankie Schist* is widely recognised east of the Loch Tay Fault but west of the fault it is found only in the Schiehallion area. It consists of semipelitic and pelitic garnetiferous mica-schists with abundant intercalated quartzose schists and pebbly quartzites. Minor amounts of graphitic schist are present locally and many concordant bands of amphibolitic metabasalt and thin beds of tuff occur towards the top of the formation. The *Carn Mairg Quartzite* crops out around Loch Lyon, in the Schiehallion area and can be traced eastwards to Ben Vrackie. It ranges from feldspathic, pebbly quartzite, to a psammitic greywacke and typically shows graded bedding. Locally it is absent and may, together with the quartzites in the underlying Killiecrankie Schist, represent sandy turbidites which periodically swept into the fine-grained basin sediments now represented by the Killiecrankie and Ben Eagach schists. The *Ben Eagach Schist* can be traced from north of Tyndrum eastwards to north of Loch Lyon and then continuously from north of Schiehallion and lower Glen Lyon to Ben Vrackie. It consists predominantly of distinctive dark grey to black graphitic schists, with impersistent thin limestones and amphibolites. Most of the stratabound mineralisation of the Argyll Group occurs in this formation, notably the bedded baryte and celsian with sulphides, up to 50 m thick, which extends intermittently for 7 km along strike around Ben Eagach and Farragon Hill, north of Aberfeldy (see Chapter 17). The *Ben Lawers Schist* forms a wide continuous outcrop extending from Tyndrum to Glen Lyon, Schiehallion and Pitlochry to the Glen Shee area. North of Loch Tay it occupies the core of the Ben Lawers Synform. The dominant lithology is a calcareous, pelitic mica-schist with some thin quartzite and limestone beds. Hornblende-schists of metasedimentary origin are common, but chloritic 'green beds' of volcanogenic origin and pods of basic and ultramafic igneous material are also recorded. A persistent zone of stratabound sulphide, mainly pyrite, occurs near the top of the formation. Above the Ben Lawers Schist, the top of the subgroup is marked by a variety of lithologies. Around Tyndrum, the *Ben Challum Quartzite Formation* consists of up to 500 m of quartzites and micaceous semipelites with minor amphibolites of possible volcanogenic origin and two bands of low-grade stratabound sulphide mineralisation (Fortey and Smith, 1986). This passes north-eastwards into the *Sron Bheag Schist* of the Ben Lawers area, consisting mainly of quartzose schists and hornblende-schists with various calcareous schists and limestones (Elles, 1926; Johnstone and Smith, 1965). East of the Loch Tay Fault the *Farragon Beds*, a complex series of quartzites, mica-schists, hornblende-schists and 'green beds' represent the earliest major volcanic episode in the Dalradian (Sturt, 1961; Harris, 1963).

The Killiecrankie Schist and Carn Mairg Quartzite cannot be traced with any confidence north-east of the Pitlochry area. In the Glen Shee–Braemar area, the Creag Leacach Quartzite of the Islay Subgroup passes up through a transition formation of interbedded quartzite and black schist into the graphitic *Glas Maol Schist* (Upton, 1986). The immediately overlying succession is dominated by metabasic intrusions with only small outcrops of limestone and calcareous schist.

North-east of the Lochnagar Granite in the Ballater area, the Easdale subgroup consists mainly of psammites and semipelites with calc-silicate lenses and thin impure limestones in the upper part. A sequence of banded amphibolites at the top of the subgroup represents basic lavas in an equivalent position to the Farragon Beds near Pitlochry. Farther north the more typical Easdale sequence is re-established in a belt extending from south of the River Don to north of Glenbuchat. Here, the Kymah Quartzite is overlain by the *Culchavie Striped Formation*, a thick sequence of striped semipelites and psammites with a distinctive pebbly quartzite. These are succeeded in turn by the *Glenbuchat Graphitic Schist* and by calcareous semipelites and minor psammites with limestone and calc-silicate beds, termed the *Badenyon Calcareous Schist*. Graphitic schists are also present around the headwaters of the River Don, where they are overlain by a prominent metavolcanic unit consisting of amphibolites, vesicular basic pillow lavas and tuffaceous beds.

To the north of these outcrops, structural complexities, major dislocations and facies changes preclude any firm correlation with the established Easdale

Subgroup successions. However, various local successions, mostly bound by faults and major shear-zones, occupy a similar stratigraphical position within the Argyll Group sequence. On the east side of the Portsoy Lineament, in the Cabrach area, a sequence of black pelites, semipelites, psammites, pebbly greywackes and metavolcanic rocks is termed the *Blackwater Formation* (Fettes et al., 1991). The metavolcanic rocks can be divided into three members which are composed of aphyric, pyroxene-phyric and pillowed metabasalts and both massive and autobrecciated ultramafic lavas (MacGregor and Roberts, 1963). The overall lithological assemblage of the Blackwater Formation is similar to that found elsewhere in the Easdale Subgroup but, since the formation appears to pass upwards into Southern Highland Group lithologies, it must be regarded as 'Argyll Group–undivided'. Semipelites and graphitic pelites with minor limestones occur above the Kymah Quartzite in the Drumdelgie area, west of Huntly. On the north coast, a thick sequence of semipelitic, pelitic and graphitic schists with limestone and quartzite in its upper part, termed the *Castle Point Pelite* and *Portsoy Limestone* formations (the 'Portsoy group' of Read, 1923), is widely sheared and attenuated within the Portsoy Lineament.

Crinan Subgroup

Throughout its outcrop, this subgroup exhibits a relatively simple stratigraphy in which local successions are dominated by single thick formations. Sedimentation was generally of deep-water, turbiditic type with marked variations in thickness and diachronous facies changes. The Crinan Grit of the South-west Highlands passes laterally into the generally thinner-bedded and finer-grained Ben Lui Schist of the Southern Highlands. In most areas of the North-east Highlands it has not so far been possible to differentiate the Crinan and Tayvallich subgroups; they crop out in a broad horseshoe of migmatitic and gneissose psammites and semipelites around the Turriff Syncline.

On the south-eastern limb of the Islay Anticline the *Laphroaig Quartz-schist* and *Ardmore Quartzite*, which crop out along the south-eastern coast of Islay, are correlated with the Crinan Grit (Borradaile, 1979). On both limbs of the Loch Awe Syncline, the subgroup is composed of the *Crinan Grit* (Knill, 1959; Roberts, 1966; Borradaile, 1973). The base is probably diachronous and the basal psammites and quartzites are locally interbedded with lithologies indistinguishable from the underlying Craignish Phyllite. Elsewhere an unconformity is implied by pebbly beds near the base of the Crinan Grit and by a basal conglomerate. Around the head of Loch Awe and on the north-western limb of the syncline the subgroup consists of 100 to 550 m of fine- and medium-grained, white quartzites with thin bands of quartzose gritty psammites and interbedded green or grey phyllites. Locally, thin limestones and black slate units are present. A lenticular zone of pale green tuffs occurs near Craignish, and the overlying psammites have a more chloritic matrix, reflecting a volcaniclastic component. On the south-eastern limb of the Loch Awe Syncline the thickness of the Crinan Grit increases to over 3000 m with the incoming of many thick-bedded coarse-grained psammites containing angular fragments of detrital feldspar, mica and carbonate. Pebbly and locally conglomeratic quartzites increase towards the top of the formation and in places become the dominant lithology. Well-developed grading, channelling and large-scale slump folding are common. The interbedded black limestones and slates are typically gritty and contain angular, slumped blocks of limestone. These features are indicative of proximal turbidite deposits which Anderton (1985) considers were deposited in submarine fan channels flowing axially in a major NE–SW-trending basin along the line of what is now the Loch Awe Syncline.

Farther to the south-east, across the Ardrishaig Anticline, the subgroup crops out in a continuous strip along the west coast of Kintyre and through Loch Fyne to the Southern Highlands. In south Knapdale the subgroup is represented by the fine-grained upper part of the *Erins Quartzite*. Grey-green, phyllitic pelites and semipelites occur locally and pebbly quartzites, usually graded, become prevalent towards the top of the formation (Roberts, 1966). A zone of sparse, stratabound pyrite-chalcopyrite mineralisation in the Upper Erins Quartzite in the Meall Mor area suggests a possible correlation with the Ben Challum Quartzite at the top of the Easdale Subgroup in the Tyndrum area. In Kintyre, the Erins Quartzite crops out in a thin strip along the west coast where locally it contains several bands of limestone (McCallien, 1929). Above the Erins Quartzite, the *Stonefield Schist* consists of garnetiferous mica-schists and chlorite-albite-schists with beds of quartzose schist, schistose psammite and lenticular sandy limestones. When traced north-eastwards, across Loch Fyne, this formation has been termed the '*Garnetiferous Mica-Schist*'. North-east of Lochgair the Erins Quartzite lenses out and towards upper Loch Fyne, the Garnetiferous Mica-Schist becomes dominantly pelitic, before passing north-eastwards into the Ben Lui Schist of the Southern Highlands. The Erins Quartzite and the Stonefield Schist/Garnetiferous Mica-Schist/Ben Lui Schist are finer-grained than the Crinan Grit. Many authors have regarded these beds as more distal turbidites, deposited in the same basin as the Crinan Grit (e.g. Harris et al., 1978), but Anderton (1985) suggests that they were deposited in a separate parallel basin which exhibits a general fining from south-west to north-east.

The *Ben Lui Schist* crops out continuously in the Southern Highlands from Ben Lui to Ben Vuirich, notably over wide areas on both limbs of the Ben Lawers Synform and including structural outliers on the lower limb of the Tay Nappe around Lochearnhead. The formation consists mainly of garnetiferous mica-schists and turbiditic, graded schistose psammites, although sparse hornblende-schists and impersistent bands of limestone have been recorded in the Killin area (Johnstone and Smith, 1965) and near Pitlochry (Sturt, 1961). At the base of the formation, between Tyndrum and Glen Lyon, is a zone up to 50 m thick containing lenticular bands of talcose or chloritic soft green schists which are dolomitised locally. These schists contain lenses and pods of dolomite and magnesite with minor amounts of chromium, copper and nickel minerals. They probably represent sediments derived from the erosion of local ultramafic rocks (Fortey and Smith, 1986).

North-eastwards from Ben Vrackie, the metamorphic grade of the Ben Lui Schist increases. Abundant concordant quartzofeldspathic segregations occur together with many thick pegmatite veins, so that the rock takes on a migmatitic appearance. In the Glen Shee area and through to the headwaters of Glen Isla and Glen Clova, the pelitic and semipelitic *Caenlochan Schist* is considered to be equivalent to the Ben Lui Schist on account of the lack of calcareous lithologies. The schists grade into the *lit-par-lit* migmatites of the *Duchray Hill Gneiss* (Williamson, 1935) in which a variety of metasedimentary lithologies can be recognised, including distinctive bands of coarse-grained staurolite- and/or kyanite-bearing rock. The equivalent *Queen's Hill Gneiss* can be traced north-eastwards into Glen Muick and to Queen's Hill, near Aboyne (Read, 1927; 1928). Here the dominant migmatitic, semipelitic to pelitic gneisses are interbedded with psammites, rare quartzites and thin calc-silicate bands comprising the Queen's Hill Gneiss Formation. The formation also includes numerous bands of hornblende-gneiss, implying that it was a preferred horizon for the intrusion of basic sheets. Similar migmatitic and gneissose lithologies occur in lower Deeside in a belt between the Hill of Fare and Mount Battock granites.

To the north of Deeside, and to the east of the Portsoy Lineament, rocks which are generally ascribed to the upper part of the Argyll Group occur in a broad arc from middle Donside to Fraserburgh and in a narrower zone from Huntly to Portsoy. Within these poorly exposed areas thick, mixed sequences of psammites, semipelites and pelites show little mappable variation and no consistent detailed stratigraphy has been established. The metamorphic grade is generally high and most of the rocks are migmatised, so that over considerable areas the rocks are predominantly gneissose. The migmatitic and gneissose textures clearly transgress primary lithological boundaries but, by analogy with the development of the Queen's Hill and Duchray Hill gneisses to the south, it has been traditional to assign all of the gneisses broadly to the Crinan Subgroup (Read, 1955; Harris and Pitcher, 1975). However, some probably belong to the Tayvallich Subgroup and it is possible that minor units of the Easdale Subgroup are included in some areas (e.g. near Portsoy).

In Middle Donside, from the eastern margin of the Movern–Cabrach basic mass to the Tillyfourie area, lithologies are comparable to those of the Queen's Hill Gneiss. These constitute the *Craigievar Formation (Donside Gneiss* of Read, 1955) which consists mainly of finely interlayered, schistose and gneissose psammites and pelites. Major developments of pelitic gneiss, concordant amphibolite and thin developments of limestone and calc-silicate-bearing rock occur locally. Relationships are further complicated by hornfelsing and partial melting close to the major basic intrusions. East and north-east of the Bennachie Granite, equivalent rocks are known as the *Aberdeen Formation* (Munro, 1986). The dominant lithologies are less pelitic than those to the west, consisting mainly of psammites and semipelites and characterised by conspicuous small-scale compositional banding. In the south, around Aberdeen and lower Deeside, the formation shows little variety, with lithological alternations rarely more than a few metres thick. To the north of Aberdeen, major sub-units of relatively more psammitic or pelitic rocks are up to 1 km thick and minor bands of calc-silicate rock and amphibolite are common.

The gneisses of the *Ellon Formation* crop out around the lower Ythan valley. (Read, 1952; Munro, 1986). They are derived mainly from semipelitic and psammitic metasedimentary rocks, although amphibolites are locally dominant. Calc-silicate rocks are rare. The gneisses are distinguished from those of the Aberdeen Formation by their lack of regular lithological banding, their poor fissility and a foliated, streaky appearance. Bodies of migmatitic 'granite' are widespread. The boundary with the Aberdeen Formation is transitional in places but elsewhere it is highly sheared. To the north and east of

Ellon, the Ellon gneisses grade into the structurally overlying *Stuartfield 'division'* of semipelites, pelites, psammites and metagreywackes. The upper part of this division has a more coherent stratigraphy and is termed the *Strichen Formation*. To the north this may be further subdivided into a lower part containing massive channel quartzites up to 500 m thick (e.g. the *Mormond Hill Quartzite*) and an upper part containing calcareous beds; the latter have been taken to indicate that the Strichen Formation spans the boundary between the Crinan and Tayvallich subgroups (Kneller, 1988).

To the north of Peterhead is the *Inzie Head 'group'* (Read and Farquhar, 1956). This mixed assemblage of rocks has a general migmatitic appearance with more homogeneous quartzofeldspathic gneisses alternating with xeno-lithic gneisses. The xenoliths consist of a wide range of metasedimentary lithologies, including calcareous schists, and are locally arranged in trails resembling dismembered sedimentary units. More coherent bands of amphi-bolite, psammite and calcareous schist with impure marble have been mapped locally. The *Cowhythe Psammite Formation* (formerly the *Cowhythe Gneiss*) crops out in a north–south belt from the coast east of Portsoy to the north-east of Huntly (Read, 1923). It is essentially a psammitic to semipelitic schist with rare limestone and pelite bands. Streaky *lit-par-lit* migmatites and feldspathised rocks occur, particularly in the semipelitic units, but for the most part the original compositional banding can still be discerned.

Tayvallich Subgroup

This subgroup is characterised by carbonate sequences; the Tayvallich Limestone and its lateral equivalents, the Loch Tay Limestone and Deeside Limestone, together constitute one of the most persistent marker bands of the Dalradian Supergroup, stretching from Donegal to the Banchory area (Gower, 1973). In Scotland the limestones are mainly turbiditic and were probably deposited in pre-existing Crinan Subgroup basins following a reduction in the supply of clastic sediments to the fringing shelves (Anderton, 1985). In the South-west Highlands an episode of basic volcanic-ity, which commenced in early Tayvallich Subgroup times, resulted in the most extensive development of volcanic and subvolcanic rocks seen in the Dalradian succession (Chapter 8). Previous accounts have included the majority of these volcanic rocks in the overlying Southern Highland Group, following Harris and Pitcher (1975), who correlate them all with the Green Beds. However, since the eruptions started in the Tayvallich Subgroup and since volcanic events generally occupy a short time interval in relative strati-graphical terms it is appropriate to include the majority of the products in the Tayvallich Subgroup (Anderton, 1985).

In the Loch Awe Syncline and the subsidiary Tayvallich Syncline, the *Tayvallich Slate and Limestone Formation* exhibits marked facies changes from north-west to south-east comparable to those in the underlying Crinan Subgroup. The overall thickness varies considerably, mainly due to variations in the amounts of slate and volcanic material. The formation reaches a maximum of 1200 m, but the total thickness of limestones is relatively constant, at about 100 m. On the north-western limb of the Loch Awe Syncline fine-grained, dark, partly oolitic limestones and coarse-grained, graded, gritty limestones are interbedded with dark blue-grey phyllites or slates. These lithologies probably represent shelf sedimentation. Coarse, slumped limestone breccias, conglomerates and gritty psammites, all suggestive of an unstable shelf margin, attain maximum development along the axial zone of the syncline. On the south-eastern limb thinner lenses of limestone-breccias and conglomerates are interbedded with turbiditic psammites (Knill, 1963). Around the north-eastern closure of the Loch Awe Syncline, the Tayvallich Limestones are succeeded by the *Kilchrenan Grit* and the *Kilchrenan Conglomerate* (Borradaile, 1973). The former is a poorly graded, feldspathic psammite with mudstone flakes, whereas the latter is a matrix-supported 'boulder bed', up to 30 m thick, consisting of well-rounded quartzite boulders in a gritty black slate matrix. It has been interpreted as a slump conglomerate (Kilburn et al., 1965).

Volcanic rocks occupy much of the core of the Loch Awe syncline between the Pass of Brander in the north-east and Loch Crinan in the south-west. Smaller outcrops occur farther to the south-west, notably in the core of the Tayvallich Syncline. The *Tayvallich Volcanic Formation* has a very sharp base in its more northern outcrops, just above the Kilchrenan Conglomerate, but in the south the lower parts consist of a complex interdigitation of limestones, clastic sediments and volcanic rocks. The volcanic formation consists of up to 2000 m of commonly pillowed basic lavas (Plate 7), hyaloclastites and a variety of epiclastic volcanic rocks (Borradaile, 1973; Graham, 1976). The epiclastic rocks include breccias and waterlain pebbly deposits such as the *Loch na Cille 'Boulder Bed'* of the Tayvallich Peninsula (Gower, 1977). Extrusion of the main volcanic pile was clearly submarine but away from the main centre of volcanicity, which corresponds to the axis of the Loch Awe Syncline, air-fall tuffs have been recognised. A suite of metabasic sills with a similar geographical distribution to the volcanic rocks intrudes the whole succession from the Craignish and Ardrishaig phyllites upwards; their total thickness attains 3000 m in places. These were originally thought to be contemporaneous with the lavas (Borradaile, 1973; Graham, 1976), but they may well be near-contemporaneous with their host sediments, building upwards with time as shallow intrusions into soft sediments (Wilson and Leake, 1972; Graham and Borradaile, 1984; Graham, 1986). A suite of NW-trending metabasalt dykes on Jura may represent feeders for the lavas and sills (Graham and Borradaile, 1984).

To the south-east of the Ardrishaig Anticline the Tayvallich Subgroup is represented by the *Loch Tay Limestone* which maintains a thickness of around 100 m along strike from Campbeltown to Glen Shee (Bailey, 1925; Elles, 1926; Johnstone and Smith, 1965; McCallien, 1929; Roberts, 1966). Thinner developments are also present to the south-east of the main outcrop, at the base of structural outliers of Ben Lui Schist in the inverted limb of the Tay Nappe around Lochearnhead (Johnstone and Smith, 1965; Watkins, 1984; Mendum and Fettes, 1985). Throughout its strike length the formation consists of thick beds of crystalline limestone, locally with various calcareous schists and semipelitic mica-schists. Thin, black graphitic schists and metagreywackes are present locally and grading in all lithologies suggests a distal turbidite origin. Hornblende-schists and amphibolites, representing mainly intrusive basic igneous rocks, are present throughout the outcrop, although there is no continuous volcanic sequence comparable with the Tayvallich Volcanic Formation.

To the north-east of Glen Isla the limestone can be traced as a 10 m-thick unit in Glen Doll, but it is absent east of the Glen Doll Fault. Here, calcareous

Plate 7   Basaltic pillow lavas, Tayvallich Volcanic Formation, Argyll Group, coast south-west of Tayvallich, Argyll (D 2274).

semipelites around the head of Glen Mark pass NNE into ribbed calc-silicate rocks and limestones of the *Water of Tanar Limestone*. In middle Deeside, this becomes the *Deeside Limestone* (Read 1927; 1928) which consists mainly of greenish calc-silicate rocks with calcareous psammite, hornblende-schist and thin layers of impure limestone. A discontinuous band of crystalline limestone, containing abundant pyrite and up to 15 m thick, has been quarried extensively. These calcareous rocks lie at the base of the Tarfside 'group' of Harte (1979), in which they are overlain by a diverse but dominantly psammitic unit, the *Tarfside Psammite Formation*, consisting of quartzites, psammites, semipelites and pelites with locally abundant calc-silicate and amphibolite bands. Parts of the unit are gneissose and between Glen Clova and Glen Lee pelitic and semipelitic gneisses predominate.

To the north of Deeside most of the gneissose units already described probably include Tayvallich Subgroup rocks, particularly in those areas where

calc-silicate and limestone bands are common. Notable examples are the calcareous part of the *Strichen Formation* at the top of the Stuartfield division and
its lateral equivalent, the *Kinnairds Head 'group'* which is well exposed on the
north coast at Fraserburgh (Read and Farquhar, 1956). In these units finely
banded calc-silicate beds occur in a sequence of pelites, semipelites and
psammites. The Strichen Formation also contains limestone bands up to 20 m
thick. Grading in the coarser-grained lithologies, which include metagreywacke, suggests a turbiditic origin and the units are transitional upwards
into the more persistent turbidites of the Southern Highland Group.

On the Banffshire coast the Tayvallich Subgroup comprises a 1200 m-thick
sequence, termed the *Boyne Limestone Formation*, which includes the *Boyne Castle
Limestone*, a thickly bedded but finely banded limestone, some 200 m thick
(Plate 8). Beneath the main limestone are subordinate purple phyllites, mica-
schists and several thin bands of white limestone (Read, 1923; Sutton and

Plate 8   Minor folding in the metamorphosed Boyne Limestone Formation, Argyll
Group, Boyne Bay, Banffshire (D 1905).

Watson, 1955). Inland the limestone can be traced for only 2.5 km. The sequence above the Boyne Castle Limestone consists of semipelites, laminated and striped calc-silicate rocks, minor thin limestones and lenses of reworked carbonate material. These rocks were formerly described as the lower part of the Whitehills 'group' (Read, 1923; Harris and Pitcher, 1975). Concordant sheets and lenses of amphibolite probably represent metabasaltic sills.

SOUTHERN HIGHLAND GROUP

The group is made up mainly of turbiditic rocks, typically coarse-grained, poorly sorted metagreywackes with subordinate fine-grained slates and phyllites. Albite-schists and minor intercalations of pelite with limestone occur locally. Volcaniclastic 'green beds' are widespread at several levels, notably in the basal part of the group, and there is one thick local development of basic lavas. In general character the turbidites are similar to those of the preceding Argyll Group. Sedimentary structures are common and environments such as fan channels, overlapping fan lobes, overbank deposits and basin plains have been inferred (Harris et al., 1978; Anderton, 1985). The sedimentary rocks are markedly more chloritic than those of the underlying Argyll Group, partly due to the generally lower metamorphic grade, but probably also reflecting the volcanic input. They are also more feldspathic, with high-grade metamorphic and granitic rock fragments, suggesting a less-mature source area.

Although several local successions have been established, detailed correlations are seldom possible over any distance. The general persistence of turbidite facies, both laterally and vertically throughout the group, means that there are few reliable stratigraphical marker bands. Green beds are useful locally, but the slate facies is probably highly diachronous. Correlations are further complicated by across-strike changes in metamorphic grade which significantly alter the appearance of the rocks and have led to different names for units which are probably equivalent stratigraphically. Green beds and slate/phyllite units are typically concentrated in the lower part of the group and coarse-grained turbidites predominate in the upper part. However, it is not at present practical to divide the group, other than locally, and consequently no subgroups are recognised. Indeed, in terms of thickness and uniformity of facies the whole group is comparable to a subgroup in other parts of the Dalradian succession.

The Southern Highland Group can be traced from County Mayo in western Ireland to the north-east Grampians. Although the group is probably no more than 7000 m thick in Scotland, much of its outcrop occurs in areas of gentle regional dip such as the Flat Belt of the Tay Nappe (see Chapter 6). Consequently its outcrop covers a wide area of some 4900 km^2. A small outlier occurs in the core of the Loch Awe Syncline, but the main outcrop is a belt, up to 34 km wide, extending along the whole south-eastern edge of the Grampian Highlands from the Mull of Kintyre to Stonehaven and Aberdeen. In the north-east the group occupies the broad core of the Turriff Syncline and a small outlier on the east coast around Collieston.

In the core of the Loch Awe Syncline, the Tayvallich Volcanic Formation is overlain by up to 1100 m of chloritic graded gritty psammites, green slates and subordinate black slates, calcareous in parts, which together comprise the *Loch Avich Grit*. The succeeding *Loch Avich Lavas* consist of 300 to 500 m of basaltic pillow lavas with no significant sedimentary intercalations (Borradaile, 1973).

In the part of the main outcrop of the Southern Highland Group to the south-west of Loch Lomond, the Loch Tay Limestone is succeeded to the south-east by a typical, predominantly metagreywacke sequence (McCallien, 1929; Roberts, 1966). The basal part of this sequence is known locally as the *Glen Sluan Schist* and consists of up to 500 m of schistose pelitic turbidites, lithologically similar to the main part of the succession but separated from it by the *Green Beds*. The latter form a well-defined stratigraphical unit up to 1000 m thick in the South-west Highlands. Metagreywackes, siliceous psammites and fine-grained quartzites are interbedded with the predominant well-foliated green schists containing abundant chlorite, epidote, biotite and albite porphyroblasts. These lithologies become hornblende-schists at higher metamorphic grades. Within the green schists, lenses of obvious detrital material, pebbles of quartz and graded bedding indicate a sedimentary origin. The beds probably represent an influx of detrital basic volcanic material, a feature reflected in their chemistry and mineralogy (Phillips, 1930; van de Kamp, 1970). Metabasic sheets within the unit represent shallow intrusions and the whole assemblage may be contemporaneous with the Loch Avich Lavas. The Green Beds are best regarded as a persistent volcaniclastic development in the lower part of the main sequence of turbidites which are known in this area as the *Beinn Bheula Schist*. This unit consists predominantly of fine-grained metagreywackes with lenticular developments of schistose psammites, commonly pebbly. Green slates and siltstones, and albite-schists occur locally, but black slates are rare.

Much of the Southern Highland Group outcrop lies on the inverted limb of the Tay Nappe which, in the South-west Highlands, is folded into the broad late Cowal Antiform. Consequently, the lower parts of the group are again exposed on the south-eastern limb of this fold. On the Cowal Peninsula the Green Beds are represented mainly by metabasic igneous bodies, but they are not present elsewhere on this limb. The Beinn Bheula Schist passes stratigraphically downwards into the *Dunoon Phyllite* (equivalent to the Glen Sluan Schist) which extends from Bute to Loch Lomond. This unit consists of dark bluish grey and greenish grey slates and phyllites with subsidiary graded metagreywacke units. Thin limestone beds and lenses are locally associated with the black slates. Near Luss, the slates have been quarried for roofing tiles. On the eastern side of Kintyre, the lower parts of the group are represented by the *Skipness Schist*. This comprises siliceous schists, many gritty and conglomeratic, with only subordinate bands of slate and phyllite so that they are difficult to distinguish from the Beinn Bheula Schist.

The Dunoon Phyllite lies within the complex hinge zone of the Tay Nappe which takes the form of a downward-facing (i.e. synformal) anticline known as the Aberfoyle Anticline farther to the north-east. Consequently the rocks which crop out to the south-east, between the phyllites and the Highland Boundary Fault Zone, are regarded as younger than the phyllites and broadly equivalent to the Beinn Bheula Schist (Anderson, 1947; Roberts, 1966; Paterson et al., 1990). Marked decreases in structural complexity, development of cleavage and grade of metamorphism occur towards the Highland

Border. It is thus difficult to correlate lithologies across the regional strike. The *Bullrock Greywacke* succeeds the Dunoon Phyllite to the south-east and hence must be broadly equivalent to the Beinn Bheula Schist. It consists mainly of pink-weathering pebbly and gritty metagreywackes with greenish grey siltstone interbeds. Black slates and dark grey limestones occur locally. In Cowal and Bute, coarse pebbly psammites with conglomerates and more persistent limestones and black slates have been placed in a separate lithostratigraphical unit, the *Inellan 'group'*, broadly equivalent to the Bullrock Greywacke. In north Arran (Richey, 1961) a similar synformal anticline structure exists, with the *Loch Ranza Slate* occupying the core of the anticline and the *North Sannox Grit* on each limb.

In the Southern Highlands to the north-east of Loch Lomond, the Southern Highland Group consists predominantly of rocks of turbiditic greywacke facies which, throughout most of this area, are referred to as the *Ben Ledi Grit* (Mendum and Fettes, 1985). Green beds and metabasic intrusions are widespread at various horizons in the lower part of the succession. Detrital sodic feldspar is common in the psammites and a redistribution of sodium during metamorphism has given rise to many porphyroblastic albite-schists in the pelitic units (Bowes and Convery, 1966; Watkins, 1983). North-eastwards from the Killin area the basal part of the Southern Highland Group is dominantly pelitic and is termed the *Pitlochry Schist*. Garnetiferous mica-schists are the characteristic lithology, but quartzose psammites and graded metagreywackes are common (Johnstone and Smith, 1965). The Pitlochry Schist grades stratigraphically upwards and laterally into the Ben Ledi Grit.

Correlations of grit and slate/phyllite units of the upper part of the Southern Highland Group in the 'steep belt', which forms the south-eastern part of the outcrop near the Highland Boundary Fault, are more complex than previously supposed (Harris, 1962; 1972; Harris and Fettes, 1972). Boundaries between grit and slate units are demonstrably diachronous. Hence, although the slates generally occur in the cores of downward-facing anticlines, neither the local slate units nor the fold axes are likely to be continuous along the whole length of the Highland Border. In both the Aberfoyle and Callander areas, fold hinges are occupied by dark bluish grey to grey-green and purple slates with subsidiary laminated siltstones, silty greywackes and bands of coarse gritty greywackes. In both areas these slate sequences are referred to as the *Aberfoyle Slate*. On the south-east limb of the major synform the *Leny Grit* is equated with the Ben Ledi Grit on the north-west limb. Adjoining the Leny Grit on its south-eastern margin is a unit of blue-grey slates and phyllites which are commonly calcareous and contain thin limestones (Harris, 1962; Francis et al., 1970). These have been referred to as the *Leny Limestone and Shales*. The fossiliferous limestone of the nearby Leny Quarry is probably part of this unit (Harris, 1969) but outcrops of both limestone units may be fault-bound and their stratigraphical affinity is uncertain (see Chapter 9). Around Dunkeld the hinge of the Tay Nappe is truncated by the Highland Boundary Fault and consequently there is no equivalent of the Aberfoyle Anticline. The whole succession in this part of the steep belt consists of *Birnam Grit* (oldest), *Birnam Slate* and *Dunkeld Grit* (youngest) and youngs to the north-west.

North-east from the Dunkeld area the general stratigraphical relationship of a lower, more pelitic metagreywacke sequence to the north-west passing into an upper, more psammitic metagreywacke sequence to the south-east persists to the east coast between Aberdeen and Stonehaven (Anderson, 1942). The Pelitic 'group' of Anderson forms an outcrop continuous with the Pitlochry Schist of the Glen Shee area and contains similar lithologies, dominated by garnet-mica-schists but passing into slates or sillimanite-gneisses depending upon metamorphic grade. In the Glen Clova area recent mapping has recognised a basal unit termed the *Longshank Gneiss*. This unit shows a marked facies change from dominantly pelitic in the south-west to dominantly psammitic in the north-east and is characterised by strongly magnetic beds throughout (Chinner, 1960). Green beds are abundant between Kirkmichael and Glen Clova but are notably absent north-east from Glen Clova. The younger graded metagreywackes and schistose psammites are continuous with the Ben Ledi Grit to the south-west and this name has been retained throughout the area by Anderson (1942).

Around upper Glen Esk, an extensive right-way-up sequence was assigned by Harte (1979) to the Tarfside Nappe, a major recumbent structure below the Tay Nappe. Alternative interpretations involving later large-scale tight folding of the Tay Nappe on gently dipping axial planes can also explain the distribution of units. Within the Tarfside Nappe the sequence consists of an intimate association of pelites, semipelites and psammites of typical Southern Highland Group metagreywacke type. A lower unit, the *Glen Effock Schist*, passes upwards into the higher *Glen Lethnot Grit*, characterised by bands of pebbly psammite, and hence a broad correlation is suggested with the Pitlochry Schist and Ben Ledi Grit of the Tay Nappe (Harte, 1979). Munro (1986) has suggested that part of the Aberdeen Formation to the south of Aberdeen may be equivalent to the Glen Effock Schist.

To the north of Aberdeen, low-grade turbiditic rocks occur in an eastward-younging sequence on the coast between Collieston and Cruden Bay and for a few kilometres inland (Read and Farquhar, 1956; Munro, 1986). These rocks are assigned to the *Collieston Formation* which is predominantly a psammitic sequence of graded metagreywackes with characteristic 'knotted' pelites containing andalusite and cordierite (Plate 9). Thin lenses of calc-silicate rock are common and minor thin limestones occur south of Collieston. Metadolerites and amphibolites, which represent intrusive sheets, also occur locally. Contacts with adjoining units are not exposed, but the deformational pattern has much in common with the older, higher-grade Stuartfield division and Ellon Formation to the west.

In the Turriff Syncline a sedimentological transition from the Argyll Group into the Southern Highland Group is well seen. On its western limb, the base of the Southern Highland Group is drawn at the base of the first massive meta-greywacke above the calcareous beds of the Boyne Limestone Formation, but below the horizon in the middle of the former Whitehills 'group' chosen by Harris and Pitcher (1975). The succession consists of some 2000 m of siliceous psammites with subordinate pelites (Read, 1923; Sutton and Watson, 1955), now renamed the *Whitehills Grit Formation*. On the eastern limb a similiar tran-sition is observed from the calcareous succession noted in the Kinnairds Head 'group' and the Strichen Formation into the non-calcareous metagreywackes and pelites of the *Rosehearty 'group'* and *Methlick Formation* (Read and Farquhar, 1956). In the core of the syncline the Southern Highland Group is repre-sented by the *Macduff Slate Formation* (1700 m), a finer-grained, more distal turbidite facies with slump deposits, clean channel sandstones and subsidiary greywackes (Sutton and Watson, 1955). A more persistent semipelitic facies to the south-west has been termed the *Clashindarroch Formation*, which has been quarried extensively in the past for roofing slate in an east–west-trending belt to the north of the Insch and Bogancloch intrusions. Boulders and pebbles of igneous and metamorphic rocks, some of extrabasinal origin, occur in some of the highest exposed parts of the Macduff Slate Formation immediately to the east of Macduff. These were interpreted as a product of ice rafting by Sutton and Watson (1954). Hambrey and Waddams (1981) have confirmed this interpretation and have described a 20 m-thick sequence of tillites culminating in a possible lodgement till deposited from a grounded ice sheet. However, the evidence is not conclusive and other workers prefer a less contentious mudflow hypothesis.

Considerable changes in metamorphic grade occur across the Turriff Syncline and several lithological units were originally defined using criteria which have subsequently been recognised as metamorphic features. Thus the *Boyndie Bay 'group'* on the western limb of the syncline, which is characterised by andalusite schists, is now regarded as equivalent to the lower part of the lower grade Macduff Slate Formation. Similarly, to the south-east, the Macduff Slate and Rosehearty 'group' grade downward into the *Fyvie Schist*, which is characterised by andalusite and cordierite porphyroblasts giving a 'knotted' appearance; it is now placed in the Methlick Formation. The closure of the syncline can be traced to the south of the Insch basic mass, in the Correen Hills. Here, the Southern Highland Group is represented by the *Suie Hill Formation* which consists dominantly of semipelites and impure gritty psammites with prominent pelite units. The base is taken at a distinctive, magnetite-bearing pelitic schist which forms a regional magnetic anomaly. Similar magnetic units occur on the western limb of the syncline and elsewhere in the basal part of the group and may indicate an influx of detrital magnetite due to a change in provenance or the unroofing of a new source.

DALRADIAN SEDIMENTATION

Recent reviews of the Dalradian Supergroup conclude that the preserved outcrop, extending across the Grampian Highlands to the west coast of Ireland, represents a segment of a former very extensive continental shelf sequence (Anderton, 1985; Winchester, 1988). Sedimentation is thought to have commenced in a broad ensialic rift which opened north-eastwards to form a marine gulf. The Grampian Group was laid down on a relatively passive continental shelf on the north-west side of the rift (Winchester and Glover, 1988). Continued extension widened and deepened the rift and the sediments of the Appin and Argyll groups represent continuing deposition on the north-western shelf margin (Figure 16). Actual rupturing of the continental crust within the widening rift is indicated by the appearance of large volumes of volcanic material towards the top of the Argyll Group, with eruption continuing during deposition of the overlying Southern Highland Group. Faulting, related to the progressive lithospheric stretching across the rift, has been recognised as a principal control on Dalradian sedimentation.

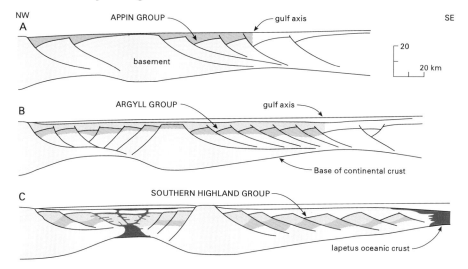

Figure 16   Schematic evolution of the Dalradian terrane in the South-west Highlands (from Anderton, 1985). **A**. late Appin Group times, **B**. Crinan Subgroup (Argyll Group) times and **C**. Southern Highland Group times. Basaltic intrusions and volcanic rocks shown (in red) in **C** only.

Anderton (1985) envisaged the area of deposition to be divided into a series of developing, SE-dipping fault blocks bounded by 'scoop-shaped' listric faults which delimited individual sedimentary basins on the continental shelf (Figure 17). These early fault-systems thus had a direct influence on both the composition and shape of the sediment pile, and ultimately affected the morphology of the regional folds. The faults also acted as controls for submarine volcanic activity, and perhaps also for the subsequent emplacement of late igneous intrusions.

Grampian Group

The Grantown Formation, at the base of the Grampian Group, is regarded as a sequence of shallow-marine shelf sediments which were deposited in local depressions on the continental shelf. Accelerated, uneven subsidence along the shelf was accompanied by an increase in the supply of immature sediment leading to the deposition of fining-upward turbidite sequences which thicken to the east and south-east where they dominate the Grampian Group succession.

The lithological variation in the upper part of the Grampian Group is attributed to regional shallowing with intertidal and estuarine sedimentation (Glover and Winchester, 1989). The deposited sediments became more mature but are chemically distinct from those of the overlying Appin Group. A marine regression of the shelf locally exposed the Grampian Group sediments prior to renewed subsidence at the start of the Appin Group. Throughout the sedimentation of the Grampian Group there was a constant supply of detritus from a 'granitic' source in the northern land mass (Hickman, 1975; Glover and Winchester, 1989).

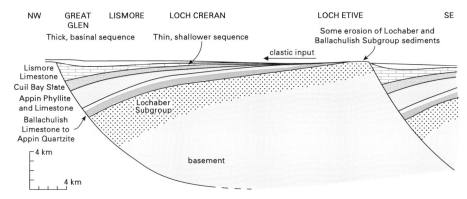

Figure 17 Deposition of the Appin Group in a 'trapdoor' basin defined by listric faults between the Great Glen and Loch Etive. Situation during deposition of the Lismore Limestone (after Anderton, 1985).

Appin Group and lower Argyll Group

Wright (1988) concluded that during Appin Group times sedimentation occurred on a tidal shelf overlying a gently subsiding crust. As in the preceding Grampian Group the supply of sediment was from a north-western landmass, although palaeocurrent indicators show that the sediment was distributed along the shelf by tidal longshore currents. According to Anderton (1985) conditions fluctuated between an open-marine oxidising environment and stagnant euxinic lagoons to produce quartzite/black pyritic shale sequences (Figure 18). In south-western parts of the Lochaber Subgroup outcrop, deltaic deposits occur over a distance of about 40 km with facies varying from proximal at Appin to distal at Glen Roy (Hickman, 1975). The various quartzite formations (the Eilde, Binnein and Glencoe quartzites) are interpreted as tidal sand bodies (Wright, 1988).

The blue limestones and black shales of the overlying Ballachulish Subgroup extend from Banff to Donegal and indicate that deposition took place on an extensive, shallow, probably lagoonal shelf with considerable along-strike continuity of facies (Wright, 1988). Nevertheless the shelf must also have been relatively narrow to explain the rapid down-dip facies changes seen, for example, between Lismore and Glen Creran (Figure 12) (Litherland, 1980). The Ballachulish Slate is interpreted as a prodelta clay deposit and encroachment of fine quartz sands from the delta into deeper water produced the overlying quartzites such as the Appin Quartzite (Wright, 1988). At intervals during Blair Atholl Subgroup times sediment supply exceeded shelf subsidence so that local nondeposition due to emergence is marked by incomplete stratigraphical sequences.

A dramatic climatic change took place at the beginning of Argyll Group times to produce the widely distributed Port Askaig Tillite Formation (and contemporaneous glacial deposits recognised around the world). The tillites, at the base of the Islay Subgroup, were probably deposited on a shallow-marine shelf by successive pulses of grounded ice, possibly advancing from the south-east. The tillites are thickest on Islay but show similar features, where present, throughout the Dalradian outcrop. Intrabasinal sedimentary rocks were first eroded by the glaciers and then covered by marine tills in which extrabasinal granitoid debris becomes increasingly common upwards. Evidence for 17 separate glacial advances is recognised from interbedded marine sediments within the tillite formation (Spencer, 1971). Isolated drop-stones in varved siltstones infer local flotation of the ice sheet but large-scale deposition from ice rafts and reworking by downslope mass-flow, as suggested by Eyles and Eyles (1983), is rejected by Spencer (1971). Periods of emergence and periglacial weathering between the glacial cycles are suggested by polygonal sandstone wedges, interpreted as ice wedges (Eyles and Clark, 1985). Beach conglomerates then herald the start of a marine transgression followed by the next glacial advance. Dolomitic limestones with

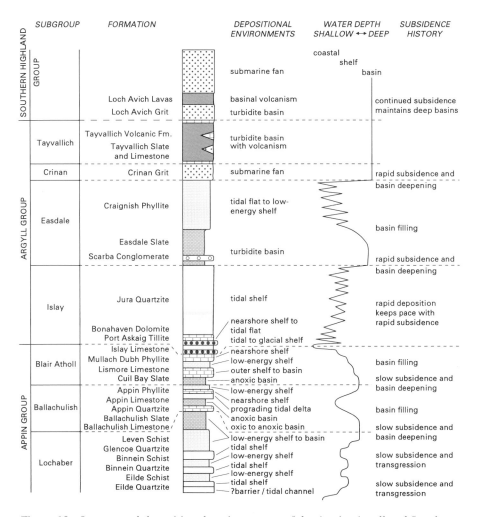

Figure 18   Interpreted depositional environments of the Appin, Argyll and Southern Highland groups (after Anderton, 1985). Approximate vertical scale 5 mm=1 km.

stromatolites, suggestive of warm water, were deposited during some of the interglacial periods and palaeomagnetic results indicate that the Port Askaig Tillite Formation was deposited in low latitudes (Tarling, 1974). This evidence would seem to be in conflict with having an ice sheet at sea level, but as yet there has been no satisfactory explanation for this paradox.

Shelf sedimentation resumed after the final retreat of the ice, resulting in the deposition of the Islay, Jura and Schiehallion quartzites. The closest shoreline remained north-west of Islay and Jura with at least 100 km of open sea to the south-east. The local absence of the Schiehallion Quartzite from the shelf sequence may be due to contemporaneous erosion rather than non-deposition (Pantin, 1961). A change to a tidal, shallow-water environment at the top of the Islay Subgroup was probably caused by a combination of source area uplift and a tectonically induced marine transgression.

Upper Argyll Group

An initial shelf-deepening event at the start of the Easdale Subgroup resulted in fault-controlled, steep shelf-slope sedimentation. The base of the subgroup is generally marked by a rapid change to fine-grained sediments showing features of deep-water sedimentation and turbidity currents, with local incursions of very coarse-grained, mass-flow deposits (as seen in the Scarba Conglomerate Formation). The sediments were probably deposited in a series of fault-controlled marginal basins having a general NE–SW trend (Anderton, 1985; 1988). The interbasinal highs are marked by local thinning, facies changes, erosion or non-deposition; they commonly coincide with long-lasting major lineaments trending north-east or north-west (Fettes et al., 1986; Graham, 1987). Syngenetic barium and base metal mineralisation is attributed to ponding of exhalative brines in local basins adjacent to active faults, which in turn controlled sea-water infiltration into buried sediments (Coats et al., 1984). A variety of lower Easdale Subgroup facies are recognised along the strike length with, for example, volcanic rocks and turbiditic quartzites exposed in the Perthshire area.

Shallow-water shelf and tidal-flat sedimentation returned during upper Easdale Subgroup times due to infilling of the local basins with fine-grained sediments to produce, for example, the Craignish and Ardrishaig phyllites in the South-west Highlands. Minor volcanism is recorded by various tuff layers such as the Farragon Beds.

A further rapid shelf-deepening and basin-forming event at the start of the Crinan Subgroup induced turbidite deposition, with soft-sediment structures preserved in the Crinan Grit, attributed to contemporaneous earthquakes. Major faulting caused slope failure and reworking of buried but poorly lithified sediments.

The sediments of the shelf became more distal in character during the interval between the Islay and Crinan subgroups; the deltaic or shallow-marine sands, which probably supplied the clastic sediments involved in the turbidite flows, being more distant from the axis of deposition (Harris and Pitcher, 1975). The marked thickening of the Crinan Grit from 600 m to over 3000 m along a line running along Loch Awe (the Loch Awe axis) is parallel to, but displaced several kilometres south-east from, the underlying major thickening seen in the Islay Subgroup (Figure 15). As the sediments in the

Crinan Subgroup fine overall from south-west to north-east across Scotland, a major input from the south-west seems likely.

The change to carbonate sedimentation at the start of the Tayvallich Subgroup to form the major limestone sequences such as the Loch Tay Limestone, implies that either the shelf became starved of clastic detritus or that off-shelf carbonate supply became dominant. Facies and thickness changes along and across-strike have been used to identify volcanic centres, such as in the Tayvallich–Loch Awe area, as well as major morphological features, such as submarine canyons.

Southern Highland Group

Preserved sedimentary structures within the dominantly psammitic turbidites of the Southern Highland Group are consistent with deposition within major submarine fans on a subsiding continental shelf (Figure 16). The turbidites are relatively enriched in feldspar and high-grade metamorphic and granitic rock fragments when compared to earlier Dalradian turbidites, features that were originally taken to indicate a newly emerged, immature source area to the south-east, the north-western continent being by this time too distant beyond a wide shelf (Harris and Pitcher, 1975; Harris et al., 1978). However, palaeocurrent directions and facies changes in the sequence all suggest a dominant flow towards the south-east with minor north-eastward and south-westward axial flows. Isotopic studies do not reveal any significant differences between the Southern Highland Group turbidites and the lower units in the Dalradian (O'Nions et al., 1983). Hence, the composition of the Southern Highland Group turbidites may reflect an erosional evolution of the north-western source area caused by stripping of a cover sequence to reveal a Lewisian-type granitoid basement (Plant et al., 1984; Anderton, 1985). The Loch Avich Lavas represent a final phase of lithospheric stretching in the Dalradian Basin; both the Tayvallich and Loch Avich lavas are thickest in the Loch Awe area.

AGE OF THE DALRADIAN SUPERGROUP

The age of the sediments which became the Dalradian Supergroup is poorly constrained. Fossils are rare, are poorly preserved and, where species have been identified, they invariably have a wide stratigraphical range. As yet few radiometric age determinations have been made on penecontemporaneous igneous rocks and most available dates relate to later tectonothermal events. Current estimates of age are based mainly upon attempts to fit this imprecise data into a framework of stratigraphical events established elsewhere in the Caledonides.

Palaeontology

The palaeontology of the Dalradian has been reviewed by Downie et al. (1971) and Downie (1975).

In the Appin Group, worm burrows have been recorded from quartzites (Peach and Horne, 1930) and algal stromatolites have been recorded from the Lismore and Islay limestones (Spencer and Spencer, 1972). The Islay Limestone has also yielded oncoliths. None are of biostratigraphical value.

In the Argyll Group, the Bonahaven Dolomite has yielded acritarchs and algal stromatolites (Hackman and Knill, 1962; Spencer and Spencer, 1972), including some which suggest a late Precambrian (Vendian, or Upper Riphean to Vendian) age (Downie, 1975). The Islay Quartzite contains worm burrows which indicate an age no older than Vendian, and the Easdale Slate contains long-ranging, Vendian to Cambrian acritarchs (Downie, 1975). Limestone clasts in the Selma Breccia Formation contain oncoliths, catagraphs and other calcareous fossils which have been assigned to the Vendian (Litherland, 1975). However, calcareous and burrowing algae which resemble Lower Cambrian and younger forms are also present (Downie, 1975). Acritarchs from the Tayvallich Limestone, which were originally thought to be Lower Cambrian (Downie et al., 1971), are now known to have a longer range, extending back into the Vendian.

In the Southern Highland Group the search for fossils has concentrated upon the weakly deformed, low-grade slates and limestones of the Highland Border area and Turriff Syncline. The Macduff Slates have yielded rare burrows, a few acritarchs and more widespread microfossils which resemble highly altered chitinozoa. Downie et al. (1971) tentatively identified one of the acritarchs as being of post-Cambrian age and possible chitinozoa which would suggest an early Ordovician age, probably Llanvirn or Llandeilo. These identifications remain highly controversial and have not gained general acceptance. An early record of Silurian graptolites is now attributed to a sample labelling error. Possibly one of the most striking features of these generally low-grade rocks is their lack of macrofossils and trace fossils which has led several workers to suggest that they must all be of Precambrian age. The 'glacial' dropstones and pebble beds at the top of the Macduff Slate Formation remain a problem since the next major global glacial period recorded after the basal Vendian (cf. the basal Argyll Group) was in late Ordovician times (Harland, 1972).

The fossiliferous limestone of Leny Quarry near Callander, which contains Lower Cambrian trilobites, was for many years regarded as the most reliable palaeontological indicator of the age of the youngest Dalradian strata in Scotland. This outcrop is now regarded as part of the Highland Border Complex (e.g. Curry et al., 1984), as are rocks reported as containing Middle Cambrian fossils in Clew Bay, Co. Mayo (Rushton and Philips, 1973; Harper et al., 1989). The structural relationships of these Cambrian rocks with the adjacent Dalradian and younger Highland Border Complex rocks, and hence their stratigraphical affinities, are still a matter of debate (see Chapter 9).

Radiometric dating *

The oldest generally accepted radiometric dates from rocks of the Grampian Highlands are from pegmatite veins within the Grampian Slide Zone and other slides above and below the junction between the Central Highland Migmatite Complex and the Grampian Group. It is believed that these veins formed during early ductile shearing along the slides. Several dates have been

---

* Rb-Sr dates quoted in this section have all been recalculated using currently accepted decay constants and hence differ slightly from values quoted in the original publications.

obtained, by Rb-Sr methods on combined muscovite and whole-rock, which cluster quite tightly around 750 Ma (Piasecki and van Breemen, 1979; 1983) suggesting a correlation with the latest tectonothermal event to affect the Moine rocks north-west of the Great Glen in Knoydart and Morar (Fettes et al., 1986). Irrespective of whether the migmatites are regarded as stratigraphically and structurally contiguous with the Grampian Group (Lindsay et al., 1989) or as an earlier basement (Piasecki, 1980), this means that the migmatite complex and the lowest Grampian Group rocks must be more than 750 Ma old. There is no good evidence to indicate how much older than 750 Ma either unit may be but Soper and Anderton (1984) have speculated that the slides may be the result of lithospheric stretching in the early stages of development of the Dalradian basin. If this is so, then the earliest Grampian Group sediments may have been deposited only shortly prior to 750 Ma and hence may be assigned to the Upper Riphean Epoch in the late Precambrian.

Later parts of the Grampian Group have been shown to be stratigraphically continuous with the Lochaber Subgroup in several areas and in general there would then appear to be stratigraphical continuity throughout the Appin Group. There are no radiometric dates which cover this interval, and it is difficult to reconcile the deformation of the inferred lower part of the succession at 750 Ma with the ongoing sedimentation, with no obvious major tectonic break, which continued at least until the next available radiometric dates at the base of the Argyll Group.

It has been generally accepted, though by no means proven, that the tillites at the base of the Argyll Group can be correlated with the Varanger Tillite, which defines the base of the Vendian Epoch in Scandinavia. Metasedimentary rocks associated with the Varanger Tillite have been dated by Rb-Sr methods at 653 Ma (Pringle, 1972), but doubts have been cast on their validity and attempted correlations with the Scottish Dalradian should be regarded as extremely tenuous.

At the top of the Argyll Group, the Tayvallich Lavas have been dated by U-Pb on zircons at 595 Ma (Halliday et al., 1989) which is the latest Precambrian on all proposed time scales (Snelling, 1985).

Estimates of a minimum age for the top of the Dalradian depend upon the dating of post-sedimentary tectonothermal events, in particular early-metamorphic granites which should theoretically give ages nearest to the age of the sediments. The Portsoy Granite, intruded into the Argyll Group of the Northeast Highlands, has been dated by Rb-Sr whole-rock methods at 655 Ma (Pankhurst, 1974). This is difficult to reconcile with the suggested 653 Ma depositional age of the base of the Argyll Group and, although it is possible that the granite was intruded shortly after deposition of the sediments (Pankhurst *in* Harris and Pitcher, 1975), both dates must now be regarded with caution. Much attention has centred upon the Ben Vuirich Granite near Pitlochry. Its host Appin and Argyll group rocks were considered to have been affected by both the D_1 and D_2 episodes of deformation prior to granite emplacement (Bradbury et al., 1976; Rogers et al., 1989). However, Tanner and Leslie (1994) have shown that the granite was emplaced between D_1 and D_2 and then subsequently deformed and metamorphosed during the D_2 and D_3 episodes. An age of intrusion of 590 Ma, obtained by U-Pb dating of carefully selected zircons from the granite (Rogers et al., 1989) and confirmed by ion-microprobe dates on individual zircons (Pidgeon and

Compston, 1992), clearly has far-reaching implications for the history of Dalradian sedimentation and deformation. However, the fact that this granite had previously been dated at 514 Ma by a less precise U-Pb zircon study, and at 552 Ma by the Rb-Sr whole-rock method, emphasises the need for caution when comparing radiometric dates, in particular those obtained by different methods (Rogers and Pankhurst, 1993). Ideally more high-precision U-Pb dates are needed, both to check the currently available Rb-Sr dates of crucial pre-metamorphic intrusions and to provide new data points.

Conclusions

The high-precision U-Pb zircon age obtained from the Ben Vuirich Granite implies that most, if not all, of the Dalradian sedimentation occurred prior to 590 Ma, entirely during the Precambrian Era. All other radiometric and palaeontological data, although less precise, are compatible with at least this broad statement, although detailed conclusions are commonly confusing and contradictory.

The detailed chronological interpretation is constantly changing as more and better dates become available, but Table 1 provides a useful 'best-fit' model at the time of writing. The Grampian and Appin groups together cover a period of at least 100 Ma in the upper Riphean from pre-750 Ma to 653 Ma but as yet there is no evidence to enable any chronological subdivision. If the tillites at the base of the Argyll Group are correlated with the Varanger Tillite at 653 Ma (itself not a particularly good quality date) then they may be considered to mark the base of the Vendian. The top of the Argyll Group is apparently well defined by the Tayvallich Lavas at about 595 Ma, implying that the group spans some 60 Ma, all well within the Vendian. The base of the Cambrian could lie within the Southern Highland Group, although there is no obvious stratigraphical horizon where it may be located. In fact, given the thickness of the group and the nature of the sediments, which suggest relatively rapid accumulation, it is possible that the whole of the Southern Highland Group, and hence the whole of the Dalradian, is Precambrian. Current interpretations of the sparse and imprecise palaeontological data are compatible with this overall interpretation.

The above model allows ample time for the deformation and metamorphism to reach a peak at 520 to 490 Ma in the early Ordovician (Chapter 7). However, problems arise over the timing of the earlier phases of deformation. For example, the Ben Vuirich Granite zircon age of 590 Ma implies that the granite was intruded only 5 Ma after the eruption of the Tayvallich Lavas and yet postdates at least one phase of deformation. If the analytical errors on the 595 and 590 Ma determinations are taken into account the two dates could even overlap and the suggestion by Rogers et al. (1989) that the true age of the Tayvallich Lavas may be slightly older than that published, eases this problem only slightly. Also, if the Tayvallich date is accepted, the D_1 and D_2 phases of deformation in the Southern Highlands must be younger than 595 Ma. However, in the Central Highlands D_2 is dated by the pegmatites at 750 Ma, implying a gap of at least 150 Ma between the early phases of deformation in the two areas. Acceptance of these dates clearly requires that more careful consideration may have to be given to the possiblity of structural and metamorphic, as well as stratigraphical, diachronism between different areas or different levels in the sedimentary pile.

6 Structure of the Grampian Caledonides

The general structure of the Grampian Caledonides has been described in Chapter 4. In that account the main fold structures and dislocations were outlined briefly in order to provide a background for the outcrop distribution and stratigraphy described in detail in Chapter 5. The more detailed structural accounts presented here involve careful consideration of the relative ages of the various structures, their relationships with each other, and hence the overall structural history of the area, which involves several phases of deformation. This history has been painstakingly pieced together and refined by a multitude of workers from the early days of C T Clough and E B Bailey to the present day. It is very much an ongoing process and ideas are constantly changing. No single hypothesis can account satisfactorily for all of the observed features but, although various controversies have raged over the years, there has also been a remarkable amount of agreement and consensus on many aspects.

In individual areas of study, separate identifiable episodes of deformation have been termed D_1, D_2, D_3 ...... etc. in order of decreasing age. However, not all of these episodes are necessarily developed in all areas and, even within individual areas, the multiplicity of workers has resulted in differences in numbering of structural events. Consequently there is little consistency in the nomenclature of structural phases when comparing detailed studies of separate areas. Authors undertaking regional syntheses have attempted to solve this problem by disregarding phases of deformation whose effects can be shown to be of local extent only. The regional phases are then renumbered to produce a sequence of major events recognisable over wide areas and accepted by most authors. Thus, for example, the sequence D_1 to D_3 identified in the regional synthesis of the South-west Highlands by Roberts and Treagus (1977c), differs numerically from that used in detailed, more local studies within the same area by Roberts (1974a; 1976) and Treagus (1974). Even with this rationalisation, problems still exist on a regional scale with the result that different nomenclatures are still being perpetuated by various authors. One major problem arises from a variation in the number of recognisable major phases across the Tay Nappe. D_1 and D_2 are widespread events recognised by most authors. However, a D_3 event associated with the development of the Tay Nappe in the Southern Highlands, becomes difficult to distinguish from D_2 farther to the north-west. Consequently, workers in the Central Highlands such as Roberts and Treagus (1977c; 1979), Thomas (1979; 1980) and Treagus (1987) recognise only D_1 and D_2 and their main late-tectonic deformation is termed D_3. Workers in the Southern Highlands (e.g. Harris et al., 1976; Bradbury et al., 1979; Harte et al., 1984; Mendum and Fettes, 1985) recognise three nappe-forming or modifying events and their main late-tectonic phase is D_4. The latter nomenclature is more generally applicable and hence will be adopted in this account unless stated otherwise.

In most areas the structural development has thus been explained in terms of three or four major episodes of deformation which occurred during the Caledonian Orogeny. There is some evidence to suggest that still older events may be recognised in the Central Highland Migmatite Complex. During the first widespread (D_1) deformation the major folds, together with accompanying slides, were initiated with a NE–SW trend. During D_2 (or D_2 and D_3 of some authors) these folds were extensively modified in places to produce a complicated pattern of refolded nappes. In other places D_2 folds form separate complexes of intermediate-scale folds, accompanied by slides which may be coincident with or extensions of those of D_1 origin. Where a separate D_3 phase is identified, it is seen to be broadly coincident with the peak of regional metamorphism and associated igneous intrusion, although it may be a little later in the north-east. Later phases are late-tectonic episodes which overprint the composite foliations of both the nappes and the separate D_2 and D_3 fold complexes. They are separated from the earlier movements by a significant time gap and seem to be the result of a change in tectonic regime from ductile folding to basement fracture and block uplift (Harte et al., 1984; Mendum and Fettes, 1985).

A diagrammatic structure of the Grampian Highlands is shown in Figure 19. This is based largely on the diagram by Thomas (1979) for blocks A, B, C and D), extended to the north and north-east by incorporating results of published and unpublished work of various authors. For the purposes of the discussions in this chapter, the diagram has been divided into three complexes defined on a combination of structural and geographical criteria.

Southern Grampians Complex: the SE-facing folds to the south-east of the axis of the Loch Awe Syncline together with all the folds above and to the south-east of the Boundary Slide and its projected north-eastern continuation.

Western Grampians Complex: the apparently NW-facing folds to the north-west of the Loch Awe Syncline axis and the Ossian–Geal Charn Steep Belt, and the north-eastward continuation of these folds above the Fort William Slide through the Lochaber area.

Central Grampians Complex: the remainder of the Grampian Highlands, which comprises mainly the outcrop of the Grampian Group with those parts of the Lochaber Subgroup below the Boundary Slide. The northern part of this last area includes the Central Highland Migmatite Complex.

In the following sections, the structures within each complex are described from south-east to north-west, across the overall Caledonoid strike. Within each section, the structures are described firstly in the south-west, where the relatively simple fold geometry is essentially as described by E B Bailey, and then progressively through to the north-east with reference to the labelled blocks of Figure 19.

SOUTHERN GRAMPIANS COMPLEX

Highland Border Steep Belt and the hinge zone of the Tay Nappe

The existence of a belt of steeply dipping rocks along the Highland Border has long been known and is implicit in the early discussions of the area. However, it was not fully recognised and described until 1957 (Stringer, 1957). The *Aberfoyle Anticline* was first recognised by Henderson (1938) who used sedimentary structures to demonstrate opposing younging directions in 'grits' on either side of a core of slates. These criteria were also applied by Anderson (1947) to interpret outcrops of slate along the length of the Highland Border as the cores of similar structures, a correlation which has subsequently been shown to be erroneous in many places. However, it was Shackleton (1958) who used both sedimentary structures and cleavage-bedding relationships to demonstrate that the Aberfoyle Anticline is a downward-facing synformal anticline and interpreted it as the closure of the Tay Nappe. Subsequently, many detailed local studies have been made of key areas of the steep belt (from south-west to north-east: Simpson and Wedden, 1974; Paterson et al., 1990; Mendum and Fettes, 1985; Stone, 1957; Harris, 1962; 1972; Harris and Fettes, 1972; Harte et al., 1984).

In the Cowal and Bute area the strata steepen gradually on the south-eastern limb of the *Cowal Antiform*, but north-eastwards from here the bend becomes a sharp, monoclinal flexure, over which the *Flat Belt* of the Tay Nappe gives way to the *Steep Belt*. This flexure, which results in a downstep to the south-east of as much as 10 km, is known as the *Highland Border Downbend* (or *Monoform*). It has long been recognised as a secondary structure, either associated with the D_4 deformation or, more likely, with a slightly earlier phase of differential block uplift (Mendum and Fettes, 1985). It is possible that the downbend is sited over a reactivated basement lineament separating two major crustal blocks (Harte et al., 1984).

The hinge zone of the Tay Nappe was originally described as a single downward-facing anticline at Aberfoyle, but elsewhere it has been shown to be more complex. In the Ben Ledi area the hinge zone consists of two major synforms, the *Aberfoyle Anticline* and the *Benvane Synform*, separated by the *Ben Ledi Antiform* (Mendum and Fettes, 1985). These structures are all inferred to be of primary, D_1, age since minor D_2 structures are not seen near the Highland Border in this area and only become overprinted on D_1 farther to the north-west. In the Dunkeld area, downward-facing folds are interpreted as lying on the north-western limb of the main Aberfoyle Anticline (Treagus et al., 1972; Harris and Fettes 1972).

Minor structures in rocks of low metamorphic grade in the Steep Belt and in much of the adjacent parts of the Flat Belt are very much influenced by lithology. Pelites develop axial planar slaty cleavages, but in the meta-greywackes, which constitute most of the Southern Highland Group, spaced cleavages fan around fold closures. Increased deformation during subsequent phases flattens the spaced cleavage and induces a finely striped rock so that, where folding is tight, as in the lower structural levels of the Tay Nappe, it is difficult to recognise the multiple origin of the dominant foliation (Harris et al., 1976). Bedding is generally not discernable, except where distinct lithological changes occur, and outcrops are commonly dominated by abundant minor folds and composite cleavages. Where NE–SW-trending D_4 folds are present, they are usually accompanied by a strong crenulation cleavage, best seen in the more pelitic or strongly foliated rocks. This cleavage dips north-westwards at moderate to subvertical angles.

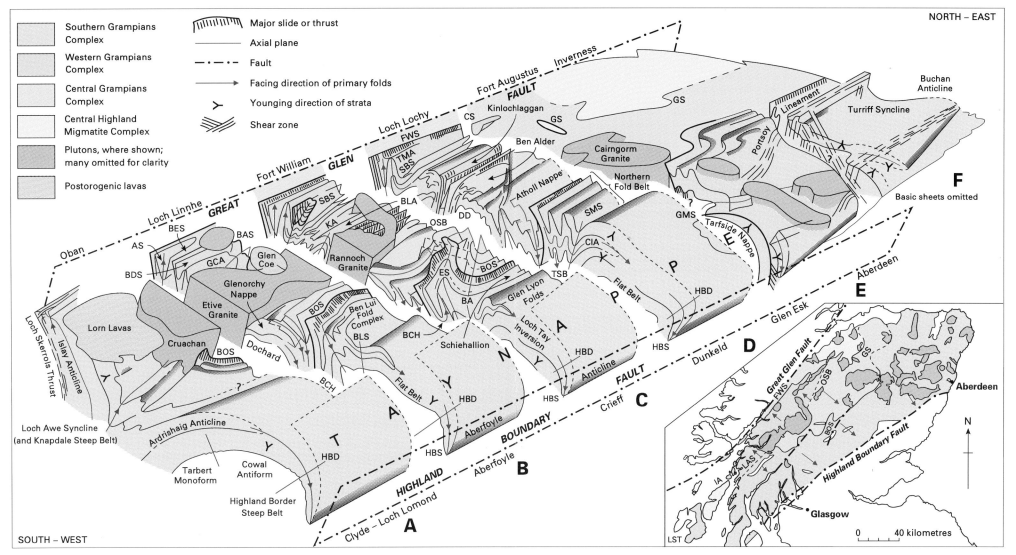

Figure 19 Block diagram of major structures in the Grampian Highlands (brittle faults not shown). Sections A, B, C and D are from Thomas (1979) and are reproduced by permission of the Geological Society of London.

AS	Appin Syncline	CS	Corrieyairack Syncline	KA	Kinlochleven Anticline
BA	Bohespic Antiform	DD	Drumochter Dome	LAS	Loch Awe Syncline
BAS	Ballachulish Slide	ES	Errochty Synform	LST	Loch Skerrols Thrust
BCH	Beinn a Chuallich Folds	FWS	Fort William Slide	OSB	Ossian–Geal Charn Steep Belt
BDS	Beinn Don Syncline	GCA	Glen Creran Anticline	SBS	Stob Ban Synform
BES	Benderloch Slide	GMS	Glen Mark Slide	SMS	Sron Mhor Synform
BLA	Beinn na Lap Antiform	GS	Grampian Slide	TMA	Tom Meadhoin Anticline
BLS	Ben Lawers Synform	HBD	Highland Border Downbend	TSB	Tummel Steep Belt
BOS	Boundary Slide	HBS	Highland Border Steep Belt		
CIA	Creag na h'Iolaire Anticline	IA	Islay Anticline		

Tay Nappe

The history of development of the Tay Nappe is a matter of continuing discussion and is central to all interpretations of the overall evolution of the Grampian Highlands. It is generally agreed that the recumbent structure now seen is not necessarily the fold as initially formed, but is the result of flattening, tightening and extension of the original D_1 anticline during D_2 and, where present, D_3 deformation. The resulting, essentially parallel-sided, flat-lying nappe has a large amplitude in the south-west, but diminishes towards the north-east (Figure 19). Bedding and the composite foliation resulting from the primary (D_1 to D_3) episodes are essentially parallel. Later NE–SW-trending secondary folds and other structures on both a major and a minor scale were subsequently imposed on the nappe. The major folds are commonly of large amplitude and open in style, for example the *Ben Lawers Synform*, but can be sharply defined locally such as in the Highland Border Downbend. Late cleavages and foliations associated with these folds cross-cut the primary structures at a high angle and trend NE–SW.

In the South-west Highlands the nappe has the form of a broad arch, known as the *Cowal Antiform* (Figure 19, block A). This is modified by two conjugate monoformal structures, probably both related to the pre-D_4 block uplift: the *Highland Border Downbend* to the south-east and the *Tarbert Monoform* to the north-west (Roberts 1974). Exposure levels are entirely within the lower, inverted limb of the nappe, except to the north-west of the Tarbert Monoform, where the core of the Tay Nappe is brought down below the level of erosion to crop out as the *Ardrishaig Anticline*. Here the fold limbs, primary axial planes and associated cleavages all dip steeply to the north-west and constitute part of the *Knapdale Steep Belt* (Roberts, 1974a). This is the only area in which the core of the Tay Nappe is exposed.

North-east from Cowal, the crest of the Cowal Antiform flattens and the inverted lower limb of the Tay Nappe forms the *Flat Belt* which dominates so much of the Southern Highlands (Figure 19, blocks B, C and D). The Flat Belt is sharply limited on its south-eastern side by the Highland Border Downbend, but the Tarbert Monoform dies out north-eastwards and the core of the nappe lies above the present level of erosion. The north-western limit of the nappe is here defined by a steeply dipping zone of downward-facing folds which must include, in part, the hinge zone of any major antiformal syncline originally underlying and complementary to the Tay Nappe. This zone is termed the *Ben Lui Fold Complex* in block B (Roberts and Treagus, 1979) but the zone widens north-eastwards in blocks C and D with the addition of many tight, upright folds of a later generation (D_2/D_3) which will be described as the *Northern Fold Belt* (Figure 19, blocks B, C and D).

Late broad, upright, NE-trending D_4 folds affect the north-western part of the Flat Belt (Figure 19, blocks B and C). Of these, the best developed are the *Ben Lawers Synform* (Treagus, 1964) and its complementary, lower amplitude *Loch Tay* (or *Ben More) Antiform* to the south-east (Watkins, 1984; Harte et al., 1984). The somewhat tighter *Sron Mhor Syncline* may be of the same generation but it is closely associated with the steeper structures of the Northern Fold Belt and was probably initiated as an earlier structure (Figure 19, block D).

Towards the north-east the amplitude of the Tay Nappe is considerably reduced (Figure 19, block E). In the Glen Esk area a broad late antiform, the

Tarfside Culmination, exposes a wide zone of non-inverted strata, apparently beneath the Tay Nappe, which Harte (1979) assigned to a separate *Tarfside Nappe.* According to Harte the axial zone of the fold separating the two nappes has been replaced by a slide, the *Glen Mark Slide* (Figure 19, block E). On the coast section north of Stonehaven, a generally flat-lying sequence is 'down-bent' (D_4) and becomes overturned to the south-east, with downward-facing D_1 structures (Booth, 1984; Harte et. al., 1987). It is not certain how this sequence, which may be right-way-up, relates to the Tay Nappe.

To the north of Aberdeen the Tay Nappe can no longer be identified with any certainty (Figure 19, block F). The rocks are mostly right-way-up and it is not clear whether the inverted limb has been cut out by major thrusting, or whether, more simply, the nappe structures have a much smaller amplitude in this area (Harte et al., 1984). However, in the coast section around Collieston the beds are regionally inverted and there are abundant small- to large-scale recumbent, isoclinal D_1 folds with a maximum amplitude of about 1 km (Plate 9; Read and Farquhar, 1956; Mendum, 1987). These folds plunge gently to the north, face eastwards and may represent a subdued equivalent of the nappe (Ashcroft et al., 1974). The folds, and their accompanying spaced cleavage, are refolded by a later set of near-coaxial, tight folds which postdate porphyroblast growth and hence are assigned to D_3.

Ben Lui Fold Complex and the Northern Fold Belt

From Dalmally north-eastwards to Strathtummel, a progressively widening zone of steeply inclined strata intervenes between the Flat Belt of the Tay Nappe and the Boundary Slide. Since the axial plane of the anticlinal Tay Nappe appears to be above the level of erosion throughout this area, early workers considered that the underlying major synclinal axial plane and fold closure (the'righting fold') must lie within this zone. In the critical area around Ben Lui the contact between the Ardrishaig Phyllite and the Ben Lui Schist can be traced around a fold closure from an inverted to a non-inverted limb. This antiformal syncline was termed the 'Ben Lui Fold' by Bailey (1922) who proposed that it is the D_1 syncline beneath the Tay Nappe.

Subsequent work by Cummins and Shackleton (1955) supported Bailey's view, but a more detailed study by Roberts and Treagus (1964; 1975) showed that the folding is more complex. Beneath the D_1 Ardrishaig Anticline, the *Ben Lui Fold Complex* consists of three recumbent elements, the *Dalmally Syncline,* the *Ra Chreag Anticline* and the *Ben Lui Syncline,* all of which were shown to be later, D_2 (or possibly D_3) structures. This tripartite structure has been traced north-eastwards to lower Glen Lyon (Roberts and Treagus, 1979), beyond which the individual folds lose their identity. In the Balquhidder area, Watkins (1984) has proposed that a large-scale recumbent fold exists, which he has tentatively correlated with the Ben Lui and lower Glen Lyon D_2 folds. Many folds in the Schiehallion area and associated folds in Strathtummel with SE-dipping axial planes (Figures 19 and 20) appear to belong to the same generation of D_2 folds, which in places are seen to fold the axial planes of major D_1 folds such as the *Beinn a Chuallich Folds* and the *Creag na h'Iolaire Anticline* (Roberts and Treagus, 1979; Treagus, 1987). It was

Plate 9 D$_1$ recumbent fold in dominantly inverted gritty and pelitic metagreywackes. Collieston Formation, Southern Highland Group, Devil's Study, near Whinnyfold, Aberdeenshire (D 3899).

these NW-overturned folds, together with the SE-facing Tay Nappe, which led to the model of a 'mushroom' structure of nappes diverging from a root zone in the *Tummel Steep Belt*, comparable to the Loch Awe Syncline (Sturt, 1961; Rast, 1963; Harris 1963). The demonstration that the D$_2$ folds to the north-west of the steep belt refold SE-facing D$_1$ structures associated with the Tay Nappe makes this interpretation untenable.

Farther to the north-east, in the Glen Shee area, a structure known as the *Kirkmichael Fold* has long been identified as a continuation of the Ben Lui Syncline (Bailey, 1925; Read, 1935; 1955). However, this too has subsequently been shown to be a complex D$_2$ structure, similar in profile to those of the Strathtummel area (Upton, 1986).

In the area around Schiehallion and Strathtummel (Figure 19, block C) the Appin and Argyll group rocks, the Boundary Slide and the underlying Grampian Group undergo a dramatic swing of strike which is a major feature of even small-scale maps (e.g. Figure 3). This displacement is caused by large, steeply plunging, north-trending late folds termed the *Errochty Synform* and the *Bohespic Antiform* (Rast, 1958; Thomas, 1980; Treagus, 1987). Thomas considers these folds to be D$_4$ structures (his D$_3$), whereas Treagus regards them as later (his D$_4$, and hence D$_5$ in the overall scheme). The two folds have contrasting geometry, consisting of a tight synform and a broad, open antiform. Such changes in fold geometry commonly occur at the junction of materials of contrasting competence, in this case the thick psammites of the Grampian Group and the multi-layered pelites and quartzites of the Appin

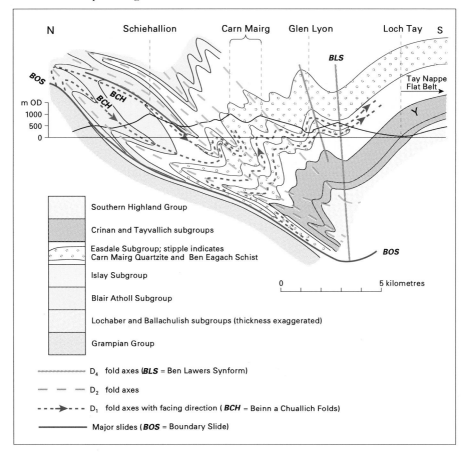

Figure 20   Section across the Northern Fold Belt in the Schiehallion–Loch Tay area showing the combined effects of D$_1$ and D$_2$ folds. Reproduced by permission of the Royal Society of Edinburgh and J E Treagus from *Transactions of the Royal Society of Edinburgh: Earth Sciences*, Vol. 78, Part 1 (1987), 1–15.

Group. A similar geometry occurs in a fold pair of similar age farther to the west, in the Loch Rannoch area. These fold pairs have a symmetrical spatial relationship to the major NNE-trending Loch Tay, Bridge of Balgie and Tyndrum faults, and Treagus (1987) considers that there may be a relationship between ductile shearing, the late folds pairs and later brittle faults. Whatever age is assigned to these fold pairs they do not represent the latest deformation in the area, since their axial traces are themselves folded locally by NW-trending folds, the most important being the *Trinafour Monoform*.

Boundary Slide

The Southern Grampians Complex is separated from the structurally underlying Central Grampians Complex by a slide which can be traced almost continuously from Dalmally to Glen Tilt. This zone of highly schistose or platy rocks, varying in thickness from 500 to 2000 m, includes several planes of dislocation and has been termed the *Boundary Slide* (Roberts and Treagus, 1977c).

A 'Boundary Slide' was first recognised in the Strathtummel area by Anderson (1923) and was described in detail in the Schiehallion area by Bailey and McCallien (1937) and by Rast (1958). It was subsequently equated with major slides in the South-west Highlands to become part of the 'Iltay Boundary Slide' (MacGregor, 1948; Rast, 1963). For most of its length the slide forms the boundary between successions with markedly different lithological associations, formerly assigned to the Moine and the Dalradian. No correlations were attempted between these successions on early maps (e.g. Bailey, 1934). Consequently the slide assumed a fundamental importance as a major dislocation separating two major tectonostratigraphical units, the Ballappel Foundation and the Iltay Nappe. Correlation across the slide involves a comparison of widely separated successions in the Appin area and in Perthshire. Unsatisfactory attempts at correlation by Anderson (1948; 1953) and Voll (1964) were followed by the more generally accepted correlation of Rast and Litherland (1970). This seemed to confirm that a considerable hiatus is represented by the slide since, in several areas, large sequences of the Appin Group and Islay Subgroup appeared to be absent. However, recent mapping has revealed that attenuated sequences of the Lochaber and Ballachulish subgroups are present in most areas (Treagus and King, 1978; Roberts and Treagus, 1979; Treagus, 1987). The attenuation may be due to a variety of factors other than sliding, and the presence of near-complete successions along much of the length of the Boundary Slide has therefore lessened its significance as a structural boundary in most current interpretations.

Roberts and Treagus (1977c) regard the Ballachulish Slide of the Western Grampians Complex as part of the Boundary Slide, although correlation between the Glen Coe and Dalmally areas is speculative on account of the intervening Etive Granite. From Dalmally eastwards the slide is well defined by a zone of strongly deformed rocks which intervenes between the Easdale Subgroup and underlying rocks of the Lochaber Subgroup (Roberts and Treagus, 1979). From Beinn Dorain eastwards through Glen Lyon, additional, structurally lower slides separate the Lochaber Subgroup from the underlying Grampian Group.

In the Schiehallion area the zone of strongest deformation is coincident generally with a porphyroblastic muscovite-schist with small quartz blebs which corresponds to the Beoil Schist of Bailey and McCallien (1937). This schist, which has a thickness of up to 95 m, was regarded as a tectonic schist by Rast (1958; 1963) and can be shown to transgress both Grampian Group and Appin Group sequences according to Thomas (1980). However, Treagus and King (1978) and Treagus (1987) argue that the Beoil Schist is a true stratigraphical unit within which the maximum deformation has been focussed. Despite this difference of opinion over local detail, there is no doubt that the slide zone does transgress the stratigraphy on a regional basis and that the total displacement may be several kilometres (Treagus, 1987).

Early isoclinal folds are truncated by the Boundary Slide and there is no doubt that it is folded extensively by later structures. Thomas (1980) considers that the dislocation was initiated during the formation of the primary nappes but that it was reactivated in places during later deformation. The strong platy or schistose fabric present throughout the zone is regarded by Treagus (1987) as an intense development of the regional schistosity developed during nappe formation, which he regards as a D_2 phase. Most of the slides within the zone occur on the short limbs of D_2 folds and result in an overall movement, in a thrust sense, towards the north-west. This fact is clearly contrary to the overall concept of the dislocation as an extensional slide related to the primary nappe formation and has not been fully explained. It may be that slides, initiated during basin development and early tectonism, were subsequently reversed during later movements (Soper and Anderton, 1984; Anderton, 1988).

A Boundary Slide has been traced north-eastwards from Glen Tilt, through the Braemar area to the eastern end of the Cairngorm Granite (Upton, 1986). However, here its overall effect may be much reduced. Farther north, zones of high strain, locally accompanied by slides, are common at or below the Grampian Group/Appin Group junction in Strath Avon and the Hills of Cromdale (McIntyre, 1951). A major shear zone, to the west of and structurally below the better-known shear zones that mark the Portsoy Lineament, has also been identified recently; this *Keith Shear Zone* may continue the zone of dislocation to the north coast at Sandend Bay.

Banff and Buchan area

In the North-east Grampians the outcrop pattern is dominated by late, open, broad, upright folds with NNE-trending axes, principally the Turriff Syncline and Buchan Anticline (Read, 1955; Read and Farquhar, 1956). Exposure is generally poor and detailed sections are to be found only on the coast.

Structures on the east coast around Collieston have been described in relation to the Tay Nappe. Here the beds are inverted regionally and a major early fold closure must therefore occur between this section and the coast section at Fraserburgh, which is regionally the right way up. The axis of this fold cannot readily be traced; it lies in a complex sheared zone with high-grade metamorphism between Ellon and Inverallochy.

Along the north coast the rocks exhibit locally complex folding with steep dips over much of the section, but regionally they have been shown to be disposed in a broad, gentle syncline (Figure 21). This *Turriff Syncline* was recognised as a relatively late fold by Read (1955), who considered that the succession, which is right way up on a regional scale, constitutes the upper limb of a major recumbent anticline which he termed the Banff Nappe. The western limb of the Turriff Syncline is steep due to a monoform, regarded as a major fold closure predating the metamorphic peak by Sutton and Watson (1956) who named the structure the *Boyndie Syncline*. Subsequent workers have regarded this as a later (D_3) structure related to the main Turriff Syncline, thereby devaluing its regional importance (Johnson and Stewart, 1960; Johnson, 1962; Fettes, 1970), although Treagus and Roberts (1981) have assigned it as an early (D_1) phase.

Despite the debate over the significance of the Boyndie Syncline, the overall structural and metamorphic sequences of this section are quite well understood due to several detailed studies of the minor structures (Johnson,

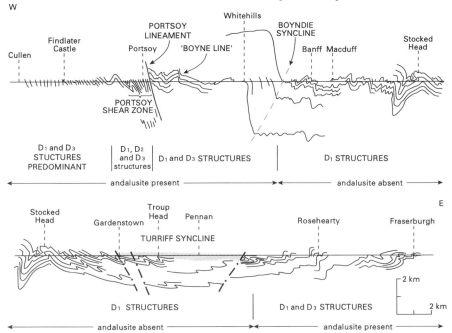

Figure 21 Cross-section along the Banffshire coast showing the main structural features and predominant deformation/fold phases (section east of Whitehills after Loudon, 1963). Red stipple = Old Red Sandstone outlier.

1962; Fettes, 1971; Treagus and Roberts, 1981). Early (D_1) structures face consistently upwards, developing a spaced cleavage in the psammitic rocks and a strong slaty axial planar cleavage in the more pelitic rocks of the Macduff Slate Formation. The attitude and style of the D_1 folds varies according to structural level, being generally upright and open to close at the highest levels in the centre of the Turriff Syncline and generally close to tight on the limbs. On both the western and eastern limbs the D_1 folds face to the north-west which is contrary to Read's (1955) model of a SE-facing Banff Nappe.

Post-D_1 structures are restricted to lower structural levels and hence are seen only on the western and eastern limbs of the Turriff Syncline. There is some confusion over the number of major phases which can be identified, and over their relationship to regional D_2 and D_3 as defined elsewhere. Both Johnson (1962) and Treagus and Roberts (1981) have identified separate D_2 and D_3 phases and their conclusions are incorporated in the following summary based upon that of Kneller (1987). D_2 folds and cleavages are recognised only to the west of the Boyne Limestone outcrop and to the east of Fraserburgh Bay. In the area around Portsoy, D_2 folds are locally the dominant folds. Here they are steeply plunging and have a very strong linear fabric attributed to thrusting associated with the Portsoy Lineament which has intensified and rotated the folds. The late, D_3 deformation, which postdates the peak of metamorphism and growth of porphyroblasts, has limits which almost coincide with the andalusite isograd (Figure 21). Large-scale D_3 folds are characteristically monoclinal. Associated finely spaced cleavages and tight crenulation cleavages occur in the more pelitic units and small-scale open to tight fold structures are well seen in the more banded units. The major Turriff Syncline and *Buchan Anticline* have been tentatively attributed to the D_3 phase by most authors, although it is difficult to make exact correlations with the observed detailed D_3 structures and their overall open and upright character is more comparable with major D_4 structures elsewhere. Numerous sets of minor brittle folds in the area are attributed to later events.

To the west of the Turriff Syncline, a number of major zones of shearing and dislocation occur and these have been used to define the western edge of a distinct structural block, the Buchan Block, possibly decoupled from the main Dalradian terrane. On the eastern edge of the Cowhythe Psammite outcrop is a zone of highly deformed rocks with a strong down-dip extension lineation and development of thin mylonite. This zone marks the position of the *Boyne Line* of Read (1955) which was interpreted as a major slide underlying his proposed Banff Nappe. In this model movement on the Boyne Line was held responsible for the removal of any calcareous lithologies equivalent to the Loch Tay and Deeside limestones over much of the north-east Grampians, apart from the Boyne Limestone which is seen only in the north coastal section.

Thrust-related fabrics at the western margin of the Cowhythe Psammite were attributed to a *Portsoy Thrust* by Elles (1931). From here a zone of sub-vertical mylonites, shear zones and faults extends westwards for over 1 km along the coast section around Portsoy. Earlier fold axes and lineaments have been rotated so that they plunge down-dip adjacent to and within the zone, and locally a pervasive down-dip stretching lineation is present. Highly sheared basic and ultramafic igneous rocks occupy the centre of the zone and cross-cutting, but lineated, sheet-like granite bodies are present near the margins. This zone of shearing and dislocation, accompanied by igneous intrusions, can be traced inland in a general SSW direction from Portsoy to the Cabrach area (Ashcroft et al., 1984) and southwards to upper Glen Clova as the *Portsoy–Duchray Hill Lineament* (Fettes et al., 1986). Major stratigraphical discontinuities occur at the lineament (Fettes et al., 1991) and marked differences in metamorphic history on each side of the lineament indicate major westward overthrusting during the regional D_3 event (Baker, 1987; Beddoe-Stephens, 1990). Some geochronological and fabric evidence suggests that the Cowhythe Psammite, along with all the other gneissose units of the north-east Grampians, represent a pre-Caledonian basement complex (Sturt et al., 1977). Ramsay and Sturt (1979) suggested that all the rocks above the Portsoy Thrust constitute an allochthonous block, and that this consists of a Dalradian metasedimentary cover separated from underlying gneissose basement by a *décollement* along Read's original Boyne Line. Another interpretation, the one followed by this handbook, regards the gneisses as migmatised equivalents of the upper parts of the Argyll Group (Chapter 5), so that an almost continuous Dalradian succession is recognised in the area. Whilst recognising the importance of the Portsoy–Duchray Hill Lineament as a major tectonic, stratigraphical and metamorphic boundary, the Dalradian of the Buchan Block is here regarded as essentially autochthonous, as argued by Ashcroft et al. (1984).

To the west of the Portsoy–Duchray Hill Lineament, large isoclinal or near-isoclinal early folds are generally upward-facing and are overturned towards the north-west. Secondary folds, which can be shown to postdate the primary metamorphism, are commonly coaxial with the early folds and give rise to a dominant penetrative cleavage which is finely spaced, or to a tight crenulation in the more pelitic lithologies. Late folds, possibly related to uplift and basement block movement, locally exert a strong control on the outcrop pattern. Such folds are typically broad, near-upright and trend north–south or NE–SW such as the *Ardonald Antiform* in the Dufftown area. Locally, penetrative spaced cleavages, minor chevron folds and kink bands are developed.

The western margin of this area, as shown in Figure 19, is marked by a major ENE- to NE-trending shear zone passing through Keith and projecting to the coast at Sandend Bay, although many of the structural features described above extend to the west of this zone into the Central Grampians Complex. This shear zone appears to be a continuation of the line of dislocation and stratigraphical attenuation at or around the Grampian/Appin group boundary to the south-west and is marked by an attenuated succession of Ballachulish Subgroup rocks. Within the shear zone are several foliated granites with a very strong ESE-plunging stretching lineation and asymmetrical fabrics that indicate overthrusting to the WNW.

One of the most striking features of both stratigraphical and structural maps of the Grampian Highlands (e.g. Figures 3, 19) is the 20 km-amplitude S-shaped 'knee-bend' in the strata in the area between Braemar and Tomintoul. The reasons for this major feature are obscure, since there are no obvious associated minor folds or regional cleavages. It is clearly a late structure, since all of the folds and dislocations described above, including the Portsoy–Duchray Hill Lineament, are folded around it. Such a large-scale structure must reflect deep crustal weaknesses and it probably results from crustal block movements in the later stages of the Caledonian Orogeny. Such movements may have been along NW–SE fractures which exerted some control on Dalradian sedimentation, although east–west-trending lines of Younger Granite intrusions appear to follow both axial traces of the 'knee-bend' and clearly indicate lines of crustal weakness.

WESTERN GRAMPIANS COMPLEX

Former subdivisions

The folds to the north-west of the axis of the Loch Awe Syncline essentially comprise a single NW-facing major recumbent nappe, the *Islay Anticline* (Bailey, 1917; Roberts, 1974). On Islay this structure overrides the Bowmore Sandstone along the *Loch Skerrols Thrust*, which was formerly equated with the Moine Thrust (Bailey 1917, 1922; Kennedy, 1946). Subsequently, the Loch Skerrols Thrust was projected into the *Benderloch Slide* at Loch Creran, which Bailey (1922) had correlated directly across the Etive Granite Complex with the Boundary Slide of the Southern Highlands, to form what was later to become known as the *'Iltay Boundary Slide'*. This continuous major dislocation was for many years regarded as a tectonic boundary between the rocks of a composite *'Iltay Nappe'*, consisting of the Islay Anticline and Tay Nappe, and those of the NW-facing recumbent folds of the *'Ballappel Foundation'*.

More recent work has shown that stratigraphical correlation is possible between these two tectonic units (Rast and Litherland, 1970) and several models have been proposed which correlate the Islay Anticline with structures to the north-east, in what was formerly regarded as part of the 'Ballappel Foundation' (Roberts and Treagus, 1977c; Hickman, 1978; Litherland, 1982). Rast and Litherland (1970) show that the Benderloch Slide continues north-eastwards and does not connect eastwards to form an 'Iltay Boundary Slide' as originally envisaged by Bailey, so that the whole of the Western Grampians Complex as shown on Figure 19 may now be regarded as a single tectono-stratigraphical unit.

Structural hypothesis of E B Bailey (1934)

The part of the Western Grampians Complex between Appin and Glen Roy is the most intensively studied area of the Scottish Dalradian, particularly in the well-exposed, cross-strike section along Loch Leven and Loch Eilde. The original interpretation of the regional structure by Bailey (1910) underwent several modifications, including a complete reversal of the order of stratigraphical succession and consequently the facing direction of the major folds. The final synthesis of this work (Bailey, 1934; 1960) has remained a sound basis for all subsequent work and interpretation (Figure 22).

Bailey recognised three NW-facing recumbent isoclinal folds, each about 15 to 20 km in amplitude. From north-west to south-east and progressing structurally upwards these are the *Appin Syncline*, the *Kinlochleven Anticline* and the *Ballachulish Syncline* (Figure 22A). A relatively minor recumbent fold, the *Aonach Beag Syncline*, was identified immediately underlying the Balluchulish Syncline to the east of Ben Nevis. The lower limbs of the two major synclines were considered to be largely replaced at an early stage in the deformation by tectonic slides: the *Fort William Slide* beneath the Appin Syncline and the *Ballachulish Slide* beneath the Ballachulish Syncline. Bailey made little attempt to establish a history of deformation in terms of a detailed sequence of fold phases, although he did recognise the existence of upright secondary folds which refold the primary recumbent folds. For example, the Kinlochleven

Figure 22   Interpretations of the structure of the Loch Leven area.

A   Bailey (1934): an attempt at an actualistic section.
B   Roberts and Treagus (1977b; c): revised stratigraphical correlations and distinction of primary and secondary folds.
C   Hickman (1978): in this interpretation the primary folds extend westwards across only part of the area.

AS	Appin Syncline		KA	Kinlochleven Anticline
BAS	Ballachulish Slide		KAF	Kinlochleven Antiform
BS	Ballachulish Syncline		MA	Mamore Anticline/Antiform
BWS	Blackwater Synform		MS	Mamore Syncline
FWS	Fort William Slide		SBS	Stob Ban Syncline/Synform
GBS	Garbh Bheinn Synform		TMA	Tom Meadhoin Anticline
			TS	Treig Syncline

Anticline was considered to be refolded by a major secondary complex termed the *Stob Ban Synform*; to the north-west of the synform axis, the anticline faces upwards to the north-west as the antiformal *Tom Meadhoin Anticline*, whereas to the south-east of the synform it faces downwards as a synformal anticline. In the latter area demonstrably inverted strata of the Lochaber Subgroup on the lower limb of the Kinlochleven Anticline crop-out over a cross-strike width of some 7 km to form what is known as the *Kinlochleven Inversion*.

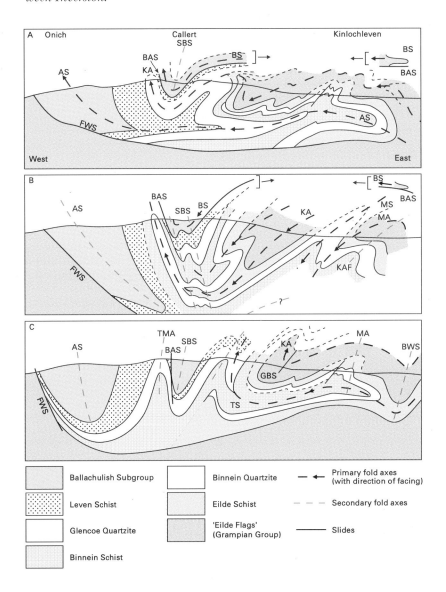

Ballachulish Subgroup		Binnein Quartzite	— ◀	Primary fold axes (with direction of facing)
Leven Schist		Eilde Schist	– – –	Secondary fold axes
Glencoe Quartzite		'Eilde Flags' (Grampian Group)	———	Slides
Binnein Schist				

Subsequent to Bailey's regional interpretation, more detailed investigations by several workers have concentrated on smaller-scale structures in an attempt to establish relative ages of the observed folds in a more complete structural sequence. Understandably, this has led to much criticism of the original, relatively simple concept of early recumbent folds and later coaxial, more upright folds. Weiss and McIntyre (1957) recognised an early NW-plunging generation of folds and a superimposed later generation trending north-east, with steeply dipping axial planes. By taking no account of the larger-scale structures or overall stratigraphy, they failed to recognise any earlier recumbent folds and their main conclusions were refuted by King and Rast (1959), by Bailey (1960) and later, following more detailed work, by Treagus (1974) and Roberts (1976). Bailey's model of nappes and slides was completely rejected by Voll (1964) who explained the observed outcrop pattern in the critical Loch Creran area in terms of major facies changes. However, in a later paper, the description of a large-scale early inversion corresponding to the Kinlochleven Inversion does imply the presence of recumbent folds (Kruhl and Voll, 1975).

Two major regional models of the structure of the Western Grampians Complex have emerged as a result of the many post-1960 detailed studies of the Lochaber–Loch Leven–Appin area. These are summarised respectively by Roberts and Treagus (1977c) and Hickman (1978).

Structural hypothesis of J L Roberts and J E Treagus (1977)

The hypothesis proposed by Roberts and Treagus (1977a; 1977b; 1977c) is based largely upon detailed studies in the Ardsheal Peninsula (Treagus and Treagus, 1971), around Ballachulish (Roberts, 1976) and around Kinlochleven (Treagus, 1974). It also draws upon the authors experience in adjoining areas of the South-west Highlands (Roberts, 1974a) and the Glen Orchy area (Thomas and Treagus, 1968; Roberts and Treagus, 1975). The geometry of the major fold structures established by Bailey is largely confirmed, but some of Bailey's stratigraphical correlations are modified and the recognition of primary and secondary structures is rationalised (Figure 22B). A problem of any regional synthesis of this area is that many structures are interrupted by the later granitic complexes of Etive and Rannoch Moor and large areas are covered by lavas.

Three major phases of deformation are recognised. Other minor phases have been identified locally, but in the overall synthesis Roberts and Treagus (1977c) ignore these local phases in renumbering the major phases, and the regional structure is interpreted in terms of four major D_1 folds, two major D_2 folds and a late 'D_3' dome structure, probably equivalent to the D_4 of other authors (Figure 22B).

D_1 folds

The Appin Syncline is confirmed as a primary fold and is traced from Ben Nevis, through a composite set of folds on the Ardsheal Peninsula, to Lismore; the Kinlochleven Anticline is correlated with the Islay Anticline, via the *Airds Hill Anticline* at Loch Creran; the Ballachulish Syncline is correlated with the *Beinn Donn Syncline* of Loch Creran and then with the Loch Awe Syncline; the NW-facing *Glen Creran Anticline* is correlated with the SE-facing Ardrishaig Anticline. Both the Islay Anticline and the Loch Awe Syncline were regarded by Bailey as secondary structures. However, Roberts and Treagus agree with subsequent workers who interpreted them as primary folds (e.g. Cummins and Shackleton, 1955; Shackleton, 1958; Rast, 1963; Borradaile, 1973; Roberts, 1974a).

D_2 folds

The Stob Ban Synform of the Ballachulish area is confirmed as a secondary fold which refolds not just the Ballachulish Syncline but also most of the D_1 folds of the Loch Creran area; a composite set of D_2 folds which deforms rocks of the Kinlochleven Inversion east of the Stob Ban Synform is termed the *Kinlochleven Antiform*.

'D_3' folds

A broad antiform in the area around Glen Orchy is shown on Figure 19 as part of the Central Grampians Complex, since it folds rocks of the Glen Orchy Nappe below the Boundary Slide. This *Glen Orchy Antiform* was recognised by Bailey (Bailey and Macgregor, 1912) and by Thomas and Treagus (1968). It was first described as a late structure by Roberts and Treagus (1975). The fold has the form of a broad dome and deforms both primary and secondary structures as seen by the deflection of strike and fold axes from the Dalmally area north-westwards and then north towards Ballachulish. Possible correlations around this deflection have been given above. Correlations across the dome have been proposed by Roberts and Treagus (1977c) and include the correlation of the D_2 upward-facing Stob Ban Synform, north-west of the dome, with the downward and SE-facing D_2 Ben Lui Antiform south-east of the dome. Both folds are underlain by major slides, beneath which the Kinlochleven Antiform in the north-west is inverted across the dome to correlate with the synformal *Beinn Chuirn Anticline* to the south-east. The *Benderloch Slide* of the Loch Creran area extends north-eastwards and is considered to continue as the Ballachulish Slide to the north of the Ballachulish Granite (Figure 19, block B). It is then folded around the closure of the Stob Ban Synform and trends back southwards to Glen Etive. Roberts and Treagus (1977c) link it across the Etive Granite Complex with a slide which is folded around the Glen Orchy Antiform in the Dalmally area, to propose a single, continuous Boundary Slide. Since it is folded by D_2 folds, the initiation and main movement on the slide clearly occurred during the D_1 deformation phase, although intense D_2 movements identified in the Dalmally area indicate local reactivation (Roberts and Treagus, 1975).

The structural interpretation proposed by Roberts and Treagus (1977c) recognises primary D_1 folds that are believed to root steeply at depth and to fan outwards at higher levels to face north-west and south-east, much as proposed in earlier interpretations. The root zone is seen as originating in Grampian Group rocks, lying at depth to the north-west of the Glen Orchy Antiform. By correlating the D_2 folds and accompanying slides across this flexure, it is possible to envisage a consistently facing set of recumbent D_2 folds, collectively referred to as the Glen Orchy Nappe. These folds affect the deeper parts of the primary structures, as seen in the north-eastern part of the area, which are thus correlated with higher level structures of the Loch Awe Syncline and Tay Nappe to the south-west.

Criticisms of the model proposed by Roberts and Treagus are founded mainly upon different interpretations of the age of individual major structures and on the correlation of folds along strike. Thomas (1979) takes a different view of the Glen Orchy Antiform, based in part upon detailed work on continuation structures to the north-east, and regards Roberts and Treagus' D_2 nappes such as the Beinn Chuirn Anticline as primary D_1 structures, albeit with a strong 'D_3' overprint. He also casts doubts upon the interpretation of the large-scale swings of strike to the south-west of the Glen Orchy Antiform, which are necessary to accommodate the proposed D_2 nappes. Litherland (1982) doubts the magnitude of the deflections of fold axes and is critical of the proposed correlations between high-level and low-level D_1 folds. His detailed work in the Loch Creran area suggests alternative correlations involving less deflection of the fold axes. Thus he correlates the Islay Anticline with the Glen Creran Anticline (rather than the Airds Hill/ Kinlochleven Anticline), and the Loch Awe Syncline with the *Beinn Sgulaird Syncline* (rather than the Beinn Don/Ballachulish Syncline). He also supports the possibility of a link between the Loch Skerrols Thrust and the Benderloch Slide, which he regards as an early D_1 reactivation of a possible syn-sedimentary fault. The Loch Creran area is described by Litherland (1982) almost entirely in terms of NE-trending upright D_1 folds, becoming recumbent to the east, in a simple 'mushroom-like' structure. No secondary folds of regional extent are recognised and the swing of strike and D_1 fold axes north-westwards towards upper Loch Creran is attributed to the presence, during D_1, of a deep-seated basement massif in the area now occupied by the Cruachan Granite.

Structural hypothesis of A H Hickman (1978)

A more radical approach to the structure of the area between Glen Roy and Lismore is taken by Hickman (1978) who interprets many of the large-scale structures, identified as primary by Bailey and as D_1 by Roberts and Treagus, as secondary (D_2) structures. (The numbering of phases is compatible with the regional scheme of Roberts and Treagus.) Thus large-scale recumbent folds are recognised only in the south-eastern part of the section, between Kinlochleven and Loch Treig, where the presence of the D_1 Kinlochleven Anticline, with its extensive inverted lower limb, is confirmed (Figure 22C). However, the underlying D_1 syncline, which Bailey took to be a south-eastern extension of the Appin syncline, is redefined as a separate *Treig Syncline*, identified during new mapping in the Loch Treig area.

In the north-western part of the section, the Appin Syncline and the Tom Meadhoin Anticline are reinterpreted as separate upright D_2 folds, rather than recumbent D_1 structures (Figure 22C). In this respect the model follows the interpretation of Bowes and Wright (1967; 1973), based upon detailed mapping in the Ardsheal Peninsula. Hickman traces the Appin Syncline from Glen Roy through Ardsheal to Lismore where its equivalent, the *Balygrundle Syncline*, refolds an early isoclinal fold. The Tom Meadhoin Anticline is an upright structure along its entire length from Glen Nevis to the equivalent Airds Hill Anticline in Benderloch. Hickman maintains that there is no evidence that either structure becomes recumbent to the south-east to connect beneath the D_2 Stob Ban Synform with the Kinlochleven Anticline and Treig Syncline. Since the Appin Syncline is regarded as a D_2 structure, it is also suggested that the underlying Fort William Slide, which cuts its north-western limb, is also D_2.

The existence of the Ballachulish Syncline and the subsidiary Aonach Beag Syncline is refuted, largely on the basis of a stratigraphical interpretation which suggests that there is no repetition of strata. Thus the Stob Ban Synform is regarded as a simple upright D_2 syncline in a continuous right-way-up sequence, a conclusion also reached by Voll (1964) and Rast and Litherland (1970) in the Loch Creran area to the south-west (Figure 22C). This D_2 structure extends north-eastwards into the Glen Roy area.

Most of Hickman's interpretations and conclusions have been contested by Roberts and Treagus (1980), essentially on the grounds that the evidence for the nature and age of the major folds is not substantiated by the small-scale evidence. Litherland (1982) accepts Hickman's proposals as compatible with his own regional 'mushroom' model. However, several of the upright folds interpreted as D_2 by Hickman are interpreted as D_1 by both Roberts and Treagus and by Litherland (e.g. the Airds Hill Anticline and Beinn Donn Syncline). Thus the matter is far from being resolved, with many mutually incompatible features present in the various models.

The two principal modern interpretations (Roberts and Treagus, 1977c; Hickman, 1978) both accept much of the near-surface fold geometry of the original structure proposed by Bailey (1934). They differ considerably in the way in which these folds are projected to depth and are correlated across strike. Consequently there are highly significant differences in the assignation of relative ages to individual folds, much of the evidence for which depends on the detailed observation and interpretation of minor structures. Correlations of structures along strike are still problematical, and these have an important bearing upon vertical sections through different levels of the nappe pile. These sections are the basis for the various theories of structural evolution of the Grampian Highlands which will be discussed in more detail later.

North-eastern Lochaber

Mapping by the BGS around Glen Spean and Glen Roy has confirmed the presence of the major folds recognised in this area by Hickman (1978). However, the upright Appin Syncline, on the west side of Glen Roy, is regarded, as in previous interpretations, as an early fold contemporaneous with recumbent D_1 folds, such as the Treig Syncline, identified farther east. Later coaxial upright folds, including the D_2 Stob Ban Synform, tighten the early upright structures. Late cross-folds and related fabrics result in local complexity. The Fort William Slide has been traced into this area and around the closure of the Appin Syncline, but it dies out eastwards, where the Lochaber Subgroup appears to rest conformably on the Grampian Group. Between Spean Bridge and Loch Leven there is no high-strain zone at the position of the slide and the geometry of sedimentological features on either side lead Glover (1993) to suggest that, here at least, the slide is an unconformity developed during Appin Group times.

CENTRAL GRAMPIANS COMPLEX

The Central Grampians Complex corresponds broadly to Bailey's 'Sub-Eilde Complex' which was defined as all rocks stratigraphically below the level of the Eilde Quartzite. It therefore consists almost entirely of rocks of the Grampian Group and Central Highland Migmatite Complex. However, since it is defined here as a structural rather than a stratigraphical unit, it also includes local infolds of Appin Group strata. Its south-eastern and eastern boundary is defined by the Boundary Slide and its northern continuation, where major slides and/or shear zones commonly occur close to the Grampian/Appin group boundary. In the west the Western Grampians Complex overlies the Central Grampians Complex and is separated from it by the Fort William Slide. This slide is not continuous around the Western Grampians Complex so that on its south-eastern edge the boundary is difficult to define. For descriptive purposes and on Figure 19 the boundary of the complex is taken as the north-western edge of the Ossian–Geal Charn Steep Belt, where slides locally separate Appin and Grampian group strata.

Folds beneath the Boundary Slide: Glen Orchy and Atholl nappes

In the area around Glen Orchy and Dalmally, Thomas and Treagus (1968) recognised three major isoclinal, recumbent primary D_1 folds beneath the Boundary Slide (Figure 19, block B), the *Glen Lochy Anticline*, the *Beinn Udlaidh Syncline* and the *Beinn Chuirn Anticline*. In later papers, Roberts and Treagus (1975, 1977c) interpreted the uppermost, Beinn Chuirn Anticline, as a secondary D_2 fold, associated with strong deformation along the Boundary Slide, and correlated it with the Kinlochleven Antiform to the north-west. Thomas (1979) subsequently reiterated his belief in a D_1 age for all three folds and added a further fold, the *Dochard Syncline*, at the base of the pile. Whatever their age, it is agreed that all of the isoclinal folds, which together constitute the *Glen Orchy Nappe*, face towards the south-east (i.e. in the same direction as the major nappes above the Boundary Slide). These folds are arched across the broad dome, of the Glen Orchy Antiform, so that they face upwards on the north-west side of the dome and downwards beneath the Boundary Slide on the south-east side.

Farther to the north-east (Figure 19, block C), in the Glen Lyon area, Roberts and Treagus (1979) have identified three D_1 closures, analogous to the isoclinal folds of the Glen Orchy Nappe. A similar structure to the Glen Orchy Antiform has also been recognised in which the dome configuration is attributed to mutual interference of D_3 and D_4 antiforms.

Still farther to the north-east, detailed studies of the stratigraphy and structure around Strathtummel and along the A9 road section (Thomas, 1979; 1980) are interpreted to show that the Grampian Group strata are disposed in a large-scale, isoclinal D_1 fold termed the *Atholl Nappe*, beneath the Boundary Slide (Figure 19, block D). The nappe has the form of a broad arch, the *Drumochter Dome*. Over most of the dome, in the Drumochter area, the level of exposure lies in the flat-lying, inverted lower limb of the nappe. To the south-east the dip of both bedding and S_1 cleavage steepens and the hinge-zone of the nappe is exposed as the downward, SE-facing *Meall Reamhar Anticline*.

On the south-eastern side of the Drumochter Dome, D_2 folds occur on similar axes to the D_1 folds. The D_2 folds are overturned, with SE-dipping axial planes and S_2 cleavages, well seen for example in the *Clunes Antiform* and *Clunes Synform*. In parts of the D_1 hinge-zone, the S_2 cleavage is subparallel to and overprints S_1. Corresponding D_1/D_2 interference on the north-west side of the dome is well seen at Crubenmore where, despite refolding by D_2 folds overturned to the south-east, the D_1 folds can be seen to have been SE-facing originally (Thomas, 1987).

The nature of the Drumochter Dome has been the subject of some controversy. Thomas (1979; 1980) originally considered that the attitude and sense of overturning of the D_2 folds change relative to their position on the dome in a manner which suggests that the dome is an early structure developed during the primary deformation. He also attributed marked variations in axial trace and plunge of major later folds such as the Bohespic Antiform and Errochty Synform to their position relative to an early dome. However, subsequent work by Lindsay et al. (1989) has shown that D_2 axial planes and cleavages are folded across the dome, which is consequently now generally accepted as a later structure, possibly D_3 (D_4 of other authors) as proposed by Roberts and Treagus (1977c; 1979) for the related domes of Glen Orchy and Glen Lyon.

The existence of large-scale, SE-facing isoclinal D_1 folds must now be considered well established. However, Treagus (1987) has suggested that it is not necessary to invoke a separate Atholl Nappe. In his view the whole of the inverted sequence to the south-east from Drumochter can be considered as the same limb of a single major D_1 fold. Such a suggestion would considerably simplify the overall structural interpretation, removing the need to identify an intervening major syncline between the Tay and Atholl nappes and consequently reducing the amount of displacement inferred on the Boundary Slide.

Ossian–Geal Charn Steep Belt

To the north-west of the Drumochter Dome is a 4 km-wide zone in which all the fold limbs and fold axial planes are near vertical, forming a complex of upward-facing isoclines. This is the *Ossian–Geal Charn Steep Belt* of Thomas (1979), which can be traced for some 40 km from south-west of Loch Ossian, through Aonach Beag and Geal Charn, to Kinlochlaggan. In the Geal Charn–Aonach Beag area, three major slide zones are recognised, which commonly form steep boundaries between Grampian Group and Appin Group strata. The steep belt includes, on its north-western side, the *Kinlochlaggan Syncline* which has long been regarded as a major isoclinal primary fold (Anderson, 1947; 1956; Smith, 1968; Treagus, 1969).

Throughout the belt, Thomas recognises a strong axial plane schistosity, cut in places by two fabrics dipping NW and SSE respectively. These later fabrics and associated minor folds are comparable in orientation to major D_2 and D_3 folds immediately adjacent to the steep belt. Consequently Thomas regards the steep isoclines as primary D_1 folds.

To the south-west the Ossian–Geal Charn Steep Belt may be aligned approximately with the axial trace of the D_1 Loch Awe Syncline, although the Rannoch and Etive granitic complexes intervene, making direct correlation difficult. Both structures appear to mark a fundamental structural divide between NW-facing D_1 folds on one side (i.e. the Islay Anticline and other primary structures of the Western Grampians Complex) and SE-facing D_1 folds on the other (i.e. the Atholl, Glen Orchy and Tay nappes). Consequently, Thomas proposed that the Ossian–Geal Charn Steep Belt constitutes a root zone, lying directly below the Loch Awe Syncline, from which all of the fundamental D_1 nappes of the Grampian Highlands have diverged.

To the north of Kinlochlaggan a continuation of the steep belt may be traced as a zone of steeply inclined strata, thrust slices and intense deformation extending in a NNE direction through the Monadhliath Mountains (Piasecki and van Breemen, 1983). However, many of the upright folds in this zone are interpreted as late, secondary structures (Smith, 1968) and this casts doubt upon the origin of the Ossian–Geal Charn Steep Belt as a fundamental D_1 structure and root zone for the primary D_1 nappes.

Strathspey and the Monadhliath Mountains

The general structure of the southern part of this area was determined by Anderson (1956). In the ground to the south-east of the continuation of the Ossian–Geal Charn Steep Belt, a detailed petrofabric study by Whitten (1959) identified two generations of minor folds trending south-east and north-east but failed to discern any early major structures. In upper Strathspey, Smith (1968) identified a series of asymmetrical overturned folds, with axial planes dipping steeply to the south-east, which refold earlier isoclines. These late folds may correlate with the D_3 folds of the Ben Alder area to the south-west, identified by Thomas (1979).

A detailed study of the area around Kincraig, coupled with reconnaissance mapping and traversing over much of the Monadhliath area between Lochindorb and Kingussie, led to the first detailed comparison of structures in the Central Highland Migmatite Complex and Grampian Group (Piasecki and van Breemen, 1979a; 1979b; Piasecki, 1980). Here rocks of the Grampian Group appear to young consistently upwards. The earliest recognisable folds are tight asymmetrical folds, overturned to the north-west and are associated with slides located near the base of the Grampian Group sequence. These folds deform an earlier schistosity (presumed D_1) and therefore the folds and the early sliding are regarded as D_2. A major D_3 event resulted in overprinting of D_2 structures by a new subparallel regional schistosity, the development of minor folds, similar in form and orientation to those of D_2, and renewed sliding. Major NW-trending open domes (D_4) have a considerable influence on the local outcrop pattern, around Kincraig for example. These were followed by three further successive sets of minor folds (Piasecki, 1980).

North-west of the projected continuation of the Ossian–Geal Charn Steep Belt, many of the major fold structures are downward continuations of those recognised in the Western Grampians Complex. Anderson (1956) recognised the *Corrieyairack Syncline*, and regarded it as a primary structure, as did Piasecki (1975) working farther to the north-east in the upper Findhorn area. Piasecki also recognised other regional-scale isoclinal folds to the south-east, including the *Loch Laggan Anticline* and the *Loch Laggan–Monadhliath Syncline Complex* which he termed D_1. Open, near-upright folds, such as the NW-trending *Glenmazeran Antiform*, were termed D_2. A more intense, more widespread phase of folding produced tight, near-upright D_3 folds, with an overall NNE trend, subparallel to the earlier major folds. Minor structures and fabrics of this phase dominate the area to the north-west of the steep belt, overprinting the earlier structures, but subsequent work has shown that they are only weakly developed or absent to the south-west.

South-west of the River Findhorn, in the area around Loch Killin and the Corrieyairack Pass, Haselock et al. (1982) recognised apparent stratigraphical repetitions, minor structures and a widespread foliation which suggested the presence of major isoclinal axes folded by the more upright major regional folds. Thus the Corrieyairack Syncline and other related near-upright, NE-trending folds became regarded as D_2 structures. No significant NW-trending folds occur in this area and D_3 structures consist of open folds with axes generally trending around north–south, possibly comparable with the D_3 structures of Piasecki (1975) farther east. A similar structural pattern was recorded in the area between the Corrieyairack Granite and Loch Laggan by Okonkwo (1988). Both studies recognised a major tectonic discontinuity and zone of high strain, attributed to the D_1 deformation and termed the *Gairbeinn Slide*, which separates the Corrieyairack Subgroup from the underlying Glenshirra succession. A similar pattern of deformation is recognised in the Glen Roy district where rocks of greenschist facies, lithologically similar to those of the Glenshirra succession, are separated from the amphibolite facies Corrieyairack Subgroup by the *Eilrig Shear Zone*, a tectonic discontinuity on which significant NW-directed transport has taken place (Phillips et al., 1993).

The phases of deformation described above are broadly comparable to those established to the south and south-west in fold complexes which affect higher structural and stratigraphical levels of the Dalradian. Structural continuity and a common deformational history have been used as an argument for including the Grampian Group in the Dalradian, which accords with the proposal of Harris et al. (1978), based largely on local stratigraphical continuity. Recent structural studies along the A9 road section, summarised by Lindsay et al. (1989), have strengthened the belief in a common structural history. Structures can be traced with some confidence from the Boundary Slide at Blair Atholl, across the Drumochter Dome to mid-Strathspey. Here the stratigraphical way-up evidence fails due to increased tectonic strain and the development of gneissose fabrics. Comparable tectonic fabric relationships have been traced somewhat more tentatively north and north-west from Aviemore into Strathnairn and the Corrieyairack area, and are recognised both in the Grampian Group and in the highly migmatised rocks.

The implications of this structural correlation are far-reaching and affect the interpretation of the overall history of the Scottish Caledonides. A major problem, based on currently available age dates, is the wide timespan involved. Pegmatites developed in the slides which underlie and cut the basal part of the Grampian Group have been dated at around 750 Ma, so sedimentation must have commenced at least prior to this. If the Grampian Group structures are to be correlated with those of higher structural levels, then at least one deformation phase had been completed by the time of intrusion of the Ben Vuirich Granite, at 590 Ma, but the peak of metamorphism at around D_3 did not occur until Ordovician time, at around 490 Ma. At present these problems are well identified, but no satisfactory explanation has yet been presented.

Central Highland Migmatite Complex

The controversial nature of the relationship between the Grampian Group and the Central Highland Migmatite Complex has been discussed in detail elsewhere. One view advocates that the migmatite complex constitutes an older basement beneath a cover of Grampian Group sedimentary rocks and consequently that it has undergone more phases of deformation and metamorphism (Piasecki and van Breemen, 1979a; 1979b; 1983; Piasecki 1980; Piasecki and Temperley, 1988). However, an alternative interpretation suggests that the earliest recognisable fabrics in the migmatite complex are coeval with those in the Grampian Group of the Atholl Nappe (Lindsay, et al., 1989). These workers, therefore, favour the view that the Grampian Group extends down stratigraphically to include the Central Highland Migmatite Complex, in which the rocks are at a higher metamorphic grade and are commonly migmatised. The descriptions which follow are essentially those of Piasecki and his co-workers; it is the interpretation of these structures which is currently a matter of great debate.

Throughout much of its length, the boundary between the Grampian Group and the Central Highland Migmatite Complex is taken at a complex zone of multiple sliding or ductile thrusting, up to 200 m in thickness, known as the *Grampian Slide* (Figure 19, blocks E and F). Minor slides occur locally both above and below this zone and the slides appear to anastomose on a regional scale. The slides generally follow lithological boundaries and movement is concentrated locally within pelitic units, although stratigraphical units are cut-out in places. A complete range of textures is observed in a transition from unsheared rocks to ultramylonites. Approaching the slide zone, foliations develop into a distinctive thin platy fabric with a striking parallel orientation of micas. Within the slides, lenticular augen or stripes of quartz and feldspar are separated by trails of mica to form tectonic schists. The intensity of this schistosity varies according to lithology. Stretching lineations related to the shearing movements trend between north and NNE, and related asymmetrical folds are overturned towards the north.

The main movement on the Grampian Slide is regarded as syn-D_2 in the structural sequence of the overlying Grampian Group, since the slide schistosity is coincident with the axial surface of near isoclinal folds which fold an earlier fabric assumed to represent D_1. A suite of podiform amphibolitised gabbros, which occurs within this slide zone and in related zones of high strain, was emplaced after the early metamorphism and migmatisation, but prior to the D_2 deformation (Highton, 1992). A suite of foliated, concordant pegmatite veins also occurs in the slide zone throughout its 60 km outcrop length. The field and petrographical relationships of the pegmatites with the foliation in the tectonic schists indicate that they segregated during the main sliding. Rb-Sr ages of around 750 Ma from whole-rock and muscovite porphyroblasts in the pegmatites are therefore considered to date the D_2 sliding event (Piasecki and van Breemen, 1979a; 1979b; 1983). These dates are similar to those obtained from pegmatites north-west of the Great Glen Fault which define a *Morarian* tectonothermal event.

Most of the outcrop of the Central Highland Migmatite Complex is characterised by a composite gneissose fabric. The gneissose banding consists typically of alternating micaceous and quartzofeldspathic layers, from millimetres to several centimetres in thickness. Broad compositional banding occurs locally and may define original bedding, generally concordant with the gneissose banding. No major D_1 folds are recognised but the disposition of rare D_1 minor folds and local low-angle discordance between the gneissose fabric and compositional banding suggest that early larger-scale recumbent structures do exist. This deformation is considered to be broadly coeval with middle-to upper-amphibolite facies metamorphism and regional migmatisation. In areas of intense migmatisation, concordant granitic (quartz-alkali feldspar) leucosomes are developed within psammitic rocks and trondhjemitic (quartz-plagioclase) segregations are prominent in semipelitic lithologies.

The disposition of lithological units within the migmatite complex is controlled largely by major and intermediate-scale second and third folds. The D_2 structures are tight to isoclinal and recumbent to reclined, with a low to moderate plunge and a well-developed axial planar crenulation cleavage. On the limbs of these structures the earlier gneissose fabric becomes transposed into a new banding which is predominant over much of the outcrop. A large fold of regional extent, which dominates the outcrop pattern in the Kincraig area, is largely due to a later phase of folding (Piasecki, 1980; Piasecki and Temperley 1988).

According to Piasecki (1980) the early structures are cut by the Grampian Slide and are overprinted locally by the characteristic fabric of the slide zone. He therefore considered that they predate the earliest structures seen in the Grampian Group which are D_2, coeval with the sliding and dated by syntectonic pegmatites at around 750 Ma. He suggested, therefore, that the early structures in the Central Highland Migmatite Complex were formed during the *Grenvillian Event*, at around 1000 Ma. Unfortunately the only pre-750 Ma dates available, from a granitic gneiss at Ord Ban, are poorly constrained within a range of 1300 to 950 Ma (Piasecki, 1980).

Several phases of near-upright folding, which affect the gneissose structures and the shear zones, were recognised by Piasecki (1980), and correlated with the later deformations in the Grampian Group. Most of these phases are local and only one, NNE-trending, is of regional extent. Folds of this regional phase have been traced into the extension of the Ossian–Geal Charn Steep Belt where they become isoclinal (Piasecki and Temperley, 1988).

The alternative interpretation of the Central Highland Migmatite Complex outlined by Lindsay et al. (1989) proposes that all of the phases of deformation identified in the Grampian Group can be traced into Central Highland Migmatite Complex rocks, regardless of their metamorphic/migmatitic state, and that no earlier structures or fabrics can be recognised. The presence and distribution of migmatitic rock is attributed to a change in original lithology and/or an increase in tectonic strain superimposed on a general gradational increase in metamorphic grade. They considered that the slides have no major regional structural significance. In areas of the migmatite complex where tectonic strain and migmatisation are less intense, lithologies are seen to be similar to those of the adjoining Grampian Group, and in some areas a continuous local stratigraphical succession has been proposed. If this interpretation is accepted the implication is that rocks of the whole of the Grampian Highlands constitute a continuous Dalradian stratigraphical succession from migmatised rocks cut by 750 Ma pegmatites, upwards into Grampian, Appin, Argyll and Southern Highland group rocks, all of which were affected by orogenesis at 600 Ma or younger and by high-grade metamorphism at around 490 Ma.

There is no doubt that slide zones exist in both the western and eastern parts of the Central Grampian Highlands, some coinciding with major changes in metamorphism, migmatisation and stratigraphy. The significance of the slide zones and in particular whether any of them separate sequences of widely differing ages and structural histories, continues to be a subject of active research. In the absence of reliable dates older than 750 Ma, the case for a separate Grenvillian basement in the Central Highland Migmatite Complex remains not proven.

STRUCTURAL DEVELOPMENT OF THE GRAMPIAN HIGHLANDS

Early mapping in the Grampian Highlands concentrated on establishing the geometrical form of folds and other structures. Later, more-detailed work led to the interpretation of temporal structural sequences in individual areas. It was only when such studies became widespread, and attention was directed towards correlating groups of structures and phases of deformation between areas, that it became possible to speculate on the overall history and mechanism of regional deformation. Even today, such interpretations of structural development are based principally on the better known areas of the South-western, Central and Southern Highlands with much emphasis on the development of the Tay Nappe.

The early regional syntheses of Bailey and others made little comment on the mechanisms of nappe development apart from an 'eddy' theory in which NW-directed movements in the lower structural levels are compensated by movements in the opposite direction at higher levels (Bailey, 1938). South-eastward gravity sliding of the higher nappes was first proposed by Cummins and Shackleton (1955) and has subsequently been invoked by several authors in conjunction with various models. The concept of a root zone, from which nappes were expelled in opposing directions, was first introduced by Sturt (1961) and divergent 'mushrooms' or 'fountains' of nappes have dominated theories for many years. More recently emphasis has been placed on the importance of NW-directed primary structures, from which SE-facing folds such as the Tay Nappe developed by back-folding. Fundamental to all these theories have been discussions on the origin and geometry of the low-angled zones of high strain and dislocation which have been variously attributed to synsedimentary growth faults, early tectonic slides, later tectonic thrusts or polyphase combinations of these. The later, post-nappe phases of deformation and uplift are more clearly defined and their history has generated less debate than the early phases.

Root-zone and 'mushroom' models

The divergence of NW- and SE-facing nappes on either side of the D_1 Loch Awe Syncline led to the suggestion that the nappes were generated by lateral compression of an underlying anticlinal root-zone. From here nappes were expressed sideways as primary recumbent folds separated by slides, giving the observed mushroom-like structure (Figure 23A). The root-zone was originally thought to be exposed in the Tummel Steep Belt, which was believed to lie beneath a north-eastern continuation of the Loch Awe Syncline (Sturt, 1961; Harris, 1963; Rast, 1963). However, this steep belt was subsequently shown to be a zone of tight D_3 folds which refold the primary recumbent structures. Furthermore the nappes on the north-west side of the steep belt, such as the Atholl Nappe, are now known to be south-east facing over much of the Central Highlands, so that the major divergence actually occurs much farther to the north-west, in the Ossian–Geal Charn Steep Belt (Thomas, 1979; 1980). Thomas therefore envisages a root-zone within the Ossian–Geal Charn Steep Belt, from which the large-scale Atholl and Tay nappes advance to the south-east, whilst comparable Mamore, Kinlochleven and Ballachulish nappes advance north-westwards. Major dislocations developed between the nappes as older rocks were brought up from below and hence slides commonly constitute the Grampian/Appin group boundary. Primary arching of the recumbent structures was particularly strong at deeper structural levels giving rise to the Drumochter and Glen Orchy domes, which had a marked influence on later phases of deformation.

In most areas the nappes were modified by a coaxial D_2 phase of folding. To the north-west of the Ossian–Geal Charn Steep Belt, this phase of folding strongly modified the attitude of the NW-facing nappes, producing upright folds such as the Stob Ban Synform as well as more asymmetrical folds such as the Kinlochleven Antiform and the Blackwater folds. However, to the south-east the Tay Nappe continued to develop. It became further separated from the Atholl Nappe by renewed movement on the Boundary Slide and was trans-ported south-eastward under the influence of deep-seated D_2 simple shear on the lower limb (Harris et al., 1976). However, the upper limb and hinge-zone of the Tay Nappe, as seen in the Aberfoyle Anticline Complex, remained little affected by the D_2 movements. The Ben Lui Syncline, which probably separated the Tay and Atholl nappes following the D_1 movements, became marked by an antiformal complex of D_2 folds overlying the Boundary Slide.

Root-zone and 'fountain'/nappe-fan models

The concept of a root-zone was adopted by J L Roberts and J E Treagus in a model, originally based upon work in the South-west Highlands (Roberts, 1974; Roberts and Treagus, 1977c), but subsequently expanded to encompass observations in the Central Highlands and Schiehallion area (Roberts and Treagus, 1979; Nell, 1986; Treagus, 1987). In this model, early nappes are considered to have originated due to lateral compression of a root-zone beneath the Loch Awe Syncline, which is deflected around the north-west of the Glen Orchy Dome into the area of the Ossian–Geal Charn Steep Belt (Roberts and Treagus, 1979). Whereas most earlier theories, and that of Thomas (1979; 1980), envisaged a primary lateral development of recumbent nappes, Roberts and Treagus consider that the first major folds were essen-tially upright structures which fanned outwards above the root-zone (the 'nappe fountain'). The major slides were initiated during this D_1 phase. As the folds tightened, the fan collapsed under gravity to produce recumbent structures (Figure 23B). New fold hinges formed by this collapse became the D_2 folds, the most important of which, the Ben Lui Fold, has the overall effect of 'righting' the inverted limb of the Tay Nappe. According to Nell (1986) and Treagus (1987), the D_2 folds were overturned towards the north-west at deep structural levels where they were associated with NW-directed simple-shear along the Boundary Slide and related dislocations. At higher levels, SE-directed simple-shear, also of D_2 age, produced the intense defor-mation of the inverted limb of the Tay Nappe in the Flat Belt as described by Harris et al. (1976) and Mendum and Fettes (1985), echoes of the 'eddy' theory of Bailey (1938).

Although Roberts and Treagus describe the nappe development as a two-stage process, recorded as separate events in the rocks of any one area, it is difficult to imagine nappes actually forming in this manner. It therefore seems more likely that the upward expulsion of upright primary folds, the tightening and the subsequent lateral gravity collapse were all part of a con-tinuous process as described by Anderton (1988).

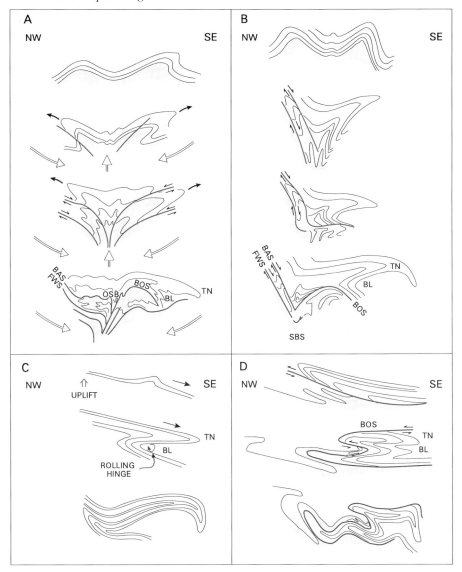

Figure 23 Alternative models for the structural development of the Grampian Highlands fold belt (see text for details). Individual stages are not numbered (D$_1$, D$_2$ etc.) to avoid confusing and unintentional time correlations between the models, but all show deformation up to and including the post-nappe D$_4$ folds.

A   Root zone and 'mushroom' models: final stage adapted from Thomas (1980)
B   Nappe fans/'fountains': adapted from Roberts and Treagus (1977 a; c) and Treagus (1987)
C   Gravity sliding of rootless nappes: adapted from Shackleton (1979)
D   North-westward movement and backfolding: from model of Hall *in* Fettes et al. (1986)

Stipple = Grampian Group; BAS = Ballachulish Slide; BLF = Ben Lui Fold Complex; BOS = Boundary Slide; FWS = Fort William Slide; OSB = Ossian–Geal Charn Steep Belt; SBS = Stob Ban Synform; TN = Tay Nappe

The most recent versions of this model, described by Nell (1986) and Treagus (1987), are based upon accurately constructed cross-sections through the Central and Southern Highlands. They include a reappraisal of all the earlier work by Roberts, Treagus, Thomas, Harris and others and are the best available syntheses at the time of writing.

Gravity sliding and 'rootless' nappe models

Although models involving root-zones have dominated the literature for many years, the tightly folded nature of the steep belts makes it very difficult to prove that they involve primary D_1 folds which can be confidently identified as roots to the flanking recumbent nappes. Furthermore, several authors are opposed to the overall concept, finding it difficult to see how such an enormous volume of nappes can be ejected from such a narrow root zone (Shackleton, 1979; Coward, 1983).

The formation and lateral movement of recumbent nappes by deep-reaching gravitational flow were first suggested by Cummins and Shackleton (1955), based upon the attitude and character of the Ben Lui and Ben Udliadh folds. Studies of the Tay Nappe revealed evidence of extensive D_2 simple shear of the lower limb (Harris, Bradbury and McGonigal, 1976). This in turn led to a model in which the nappe nucleated at a high level (with no root) and moved south-eastwards on a shear zone above a 'rotation zone' located in the Tummel Steep Belt (Bradbury, Harris and Smith, 1979). The 'rotation zone' acted as a rolling hinge which connected the inverted limb of the Tay Nappe, through the Ben Lui Fold, with the underlying right way-up succession (Figure 23C). Movement of the nappe was seen as being due to gravitational instability, possibly accentuated by a density inversion of the Dalradian succession, whereby dense metagreywackes and basic igneous rocks of the Argyll and Southern Highland groups overlie less dense carbonates, pelites and siliceous rocks of the Grampian and Appin groups. This model was extended by Shackleton (1979) who envisaged a general gravity-operated, down-plunge flow of both D_1 and D_2 folds away from a Central Highlands plunge culmination, south-westwards towards a plunge depression in the Loch Awe area. There is, however, no evidence to support this down-plunge flow and much contradictory evidence exists.

Other authors, whilst not attributing the whole structural development of the region to gravity gliding, do accept the possibility that gently dipping gravity structures may have contributed to lateral flow at high structural levels (Anderton, 1988). Indeed, the continued south-eastward translation of the Tay Nappe during D_2 deformation in the model of Thomas (1980), which accepts the deep-seated simple shear described by Harris et al. (1976), is suggestive of a gravity structure.

North-westward movement and backfolding models

Wider considerations of the tectonic evolution of the whole of the Scottish Caledonides emphasise the north-westerly overturning and movement of most major structures (Coward, 1983; Dewey and Shackleton, 1984; Watson, 1984). Some detailed structural models of the Grampian Highlands also include a strong element of NW-directed thrust movement on such structures as the

Fort William, Ballachulish and Boundary slides, possibly rising from a funda-mental basal or 'floor thrust' (Bradbury, 1985; Nell, 1986). North-westerly movement has been demonstrated in the Eilrig Shear Zone (Phillips et al., 1993) but in many areas, particularly in the Western Grampians Complex, major dislocations have the geometry of extensional faults with younger rocks emplaced on older, and it has been argued that they may have originated as growth faults during basin formation (Soper and Anderton, 1984; Anderton, 1988). It is therefore necessary to postulate reactivation and reversal during the subsequent compressional phase. Anderton (1988) argues that this need not necessarily involve significant north-westerly tectonic transport.

It is possible that the SE-facing folds, such as the Tay Nappe, originated as high level 'rootless' gravity nappes triggered by the thrusting and consequent crustal thickening. However, in many current models the distinction between these and the NW-facing folds is of lesser significance. Although they may well have originated as primary structures, their current facing direction is probably a result of backward gravitational collapse or of subsequent defor-mational events.

Both Treagus (1987) and Anderton (1988) suggest that the south-eastward translation of compressional upright folds took place at high level, above NW-directed thrusts. Other authors envisage large-scale isoclinal backfolding of an original NW-facing Tay Nappe by the D_2 Ben Lui Folds, during continued NW-thrusting (Bradbury, 1985; Nell, 1986; L M Hall, in Fettes et al., 1986) (Figure 23D).

Accepting an overall north-westerly movement enables the construction of more realistic continuous structural cross-sections from the Highland Border north-westwards across the Great Glen to the Moine Thrust Zone (Figure 24), where Caledonian structures override the foreland (Coward, 1983; Fettes et al., 1986). The structural development of the whole of the Scottish Cale-donides can be modelled as a complete entity and thereby be more easily related to even wider plate tectonic reconstructions (Garson and Plant, 1972; Lambert and McKerrow, 1976; Phillips et al., 1976; Watson, 1984; Dewey and Shackleton, 1984; Soper, 1988).

Post-nappe deformation and uplift

Following the development of the primary nappes, the later phases of defor-mation were, with notable local exceptions, generally less intense and produced more-open, near-upright structures trending between ENE and north-east. Refolding and overprinting by still later fabrics are generally only of local extent. Although the numbering of later phases is frequently incon-sistent and complicated by local variations, most authors are agreed on the overall sequence of events.

In most areas the peak of metamorphism is coeval with or immediately predates the D_3 deformation (the Grampian Event of some authors; Table 1) and later deformations are associated with retrograde effects. Most authors therefore attribute the initial D_4 deformation to the commencement of late-orogenic isostatic uplift (Watson 1984; Harte et al., 1984; Dempster, 1985). The scale, monoformal nature and lateral continuity of many of these early D_4 folds, especially the Highland Border Downbend, suggest that they are controlled by a parallel series of major basement lineaments. Between these

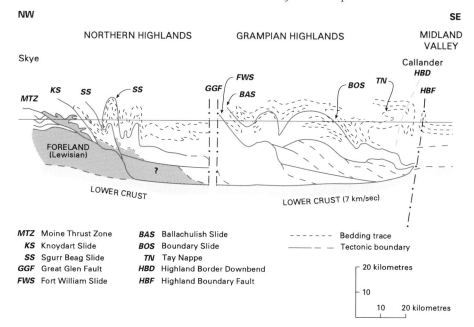

Figure 24   Diagrammatic structural cross-section of the Grampian and Northern highlands. (Based on sections by Powell, 1974; Roberts and Treagus, 1977a; Coward, 1983; Bradbury, 1985; Mendum and Fettes, 1985; Fettes, Harris and Hall, 1986; Nell, 1986; Roberts et al., 1987.) Deep crustal divisions relate to the LISPB seismic refraction profile (Bamford et al., 1978).

lineaments episodic uplift of crustal blocks at different rates in both space and time generated the pattern of contrasting flat and steep belts which now dominates the Southern Highlands. Minor folds and crenulation cleavages, which locally overprint the early D_4 major flexures, can be traced north-westwards across the flat belt, where they are seen to be related to broad, upright folds such as the Ben Lawers Synform. These are taken to indicate a late D_4 compressional phase which postdates the early D_4 uplift event. Roberts and Treagus (1977c; 1979) and Treagus (1987) also consider that the Drummochter and Glen Orchy domes are late D_4 structures (their D_3). The D_4 deformation is also probably responsible for the steepening of major D_2 (and/or D_3) fold limbs in the Knapdale and Tummel steep belts.

Other late major folds, such as the Bohespic Antiform and Errochty Synform, which have a more north–south trend, are attributed to D_4 (his D_3) by Thomas (1979; 1980), but are considered to belong to a later phase by Treagus (1987). The Turriff Syncline, Buchan Anticline and other late NE- to NNE-trending open folds in the North-east Highlands, together with renewed movements on major shear-zones such as those of the Portsoy Lineament, may also belong to this phase. Still later structures tend to be small-scale, brittle open box folds and conjugate kink-zones, although NW-trending flexures and monoforms affect the limbs of the Errochty Synform (Thomas, 1980) and Roberts (1974) describes a complex sequence of late structures from the South-west Highlands.

7   Metamorphism

Traditionally the metamorphic grade or facies in the Grampian Highlands is referred to zones defined by a set of index minerals developed in pelitic rocks. These zones in the south-east Highlands were first described by Barrow (1893; 1912). Slightly modified by Tilley (1925) the *Barrovian zones* (chlorite–> biotite–> garnet–> staurolite–> kyanite–> sillimanite) were subsequently extended across the Southern, South-west and Central Highlands, mainly by Elles and Tilley (1930) and Kennedy (1948). At the same time Read (1923; 1952) identified a different style of metamorphism in the North-east Highlands which formed the basis of the *Buchan zones* (biotite–> cordierite–> andalusite–> sillimanite). Subsequently Winchester (1974), working mainly from assemblages in calc-silicate rocks, which he was able to correlate on an empirical basis with the pelitic zones, extended the Barrovian zones across the remaining areas of the Grampians region.

The two zonal sequences define different styles of metamorphism or facies series. Read (1952) regarded the Buchan and Barrovian styles of metamorphism as the products of separate events. However, Fettes et al. (1976) showed that the two zonal sequences reflected a regional variation from an intermediate/high-pressure facies series in the South-west Highlands through to a low-pressure facies series in the North-east Highlands. This transition between the Barrovian and Buchan series was formalised by Harte and Hudson (1979) who defined four zonal sequences, namely, in order of decreasing pressure: *Barrovian* (biotite–> garnet–> staurolite–> kyanite); *Stonehavian* (biotite–> garnet–> chloritoid + biotite–> staurolite–> sillimanite); *West Buchan* (biotite–> cordierite–> andalusite–> staurolite–> kyanite); and *East Buchan* (biotite–> cordierite–> andalusite–> sillimanite).

Distribution of facies

Fettes et al. (1985) and Harte (1988) devised metamorphic facies maps of the Grampian region based on the wealth of published data on the distribution of pelitic and calc-silicate index minerals allied to data from assemblages in basic rocks (Wiseman, 1934).

The facies map (Figure 25) shows the South-west Highlands as lying almost wholly in the greenschist facies. The grade rises to the amphibolite facies in the north (at lower stratigraphical levels) and east (at similar stratigraphical levels) with the greenschist facies wedging out northwards against the Great Glen and narrowing markedly eastwards against the Highland Boundary Fault, so that on the Stonehaven coast section the metamorphic grade rises rapidly northwards from the fault. Greenschist facies assemblages reappear northwards in the Turriff Syncline of Buchan.

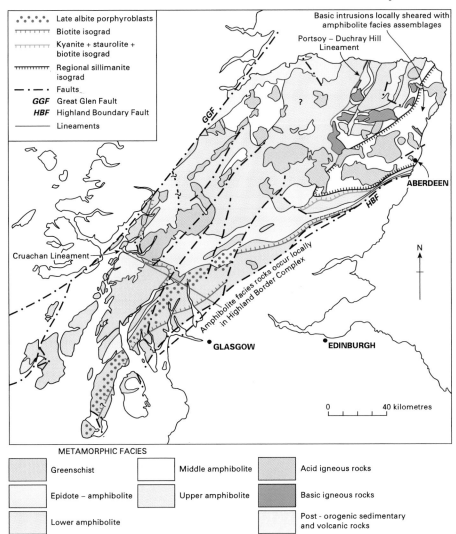

Figure 25 Metamorphic zones of the Grampian Highlands (modified after Fettes et al., 1985. Reproduced by permission of the Geological Society of London).

The greater part of the higher-grade area lies in the lower amphibolite facies, characterised by kyanite + staurolite- and andalusite + cordierite-bearing assemblages. In the North-east Highlands and in the northern Central Highlands, middle and upper amphibolite facies rocks are found. These are defined, respectively, by sillimanite + muscovite- and sillimanite + K-feldspar-bearing assemblages. The sillimanite-bearing rocks are commonly characterised by the presence of quartzofeldspathic lenses and augen. They have been referred to as migmatites (e.g. Johnstone, 1966), although they have evolved through a wide variety of processes including partial melting, metamorphic segregation and metasomatism; they are therefore better termed sillimanite-gneisses.

The facies boundaries or isograd surfaces are broadly flat lying in the Central Highlands, steepening markedly against the Highland Boundary Fault. This general disposition of facies and facies boundaries gave rise to the concept of a 'thermal anticline', the high-grade rocks of the Central Highlands forming the core and the 'axis' plunging to the south-west along the spine of epidote-amphibolite facies rocks in Knapdale and Cowal (Kennedy, 1948).

The regional facies pattern is taken to reflect the peak conditions of metamorphism. Most of the early workers regarded this metamorphic imprint as resulting from a single metamorphic event, albeit *polyphasal* with progressive phases broadly related to burial and retrogressive phases broadly related to uplift. However, the possibility exists that in the northern Central Highlands the rocks were subjected to more that one metamorphic event and are *polymetamorphic*. In these cases the facies pattern is likely to be composite, reflecting the effects of the various metamorphisms.

The nature of the chemical reactions governing the development of the various index zone assemblages and isogradic surfaces has been extensively studied. A full discussion of these is not possible here but excellent summaries are given, amongst others, by Atherton (1977), Harte and Hudson (1979), and Chinner and Heseltine (1979).

Metamorphic provinces

Harte (1988) divided the Grampian metamorphic rocks into six provinces or domains on the basis of their metamorphic characteristics. The provinces are as follows:

South-west province: covers the area south-west of the Cruachan Lineament. The province is defined by greenschist facies assemblages and high pressures. The area also includes thick accumulations of basic lavas and associated sills (Graham, 1976).

South-east province: lies east of the Portsoy–Duchray Hill Lineament and between the Highland Boundary Fault and the sillimanite zone. This area, which encompasses the type area of Barrow's zones, is characterised by intermediate/high pressure and high thermal gradients (narrow metamorphic zones) adjacent to the Highland Boundary Fault and has a distinctive uplift history.

Southern province: comprises the area bounded to the west by the Cruachan Lineament, to the east by the Portsoy–Duchray Hill Lineament, to the north by the Tummel Steep Belt and to the south by the Highland Boundary Fault. Its metamorphic characteristics are transitional between provinces 1 and 2. A zone of albite porphyroblasts of regional extent occurs between Balquidder, Cowal and Kintyre. The zone straddles the junction of the greenschist and epidote-amphibolite facies and the boundary between the Southern and South-west provinces; the porphyroblasts are believed to have developed relatively late in the metamorphic history.

Buchan province: comprises north-east Scotland east of the Portsoy–Duchray Hill Lineament and is characterised by low-pressure metamorphism. The highest-grade rocks are granulite-facies hornfelses found in the aureoles of the basic complexes in the Buchan area. These hornfelses are characterised by garnet orthopyroxene-cordierite and related assemblages (Droop and Charnley, 1985).

Central province: the area north of the Tummel Steep Belt and west of the Portsoy–Duchray Hill Lineament. The area is characterised by marked increases in pressure during the peak metamorphism.

North-west province: the remaining part of the Grampian Highlands. Part of this area may be polymetamorphic.

To these six provinces a small seventh one near the Great Glen south of Fort Augustus may now be added. Here Phillips et al. (1993) have proved greenschist facies rocks at low structural levels beneath the Eilrig Shear Zone, which separates them from amphibolite facies rocks above (Figure 25).

Metamorphic history

Early workers regarded deformation and metamorphism as single events, although there was considerable debate on the relative ages of the two. For example, Bailey (1923) regarded the two as contemporaneous, Read (1952) suggested that the metamorphism was later and Elles and Tilley (1930) believed that the deformation was later. Much of this debate was based on the spatial attitude of the isogradic surfaces and how far, if at all, this reflected deformation. However, detailed textural studies by, amongst others, Rast (1958), Sturt and Harris (1961), Johnson (1962; 1963), Harte and Johnson (1969) and Upton (1986) showed that the development of the metamorphic assemblages could be referred to stages marked by deformational phases. The results of these and similar studies have shown that progressive metamorphism occurred during the main nappe-forming movements (D_1 and D_2) and reached a peak around the time of the nappe-folding events (D_3). Retrogressive metamorphism took place during uplift and associated movements (D_4). The isograd surfaces were folded by these late structures, for example, the Boyndie and Ben Lawers synforms and the Highland Border Downbend.

In general, progressive metamorphism is assumed to have taken place along a simple curve on a pressure–temperature plot. However, in central Perthshire, Dempster and Harte (1986) document a significant increase in pressure close to the peak of metamorphism, with the replacement of chloritoid + biotite by garnet + chlorite as well as the localised growth of kyanite- and staurolite-bearing assemblages. They ascribe the pressure increase, of about 2 to 3 kb, to tectonic movements associated with the development of the D_3 Tummel Steep Belt. In the North-east Highlands, Baker (1985) and Beddoe-Stephens (1990) detailed a similar increase in pressure which they suggested is due to westward overthrusting during D_3; this resulted, locally, in the inversion of andalusite to kyanite over a restricted area lying immediately west of the Portsoy–Duchray Hill Lineament.

It has been suggested that the growth of sillimanite in the Buchan area may also depart from the simple progressive model. Chinner (1966) believed the sillimanite formed in response to a thermal overprint on a depth-controlled metamorphism. Ashworth (1976) suggested that, at least in the Huntly–Portsoy area, the sillimanite developed as a result of the thermal effects of the basic complexes, a suggestion supported by the work of Fettes (1970) and Pankhurst (1970) who believed that the basic masses were intruded close to the peak of metamorphism. Harte and Hudson (1979) recognised two phases of sillimanite growth, one regional and the other related to the basic masses, although they considered that the two phases were closely linked in time. They delineated a 'regional' sillimanite isograd within the mapped sillimanite zone (Figure 25), and regarded this isograd as the limit of the sillimanite zone prior to the intrusion of the basic complexes.

In the northern Central Highlands migmatisation and the development of metamorphic porphyroblasts may have begun before 750 Ma, that is prior to movement on the Grampian Slide. In the Southern and North-east Highlands the peak of metamorphism occurred at around 520 to 490 Ma (the Grampian Event of some authors; Table 1). The question then arises whether the metamorphism was polymetamorphic or a single event and, if the latter, whether metamorphic crystallisation was effectively suspended or interrupted for a period of over 200 Ma or whether there is marked diachroneity or differences in the tectonothermal evolution across the region. At present the geochronological and geological evidence cannot be satisfactorily reconciled and the metamorphic evolution remains equivocal.

In the Southern and North-east Highlands retrogressive metamorphism took place from 490 to 390 Ma although considerable regional variation occurred in the rate and timing of uplift phases (Dempster, 1985). Watkins (1983) has argued that the growth of the albite porphyroblasts in the South-west Highlands occurred during the retrogressive phase. He suggests that this growth of albite was facilitated by dehydration fluids trapped in D_3 structures. Dymoke (1989), however, has suggested that the growth of albite was a late prograde reaction initiated by, and spatially related to, zones of D_3 movements, the rocks being close to the temperature conditions which they achieved during the pre-D_3 peak of regional metamorphism.

Presssure–temperature estimates

A considerable body of data has been produced on the pressure and temperature (PT) conditions of metamorphism. These highlight the variation in pressure within various facies and index mineral zones. For example, peak temperatures in the kyanite + staurolite zone were between 500°C and 550°C across the region but pressures varied from 9–10 kb (c. 30km depth) in the Central Highlands (Moles, 1985; Baker, 1985), through 5–6 kb (c. 18 km) in Angus (Dempster, 1983; 1985; McLellan, 1985) to 3–4 kb (c. 11 km) in Banffshire (Hudson, 1985; Beddoe-Stephens, 1990). Also, Graham (1983) recorded pressures of 8–10 kb (c. 30 km) for epidote-amphibolite facies rocks in the South-west Highlands compared with 2–3 kb (c. 8 km) for similar grade rocks in Banffshire (Hudson, 1985). The highest temperatures occurred in the sillimanite + K-feldspar zones associated with the inner aureoles of the North-east Highlands basic complexes, with measurements of 800–850°C (Droop and Charnley, 1985). These extreme conditions are believed to have occurred in the roof zones of the basic masses (Fletcher and Rice, 1989).

The wealth of PT data allied to detached mineral cooling ages has allowed the erection of sophisticated PT evolution paths (at least for the later phases) for the South-east Highlands (Dempster, 1983; 1985); these illustrate regional variation as well as phases of rapid uplift and cooling in the periods 460 to 440 Ma and about 410 Ma.

Metamorphic models

A great variety of models has been produced to explain the metamorphic pattern. The models have included the thermal effects of older granites (Barrow, 1912), burial (Elles and Tilley, 1930), the tectogen or mountain root model of Kennedy (1948), uprising migmatite domes (Read, 1952; Read and Farquhar, 1956), the self generation of heat in a tectonically thickened crust (Richardson and Powell, 1976), and crustal overplating on a NW-directed thrust duplex analagous to the Alps (Bradbury, 1985).

The North-east Highlands were not only characterised by high geothermal gradients during metamorphism, but also by a suite of late- to post-kinematic granites and, uniquely in Scotland, basic and ultramafic complexes. However, although it is accepted that high heat flow is a factor in the metamorphic evolution, its exact cause is uncertain. It has been ascribed to the influence of deep-seated magmatic intrusions (Harte and Hudson, 1979) and lithospheric stretching (Kneller, 1985).

In the South-east Highlands the extremely steep lateral thermal gradients and possible inverted zoning have been related to some form of underthrusting by cold crust (Chinner, 1978), the junction of a Dalradian block heated and uplifted by deep-seated intrusions against a subsiding basin filled with cold sediment (Harte and Hudson, 1979), and the southward translation of a relatively hot Tay Nappe (Chinner, 1980).

The South-west Highlands contrast markedly with the Buchan region in having an intermediate- to high-pressure facies series, an absence of granite plutons and thick accumulations of basic lavas and related intrusions which may have produced a dense relatively thick crust which suppressed the geothermal gradient.

In the Balquidder area Watkins (1984) suggested that the inverted metamorphic zoning noted by Tilley (1925) is related to the southward translation of relatively hot rocks, in the core of the Tay Nappe, over cooler strata during D_2 (cf. Dempster and Bluck, 1991). However, Dymoke (1989) and C M Graham (oral communication, 1992) have established that the position of the present garnet isograd in north Kintyre is a product of retrogression. The 'original' garnet isograd, related to the progressive metamorphism, lay to the south-east of the present position with a symmetrical arrangement to the Cowal Antiform; this negates the arguments for inversion, at least in the South-west Highlands.

The various models illustrate the complex variations imposed on the metamorphic pattern during the progressive phases of metamorphism. Dempster (1985) has also shown that variable uplift and tectonism during the retrogressive phases, involving the juxtaposing of relatively warm and cold rocks, distorts and complicates the facies pattern.

These regional variations in the metamorphic history make it difficult to sustain any simple single model for the Grampian Highlands. The six provinces recognised by Harte (1988) may therefore represent domains with different metamorphic histories.

Harte (1988) regarded the boundaries to the domains or blocks as part metamorphic and part tectonic. This concept is built in part on the work of

Ashcroft et al. (1984) and Fettes et al. (1986), who recognised the long-lived influence of major lineaments such as the Cruachan and Portsoy–Duchray Hill lineaments on the evolution of the Grampians. In particular the lineaments delimit domains with different sedimentological and tectonothermal histories. Regionally, therefore, the metamorphic pattern was influenced by variations in the thickness and nature of the sedimentary pile, degrees of compression, tectonic thickening or lithospheric stretching, igneous events and rate and degree of uplift, all of which produced a complex series of pressure–temperature–time curves for the Grampian region.

8 Caledonian magmatism

In the discussion which follows, the established divisions of Older, Syntectonic and Newer Igneous Rocks used in the third edition of this handbook have largely been modified, to become a seven-fold division of the Caledonian Igneous Suite. Each division or grouping is linked to one or more episodes in the history of deformation events in the development of the Caledonian Orogen, which spanned the period from about 750 to 390 Ma. Excluding the events recorded by the tectonically emplaced Highland Border Complex (Chapter 9), the principal magmatic episodes were as follows:

Pre-tectonic basic magmatism
Syntectonic granitic intrusions
Syn- to late-tectonic basic and ultramafic intrusions
Late-tectonic granitic intrusions
Post-tectonic granitic intrusions
Late- to post-tectonic minor intrusions
Lower Old Red Sandstone volcanism

The distribution of components of these groups, except for the minor intrusions, is shown on Figure 26; each is numbered and referenced in italics in the text, e.g. *Dunfallandy Hill Granite (8)*.

The age of an intrusion with respect to the tectonic history of the surrounding rocks cannot always be determined from the local field evidence, especially in areas of poor exposure, and hence certain intrusions have been assigned to a suite by analogy with petrologically similar intrusions whose structural setting is better known. It should be noted also that the three latest suites were largely contemporaneous.

PRE-TECTONIC BASIC MAGMATISM

The Central Highland Migmatite Complex contains partially amphibolitised two-pyroxene-gabbros in two zones, structurally underlying the boundary between the migmatite complex and the Grampian Group (Highton, 1992); they are small bodies not shown on Figure 26. They were intruded after the local D_1 deformation and associated metamorphism and migmatisation, but before D_2. They are preserved as concordant sheets and larger podiform masses up to 800 m by 200 m. Primary igneous textures are preserved in the largest masses. Cumulate textures are widely developed, and large pyroxene oikocrysts are prominent in the coarser lithologies. The primary igneous assemblages were clinopyroxene + plagioclase + brown hornblende + ilmenite ± orthopyroxene ± biotite.

Igneous activity during later deposition of the Dalradian Supergroup was closely linked with crustal instability, which resulted in the commencement of high-level rift faulting in Argyll Group times (Chapter 5). Increased crustal stretching, associated with the opening of the Iapetus Ocean to the south-east of the Dalradian depositional basin (Graham, 1976; Leake, 1982), led to the development of volcanism, particularly in the South-west Highlands.

The earliest recorded evidence of volcanic activity, widespread but minor and basic in character, is in the Argyll Group. Minor tuffs and pillow-lavas occur in the Muckle Fergie and Kymah burns (Tomintoul/Glenlivet area), in the lower part of the Islay Subgroup. A few thin cross-bedded tuffaceous beds occur at a similar stratigraphical level in the Loch Creran area (Litherland, 1980). To the north of the Dee and south of Tomintoul, the Easdale Subgroup contains the *Delnadamph Volcanic Member* (*5*) lying within the local equivalent of the Ben Eagach Schist. Pillow lavas, fine-grained vesicular amphibolites, and coarse-grained actinolite-rich amphibolites are associated with pelitic to psammitic rocks. The Ardrishaig Phyllite (Easdale Subgroup) contains sporadic 'green beds' (Borradaile, 1973), while the equivalent Ben Lawers Schist contains minor basic lavas as well as some green beds.

The first major outbreak of volcanic activity is recorded close to the top of the Easdale Subgroup, in which metavolcanic rocks extend discontinuously for 120 km along strike. Several eruptive centres are indicated. The Beinn Challum Quartzite and Sron Bheag Schist, occurring between Tyndrum and the Loch Tay Fault, show a north-eastward increase in the proportion of volcanic material, with lavas and tuffaceous 'green beds' being intercalated within a metasedimentary succession. From the Loch Tay Fault to Glen Shee, the Farragon Beds (Goodman and Winchester, 1993) show an increased proportion of metavolcanic material, reaching 400 m in thickness in the Ben Vrackie area. Here, fine-grained, foliated metabasalts and tuffaceous metasedimentary rocks occur together with rubbly meta-agglomerates (Graham and Bradbury, 1981). The Farragon Beds thin to 40 m in Gleann Fearnach, and are separated by a 20 km gap from the metabasite at Meall Dubh in Glen Girnock. Slightly higher in the succession, the metabasite at *Balnacraig* (*4*) lies within the lower part of the Queen's Hill Gneiss Formation (Crinan Subgroup) in Glen Muick. It consists of fine-grained hornblende-schist intercalated with layers of psammite and calc-silicate rock. Basaltic pillow lavas together with fragmental ultramafic rocks of possible volcanic origin also occur in the Blackwater Formation of the Argyll Group (possibly Easdale Subgroup in part) between Upper Donside and Dufftown (*1*; MacGregor and Roberts, 1963; Fettes et al., 1991).

A lenticular zone of tuffs occurs within the Crinan Grit at Craignish, and a 50 m-thick band of green schists enriched in Mg, Cr, Cu and Ni at the base of the Ben Lui Schist between Tyndrum and Glen Lyon was probably derived by the erosion of local ultramafic rocks (Fortey and Smith, 1986). A major phase of sill injection took place, centred on the Knapdale–Tayvallich area, commencing roughly contemporaneously with the deposition of the Ardrishaig Phyllite and continuing until the extrusion of the Tayvallich Volcanic Formation (Wilson and Leake, 1972; Graham, 1976).

The main centre of extrusive activity in late Argyll–early Southern Highland group times was located in the *Tayvallich–Loch Awe* area of the

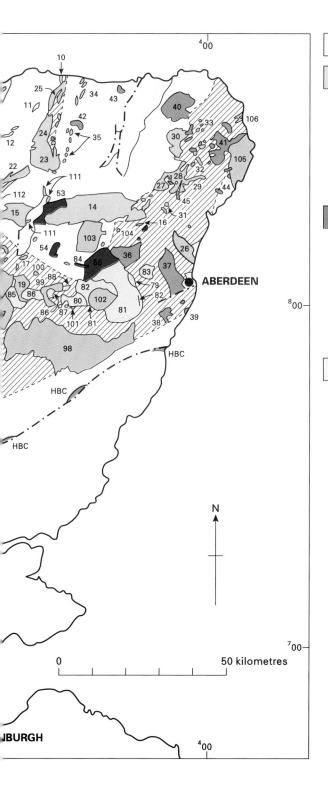

Upper Silurian and younger sedimentary rocks in the Grampian Highlands

Volcanic rocks in the Lower Old Red Sandstone

108. Lorn Plateau
109. Glencoe
110. Ben Nevis
111. Rhynie
112. Cabrach
113. Gollachy Burn
114. Forest of Alyth
115. Monzie

Post-tectonic granitoid intrusions

(i) South Grampians Suite

56. Garabal Hill–Glen Fyne
57. Arrochar
58. Doune Farm
59. Inversnaid
60. Glen Tilt
61. Glen Doll
62. Glen Shee
63. Glen Derry
64. Abergeldie
65. Comrie

(ii) Argyll Suite

66. Moor of Rannoch
67. Strath Ossian
68. Ballachulish
69. Ben Nevis
70. Mullach nan Coirean
71. Glencoe
72. Etive
73. Kilmelford
74. Foyers
75. Findhorn
76. Boat of Garten
77. Corrieyairack
78. Allt Crom
79. Gask
80. Torphins
81. Crathes
82. Balblair
83. Clinterty
85. Logie Coldstone
86. Tomnaverie
87. Kincardine O'Neil
88. Lumphanan

(iii) Cairngorm Suite

89. Auldearn
90. Ben Rinnes
91. Glenlivet
92. Dorback
93. Monadhliath
94. Cairngorm
95. Lochnagar
96. Glen Gairn
97. Ballater
98. Mount Battock
99. Cromar
100. Cushnie
101. Ord Fundlie
102. Hill of Fare
103. Bennachie
104. Middleton
105. Peterhead
106. St Fergus
107. Coilacreich

Late-tectonic granitoid intrusions

(i) North-eastern biotite - muscovite granites

9. Glen Clova-Hunt Hill and Cairn Trench
36. Kemnay
37. Aberdeen
38. Auchlee
39. Cove
40. Strichen
41. Forest of Deer
42. Aberchirder
43. Longmanhill
44. Lochlundie
45. Ardlethen

(ii) North-western biotite-muscovite granites

46. Ardclach
47. Moy
48. Grantown
49. Glen Kyllachy
50. Maol Chnoc
51. Strathspey
52. Loch Laggan

(iii) North-eastern diorites to granites

53. Kennethmont
54. Syllavethy
55. Tillyfourie
84. Corrennie

Syn-to late-tectonic basic and ultramafic intrusions

(i) Largely basic

14. Insch
15. Boganclogh
17. Kildrummy
18. Morven-Cabrach
19. Tarland
21. Blackwater
23. Huntly
24. Knock
25. Portsoy
27. Haddo House
28. Arnage
29. Arthrath-Dudwick
30. Maud
31. Udny-Pitmedden
32. Kinnadie
33. West Crichie
34. Boyndie
35. Marnoch

(ii) Basic, with large amounts of ultramafic rock

16. Lawel Hill
20. Coyles of Muick
22. Succoth - Brown Hill
26. Belhelvie

Highland Border Complex lavas and tectonically emplaced ultramafic rocks (HBC)

Syntectonic granitic intrusions

6. Ben Vuirich
7. Meall Gruaim
8. Dunfallandy Hill
9. Glen Clova-Rough Craig
10. Portsoy
11. Windyhills
12. Keith
13. Muldearie

Pretectonic basic magmatism

1. Blackwater
2. Tayvallich
3. Loch Avich
4. Balnacraig
5. Delnadamph

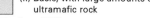 **Areas of Caledonian migmatisation**

 Areas of migmatisation in Central Highland Migmatite Complex (possibly pre-Caledonian)

Figure 26 Distribution of Caledonian igneous rocks in the Grampian Highlands.

South-west Highlands, where there are 2000 m of low-K basaltic pillow lavas and tuffs, with a few interbedded pyroclastic layers (*2*; Borradaile, 1973; Graham, 1976). The volcanic pile thins rapidly to the north-east and contains in its upper part indications of subaerial extrusion. Keratophyre, from a small laccolith cogenetic with the Tayvallich Volcanic Formation, has been dated at 595 ± 4 Ma (Halliday et al., 1989). The overlying turbiditic Loch Avich Grit is succeeded by the 300 to 500 m-thick *Loch Avich Lavas* (*3*), preserved only in the centre of the Loch Awe Syncline. The volcaniclastic Green Beds, which form a mappable unit within the Southern Highland Group from Kintyre to Glen Clova, may represent reworked debris from lateral equivalents of the Loch Avich Lavas. The metamorphosed basaltic dykes of northern Jura (Graham and Borradaile, 1984) are considered to represent feeders to the sills and lavas of mid-Argyll. These factors indicate that the Loch Awe–North Knapdale area was probably a centre of crustal instability from Easdale Subgroup times until early Southern Highland Group times.

The petrochemistry of the metamorphosed sills and lavas in mid-Argyll has been studied by Graham (1976) and Graham and Bradbury (1981). The original ophitic textures of the dolerites are commonly preserved, as are some primary skeletal ilmenite crystals, despite the recrystallisation of the rocks to typical greenschist and epidote-amphibolite facies assemblages. Patches of pegmatitic metadolerite within the sills commonly develop interstitial quartz and alkali feldspar, indicating the tholeiitic character of the dolerites, and there is evidence of a degree of cumulate settling in the thicker sills. Some of the basic rocks suffered hydrothermal spilitisation (albitisation of the plagioclase) shortly after extrusion.

The non-spilitised lavas and sills have an evolved tholeiitic basalt chemistry, showing a differentiation trend marked by considerable enrichment in Fe, Ti, Zr and P. The lavas of the Tayvallich Volcanic Formation are more differentiated than the Loch Avich lavas. The rocks are chemically similar to modern basalts produced during continental rifting, representing the first stage of oceanic spreading.

Amphibolite sills and sheets are widely developed in the upper Argyll Group from Perthshire to Deeside. In Perthshire, Pantin (1956) recognised a metamorphosed contact aureole around the Ben Vrackie sill complex, which intrudes the Ben Lawers and Ben Lui schists. The BGS resurvey during the 1980s of the area from Donside to the Banffshire coast has shown that many of the amphibolites, hornblende-schists and serpentinites previously regarded as pretectonic intrusives are actually sheared members of the syntectonic basic suite. However, the many smaller conformable bodies of amphibolite to the east of this line are probably metamorphosed sills intruded prior to the regional metamorphism.

SYNTECTONIC GRANITIC INTRUSIONS

Several discrete episodes of acid magmatism can be recognised in the Grampian Highlands (Figure 26). The earliest is represented by the *Ben Vuirich Granite* (*6*) which has yielded a U/Pb zircon age of 590 ± 2 Ma (Rogers et al., 1989). Bradbury et al. (1976) reported that the Ben Vuirich Granite was emplaced between the regional D_2 and D_3 events; current work casts doubt on this interpretation (Tanner and Leslie, 1994). South-east of the post-tectonic Glen Tilt Complex, a number of small granitic bodies, of which the *Meall Gruaim* intrusion (*7*) is the largest, show many similarities with the Ben Vuirich intrusion, and are probably of similar age.

The *Dunfallandy Hill Granite* (*8*; Bradbury et al., 1976) is a set of sheet-like bodies with a foliation parallel to the axial planes of F_3 folds; it is considered to be pre- to early-D_3 in age. The high initial Sr isotope ratios of the Ben Vuirich and Dunfallandy Hill granites suggest that they are both S-type granites.

The relationships of the granites in the vicinity of *Milton of Clova* (*9*; Harry, 1958) to the regional tectonic episodes are more complex. They are muscovite-biotite-granites which form sheet-like intrusions whose textures range from medium grained and homogeneous to strongly foliated and gneissose. Robertson (1991; 1994) has shown that the gneissose granites predate D_2, whereas other granites in the vicinity are late- to post-D_3 in age.

Small, elongate masses of granite are intruded into Appin Group and Argyll Group metasedimentary rocks at *Portsoy* (*10*), *Windyhills* (*11*), *Keith* (*12*) and *Muldearie* (*13*). They exhibit strong tectonic fabrics, although cleaved xenoliths in the Keith granite indicate some pre-intrusion deformation. The granites consist of alkali-feldspar augen in a matrix of quartz, plagioclase, biotite and secondary muscovite.

Migmatitic rocks occur in the north-west and north-east Grampians (stippled ornament in Figure 26). Those in the north-west form the Central Highland Migmatite Complex, parts of which have suffered two separate periods of migmatisation. Migmatisation in the north-east Grampians coincided with the peak of the Grampian event: regional metamorphism, the D_3 deformation, and the intrusion of large volumes of basic magma (dated at around 489 ± 17 Ma by Pankhurst, 1970). In those areas subjected to middle amphibolite facies metamorphism, migmatites were formed in rocks of susceptible compositions. Both groups of migmatites comprise lenticles and veinlets of coarse-grained quartzofeldspathic material (the neosome), frequently bordered by a biotite-rich selvedge, set in a finer-grained host rock or palaeosome. The neosome forms bands and segregations, both parallel to and, more rarely, cross-cutting the layering of the host metasedimentary rocks. Some of the migmatites, particularly those adjacent to the basic plutons, may have formed by partial melting (anatexis) of the host rock but the majority of them are now believed to have formed by subsolidus segregation (Ashworth, 1985; Ashworth and McLellan, 1985).

SYN- TO LATE-TECTONIC BASIC AND ULTRAMAFIC INTRUSIONS

A suite of basic and ultramafic rocks was intruded into the Dalradian rocks of the North-east Grampians during the D_3 tectonic episode, shortly after the peak of the regional metamorphism. The largest of these intrusions, the Insch mass, has been dated at 489 ± 17 Ma (Pankhurst, 1970), while others have yielded ages between 497 and 464 Ma (Table 2). The bodies occur in areas of intense D_3 deformation, particularly along the Portsoy–Duchray Hill Lineament, and are associated with areas of high-temperature low-pressure 'Buchan' metamorphism (Fettes et al., 1986). The contacts, where seen, are mostly tectonic, and are for the most part coincident with large-scale shear structures (Ashcroft et al., 1984), which have in places detached the aureole hornfelses from individual basic masses, and displaced them by up to several kilometres. Other shear planes are developed within several of the basic masses. The shearing commenced shortly after emplacement while the rocks were still hot, but continued for a considerable period. Basic mylonites derived from the Huntly mass are cut by granites and pegmatites belonging to a 460 Ma suite but similar pegmatites near Oldmeldrum show cataclastic textures with lensoid feldspar porphyroclasts set in a fine-grained granular quartz matrix (Munro, 1986a), confirming the syntectonic intrusion age of the basic bodies. Most of the larger bodies show local evidence of cumulate layering. The non-cumulate rocks comprise fine-grained granular gabbros and contaminated and xenolithic rocks, which include hybrid rocks derived by partial melting of the country rocks.

The Huntly, Insch–Boganclogh, Morven–Cabrach, Haddo House–Arnage, Belhelvie and other smaller masses were thought by Stewart and Johnson (1960) to be remnants of a single disrupted layered intrusion, but Munro (1970) and Weedon (1970) have suggested that they are, in fact, several sill-like bodies differing slightly in composition and conditions of crystallisation. Disruption and translocation occurred shortly after consolidation, before the rocks had acquired a permanent magnetisation.

The south-eastern portion of the *Insch* mass near Oldmeldrum (*14*; Figure 27), is much faulted but comprises layered dunites passing up into troctolites and then olivine-gabbros; all of them are assigned to a Lower Zone. They are separated by a 1 to 2 km-wide zone of sheared gabbroic rocks from the Middle and Upper zone rocks which form the remainder of the Insch mass (Clarke and Wadsworth, 1970). The Middle Zone comprises mainly coarse-grained clinopyroxene-norites showing cumulus textures, intimately associated with fine-grained clinopyroxene-norites with granular, non-cumulate textures (Wadsworth, 1988). These pass north-westwards into the Upper Zone of layered olivine-ferrogabbros, olivine-monzonites and finally syenites (Wadsworth, 1986), a sequence characteristic of the upper part of a layered tholeiitic intrusion. In the area to the west and south-west of Insch, the layering is gently folded into a broad, north-plunging syncline, cut by normal faults. On the western limb of the syncline, the equivalents of the clinopyroxene-norites are biotite-norites with interstitial quartz, similar to those of the Boganclogh intrusion. Along the south-west margin of the Insch mass a discontinuous line of serpentinite pods crops out (Read, 1956). These represent part of the lowest zones of the cumulate pile, moved to higher structural levels by the shearing event. The ultramafic rocks are extensively serpentinised and the cumulate and granular clinopyroxene-norites are widely but patchily uralitised, while the other rocks of the Insch mass are not significantly altered.

The *Boganclogh* mass (*15*; Figure 27) is the western continuation of the Insch mass beyond the late-tectonic Kennethmont Complex and the Rhynie Old Red Sandstone outlier (Busrewil et al., 1973). It consists mainly of Middle Zone quartz-biotite-norites, passing upwards into Upper Zone olivine-ferrogabbros, olivine-monzonites and syenites, as in the central and western parts of the Insch mass. An almost continuous belt of serpentinite occurs along the southern margin of the mass (Blyth, 1969). It is flanked to the north by sheared and mylonitised syenite, and contains shear zones and

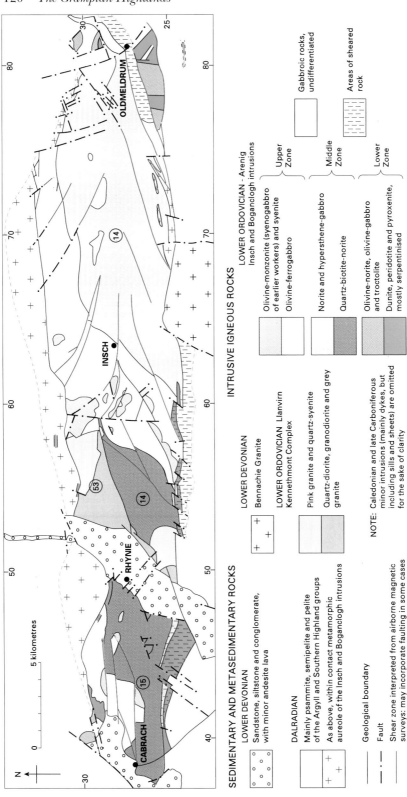

SEDIMENTARY AND METASEDIMENTARY ROCKS

LOWER DEVONIAN

Sandstone, siltstone and conglomerate, with minor andesite lava

DALRADIAN

Mainly psammite, semipelite and pelite of the Argyll and Southern Highland groups

As above, within contact metamorphic aureole of the Insch and Boganclogh intrusions

─── Geological boundary

─ · ─ Fault

──── Shear zone interpreted from airborne magnetic surveys: may incorporate faulting in some cases

INTRUSIVE IGNEOUS ROCKS

LOWER DEVONIAN

Bennachie Granite

LOWER ORDOVICIAN Llanvirn
Kennethmont Complex

Pink granite and quartz-syenite

Quartz-diorite, granodiorite and grey granite

NOTE: Caledonian and late Carboniferous minor intrusions (mainly dykes, but including sills and sheets) are omitted for the sake of clarity

LOWER ORDOVICIAN - Arenig
Insch and Boganclogh intrusions

Olivine-monzonite (syenogabbro of earlier workers) and syenite　⎫
　　　　　　　　　　　　　　　　　　　⎬ Upper Zone
Olivine-ferrogabbro　　　　　　　　　⎭

Norite and hypersthene-gabbro　⎫
　　　　　　　　　　　　　　　⎬ Middle Zone
Quartz-biotite-norite　　　　　　⎭

Olivine-norite, olivine-gabbro and troctolite　⎫
　　　　　　　　　　　　　　　　　　　　⎬ Lower Zone
Dunite, peridotite and pyroxenite, mostly serpentinised　⎭

Gabbroic rocks, undifferentiated

Areas of sheared rock

Figure 27　Geological sketch map of the Insch (*14*), Boganclogh (*15*) and Kennethmont (*53*) intrusions.

enclaves of sheared syenite and quartz-biotite-norite (Gould, in press). Narrow pods of sheared serpentinite occur along the northern contact of the Boganclogh mass, and faulted inliers of massive, unsheared, lightly serpentinised dunite occur within the mass.

Small detached bodies consisting mostly of dunite, harzburgite, troctolite and olivine-gabbro occur in the area to the south of the Insch mass. The *Lawel Hill* mass (*16*; Whittle, 1936), forms a gently dipping sheet bounded by north-dipping thrust planes, and shows lithological similarity to the Lower Zone cumulates at the south-eastern corner of the Insch mass. Three small fault-bounded bodies of basic rocks occur near *Kildrummy* (*17*; Gould, in press). Two of them consist of clinopyroxene-norite similar to the Tarland mass, but the third includes troctolites with bytownite plagioclase, more similar to Insch Lower Zone material.

The northern tip of the *Morven–Cabrach* mass (*18*; Allan, 1970) lies close to the western end of the Boganclogh mass, and the two bodies are probably connected at depth. This northern part of the 30 km-long body consists of quartz-biotite-norite (Plate 10) similar to that of Boganclogh with minor quartz-diorite and monzonite. These are succeeded southwards by layered norite and hypersthene-gabbro with minor ferrogabbro and monzogabbro. In the extreme south, minor intercalations of ultramafic material are present. The rocks of the southern part of the mass have suffered extensive uralitisation of the pyroxenes and saussuritisation of the plagioclase, but the primary gabbroic texture is preserved, except in the southern extremity and close to the highly sheared western margin of the body, where sheared gabbroic rocks have been recrystallised to foliated amphibolites. There are several large serpentinite bodies along the western margin. The cumulate

Plate 10 Layering in norites of the Morven–Cabrach mass, Hill of Allamuc, Aberdeenshire. The layers are alternately plagioclase-rich and orthopyroxene-rich (D 4320).

layering in the northern part of the body dips steeply to the east but in the southern part swings round to dip north, and the angle of dip reduces to 30° to 40°. Extensive hornfelsing and partial melting has occurred in the metasedimentary rocks along the eastern (upper) contact of the body, and xenolithic hybrid rocks are developed (Fettes, 1970). Allan (1970) suggested that Morven–Cabrach represents a feeder to the other basic masses, but the attitude of the layering, together with the shape of the related gravity anomaly, suggests that it is more likely to be a sheet-like body which has been extensively stoped by the Ballater granite at its southern end.

The adjoining *Tarland* mass (*19*; Gould, in press) is almost entirely composed of norite, with uralitised pyroxenes and saussuritised plagioclase, showing a strong resemblance to that of the southern part of the Morven–Cabrach mass.

The serpentinised ultramafic rocks of the *Coyles of Muick* (*20*) are similar to those along the western margin of the Morven–Cabrach mass, and represent a southern continuation of the zone of tectonically emplaced ultramafic bodies.

The small *Blackwater* mass (*21*; Fettes and Munro, 1989) consists of cumulate gabbroic rocks with a number of scattered ultramafic bodies, mostly serpentinised. The rocks have been strongly deformed, and the basic rocks amphibolitised, along narrow shear zones.

The *Succoth–Brown Hill* mass (*22*; Gunn et al., 1990) consists of serpentinised dunites, pyroxenites and minor amphibolitised gabbroic rocks characterised by unusually primitive mineral compositions. The ultramafic rocks are comparable to those along the southern margin of Boganclogh, but the feldspar-bearing rocks are unique to the mass.

The *Huntly* (*23*) and *Knock* (*24*) masses (Munro, 1970; 1984; Munro and Gallagher, 1984) are joined together by a narrow neck of basic rocks. The cumulate layering within each body strikes north–south, parallel to the contacts of the bodies, and is subvertical, younging eastwards. Olivine ± plagioclase ± clinopyroxene cumulates occur as layered rocks of peridotite, troctolite and olivine-gabbro composition; cumulus textures are particularly well developed in the western parts of all three masses. Isolated patches of rather more differentiated olivine-gabbro cumulates crop out in the central and eastern parts of the Huntly mass, enclosed within granular gabbros similar to those of the Insch mass. Xenolithic and contaminated gabbroic rocks are also a prominent feature of the eastern parts of the masses (Read, 1923, pp.128–135); in places they appear to have intruded both the cumulates and the granular gabbros. The masses are traversed by several N–S-trending shear zones (Munro and Gallagher, 1984).

The *Portsoy* mass (*25*; Munro, 1970; 1984; Munro and Gallagher, 1984) lies to the north of the Knock mass; it is very poorly exposed. Olivine-gabbros similar to those of Huntly are associated with serpentinised ultramafic rocks of uncertain affinity, some of which may be pretectonic.

The *Belhelvie* mass (*26*) is a layered complex of ultramafic and mafic cumulates (Wadsworth, 1991). The body is elongated in a NW–SE direction and the cumulate layering dips very steeply to the north-east, parallel to the contacts (Ashcroft and Boyd, 1976). A succession from dunite through troctolite to olivine-gabbro occurs along the south-western margin. It is overlain by a narrow, impersistent raft of metasedimentary hornfelses, which is overlain in turn by a repetition of the dunite–troctolite–olivine-gabbro succession. The rocks of the eastern part of the intrusion have been intensely sheared, and narrow belts of sheared and recrystallised rock occur along the contacts. The sheared basic rocks are cut by bodies of pegmatitic granite, dated at 472 ± 5 Ma, which have suffered only limited crushing (Boyd and Munro, 1978).

The *Haddo House (27)* and *Arnage (28)* masses present classic examples of contamination of basic igneous rocks (Gribble, 1967; 1968). Both intrusions are chiefly quartz-biotite-norites with minor olivine-norite, showing granular textures. These rocks are intimately associated with cordierite-norites which are considered to represent a partial melt derived from pelitic and impure psammitic Dalradian metasedimentary rocks, with only the most refractory components (quartzite and highly aluminous rocks) preserved as xenoliths. The *Arthrath–Dudwick* body (*29*), which is adjacent to the eastern margin of the Arnage mass, is bounded to the north and south by faults and cross-cut by N–S-trending dislocations. The rocks are mostly orthopyroxene-rich cumulates interlayered with olivine-rich cumulates. Minor olivine-gabbros and norites are present locally.

The poorly exposed *Maud* mass (*30*) comprises medium-grained quartz-biotite-clinopyroxene-norites with local coarser- and finer-grained variants. Xenoliths of locally derived metasedimentary rocks are abundant, particularly in the eastern part of the body, and show a strong north–south alignment. The foliation in the xenoliths typically dips steeply to the east. The basic rocks are bounded to the east by a linear ductile shear zone and a similar zone truncates the metamorphic aureole a short distance beyond the western contact.

Other, smaller, bodies ranging from ultrabasic rocks through gabbros and norites to diorites (*31–33*) occur in a belt running northwards from the Belhelvie mass, past the eastern margins of the Arnage and Maud masses, as far as the coast at Inzie Head. Another belt runs parallel to the eastern margin of the Huntly–Knock and Portsoy masses. The rocks are commonly amphibolitised (e.g. *Boyndie Metagabbro, 34*) and there is some uncertainty as to whether they are pre- or syntectonic.

Substantial metamorphic aureoles occur along the margins of most of the basic masses, but they are often incomplete; for example, intense D_3 shearing has removed the aureole from most of the southern margin of the Insch mass (Figure 27). Elsewhere, hornfels-like rocks occur in shear-bounded slices up to several kilometres away from the nearest outcrops of basic rock; they are interpreted as representing aureole rocks displaced by movement along shear zones propagated along the margins of the basic masses (Ashcroft et al., 1984). In places, however, the hornfelses are difficult to recognise because they bear mineral assemblages similar to those produced by high-grade regional metamorphism of the low-pressure Buchan facies.

The basic masses give rise to magnetic and gravity anomalies of varying intensity. Where fresh, the ultramafic rocks have low magnetic susceptibilities, but abundant magnetite is released during the serpentinisation process, giving rise to a high susceptibility within most of the ultramafic components. Most of the gabbroic rocks are moderately magnetic when fresh, but sheared and uralitised gabbroic rocks commonly have lower susceptibilities, due to destruction of the primary magnetite and ilmenite. The contaminated rocks generally have low susceptibilities, as do the more quartzitic hornfelses. However, pelitic hornfelses frequently have high susceptibilities, due partly to layers rich in recrystallised detrital magnetite, but mostly due to release of magnetite during the formation of fibrolitic sillimanite after biotite.

Most components of the basic masses (except for syenites and serpentinised ultramafic rocks) are considerably denser than the Dalradian country rocks and consequently give rise to large positive Bouguer gravity anomalies.

Gravity and magnetic survey data were used by McGregor and Wilson (1967) and Gallagher (1983) to interpret the overall shape of the basic intrusions of the north-east Grampians. The Portsoy, Knock and Huntly masses extend to a depth of no more than about 1 km. A thickness of about 5 km of basic rocks underlies the south-western part of the Boganclogh intrusion, the northern tip of the Morven–Cabrach mass and the intervening area (Gould, in press). The thickness of the basic rocks decreases steadily southwards within the Morven–Cabrach mass, whose southern portion is underlain at shallow depth by a continuation of the Ballater granite. The Insch mass has an interpreted thickness of 2 km to the north of Bennachie (McGregor and Wilson, 1967). Its sheared southern margin dips to the north, probably at a fairly steep angle. The Belhelvie intrusion is thought to continue south-eastwards along strike for up to 10 km under a cover of younger sedimentary rocks. The main gravity anomaly lies to the north-east of the outcrop of the mafic and ultramafic rocks, indicating a probable shallowing of the dip of the contacts in that direction. Prominent gravity anomalies with a slight NNE trend in the area between Haddo House and Fraserburgh indicate that there are probably considerable quantities of basic igneous rocks at depth throughout this area.

LATE-TECTONIC GRANITIC INTRUSIONS

Three distinct suites of late-tectonic granitic intrusions are recognised in the northern Grampian Highlands. Two of the suites consist of S-type granites, commonly foliated and garnet-bearing, which postdate the regional migmatisation, the major basic intrusions described above and the D_3 tectonic episode. One suite in the north-east (*36–45*) is dated at 475–470 Ma, and the other, in the north-west (*46–52*), is dated at 445–435 Ma (Table 2). The third suite (*53–55*) consists of I-type diorites, tonalites and granodiorites.

(i)   North-eastern biotite-muscovite-granites

Several members of this group have yielded U/Pb monazite ages of 470 to 475 Ma and Rb/Sr isochrons of 435 to 470 Ma (Pankhurst, 1970; 1974; Pidgeon and Aftalion, 1978; Kneller and Aftalion, 1987). The former has been taken as the best estimate of the time of intrusion and the latter as the age of the main cooling episode. Age determinations for individual intrusions are listed in Table 2.

The *Kemnay* (*36*) and *Aberdeen* (*37*) granites comprise grey, foliated biotite-muscovite-granite with a composition very similar to that of the minimum melt (Munro, 1986b). Some primary muscovite occurs, although most of the muscovite is secondary. Contacts with the Dalradian country rocks are gradational, passing through a zone containing partially digested xenoliths into country rocks which had previously suffered regional migmatisation. The *Auchlee* (*38*) and *Cove* (*39*) granites to the south of Aberdeen are similar in composition to the Aberdeen granite and also show gradational contacts. The *Strichen* (*40*), *Forest of Deer* (*41*), *Aberchirder* (*42*), *Longmanhill* (*43*), and *Lochlundie* (*44*) (Read, 1923) granites are all light grey, unfoliated, porphyritic biotite-granites showing marked similarities to the Kemnay and Aberdeen intrusions. However, muscovite, where present, is secondary.

The basic masses of the north-east Grampians are cut by a suite of biotite-granites, of which *Ardlethen* (*45*) is the largest, and has associated aplites and pegmatites, the latter muscovite- and tourmaline-bearing with rare beryl. This suite is roughly coeval with the Aberdeen and related granites, but its members lack a foliation and the finer-grained members are frequently reddened.

(ii)   North-western biotite-muscovite-granites

The members of this suite of late-tectonic granitic intrusions were probably intruded at 445 to 435 Ma (Table 2). Initial $^{87}Sr/^{86}Sr$ ratios of members of this suite are in the range 0.714 to 0.718, indicating that they are S-type granites with magmas involving melting of a metasedimentary protolith. No significant gravity or magnetic anomalies are associated with them.

The *Ardclach Granite* (*46*; Horne, 1923) is pink, coarse grained and biotite and muscovite bearing; megacrysts of potash feldspar are zoned, with inclusion trails. Contacts of the pluton are poorly defined, with a xenolithic medium-grained marginal granodiorite facies passing outwards into a vein complex.

The *Moy Granite* (*47*) varies from porphyritic biotite-granodiorite to rare biotite-granite (Zaleski, 1985). The abundance of feldspar megacrysts increases towards the southern part of the intrusion. Rafts and xenoliths of country rock are locally abundant close to the eastern margin of the granite. The margins of the intrusion are marked by intense granite veining. There is little evidence of hornfelsing of adjacent country rocks. The Moy Granite was subsequently intruded by the post-tectonic Saddle Hill Granite.

The *Grantown Granite* (*48*) comprises a fine- to medium-grained microcline-bearing granite with a subordinate leucogranite phase (Highton, in press), and both are weakly to moderately foliated (Mackenzie, 1958). It is locally garnetiferous, and in places extensively recrystallised, with abundant epidote porphyroblasts. Muscovite is present throughout, but appears to be secondary. The main phases are cut by sheets and veins of foliated aplitic microgranite and pegmatite. Emplacement appears to have been by passive stoping, with the preservation of a 'ghost' metasedimentary stratigraphy through trains of rafts and xenoliths. Contacts are mostly obscured, with the exception of the western margin, which is seen to be sheeted and characterised by large screens of metasedimentary material. The intrusion has no discernible metamorphic aureole.

Table 2   Radiometric ages of Caledonian igneous rocks in the Grampian Highlands.

No. on Fig. 26	Name	Age (Ma)	Type	$^{87}Sr/^{86}Sr_i$	Geochronological reference
Pre-tectonic basic magmatism					
2	Tayvallich	595 ± 4	U/Pb Zr	—	Halliday et al., 1989
Syntectonic granitic intrusions					
6	Ben Vuirich	590 ± 2	U/Pb Zr	—	Rogers et al., 1989
8	Dunfallandy Hill	481 ± 15	Rb/Sr WRI	0.7185 ± 0.0008	Pankhurst and Pidgeon, 1976
9	Glen Clova (Rough Craig)	549 ± 44	Rb/Sr WR	0.7166 ± 0.0024	Robertson, 1994
Syn- to late-tectonic basic and ultramafic intrusions					
14	Insch	489 ± 17	Rb/Sr WRI	0.7117 ± 0.0003	Pankhurst, 1970
23	Huntly	476 ± 5	K/Ar Bi	—	Brown et al., 1965
27	Haddo House	491 ± 6	K/Ar Bi	—	Brown et al., 1965
Late-tectonic granitic intrusions					
(i) North-eastern biotite-muscovite-granites					
26	Belhelvie pegmatites	472 ± 5	Rb/Sr WRI	0.7185 ± 0.0016	van Breemen and Boyd, 1972
36	Kemnay	411 ± 7	Rb/Sr Mins	0.715 ± 0.002	Bell, 1968
37	Aberdeen	470 ± 1	U/Pb Mon	—	Kneller and Aftalion, 1987
38	Auchlee	439 ± 7	Rb/Sr Mins	0.733 ± 0.004	Bell, 1968
40	Strichen	475 ± 5	U/Pb Mon	—	Pidgeon and Aftalion, 1978
		455 ± 22	Rb/Sr WRI	0.7172 ± 0.0011	Pankhurst, 1974
42	Aberchirder	444 ± 9	Rb/Sr WRI	0.7157 ± 0.0008	Pankhurst, 1974
43	Longmanhill	470 ± 50	Rb/Sr WRI	—	Pankhurst, 1974
(ii) North-western biotite-muscovite-granites					
46	Ardclach	475 ± 7	U/Pb Mon	—	Zaleski, 1983
47	Moy (main phase)	455 + 27/ -15	Rb/Sr WR + U/Pb Mon	0.7185 ± 0.0001	Zaleski, 1983
49	Glen Kyllachy	443 + 5/ -15	Rb/Sr WR/Bi	0.7176 ± 0.0004	van Breemen and Piasecki, 1983
50	Maol Chnoc	436 ± 8	Rb/Sr Mins	0.717	Clayburn, 1981
51	Strathspey	437 ± 16	K/Ar Bi	—	Miller and Brown, 1965
52	Loch Laggan	439 ± 7	Rb/Sr Mins	0.705 - 0.718	Clayburn, 1981
(iii) North-eastern diorites to granodiorites					
53	Kennethmont	453 ± 4	Rb/Sr WRI	0.7145 ± 0.0013	Pankhurst, 1974
Post-tectonic granitic intrusions					
(i) South Grampian Suite					
56	Garabal Hill– Glen Fyne	406 ± 4[1] 429 ± 2[2]	Rb/Sr WRMI U/Pb Zr	0.705 ± 0.003 —	Summerhayes, 1966 Rogers and Dunning, 1991
57	Arrochar	426 ± 3[3]	U/Pb Sph	—	Rogers and Dunning, 1991
65	Comrie	408 ± 5	Rb/Sr WRMI	0.705 - 0.707	Turnell, 1985
—	Rubha Mor (appinite pipe)	427 ± 3	U/Pb Sph	—	Rogers and Dunning, 1991

Table 2 Radiometric ages of Caledonian igneous rocks in the Grampian Highlands (continued).

No. on Fig. 26	Name	Age (Ma)	Type	$^{87}Sr/^{86}Sr_i$	Geochronological reference
(ii)	Argyll Suite				
66	Moor of Rannoch	408 ± 18	K/Ar Bi; Rb/Sr WRMI	0.7048	Miller and Brown, 1965; Clayburn, 1981
67	Strath Ossian	405 ± 9	Rb/Sr WRMI	0.7059	Clayburn, 1981
		400 ± 10	U/Pb Zr	—	Pidgeon and Aftalion, 1978
68	Ballachulish	406 ± 34	Rb/Sr WRI	0.7040 ± 0.0002	Haslam and Kimbell, 1981
72	Etive	401 ± 6[4]	Rb/Sr WRI	0.7058 ± 0.0004	Clayburn et al., 1983
		396 ± 12[5]	Rb/Sr WRI	0.7055 ± 0.0005	Clayburn et al., 1983
74	Foyers	c.415	various	—	Pankhurst, 1979
		405 ± 8[6]	Rb/Sr WRMI	c. 0.707	Clayburn, 1981
		403 ± 9[7]	Rb/Sr WRMI	c. 0.7045	Clayburn, 1981
75	Findhorn	413 ± 5	Rb/Sr WRMI	c. 0.706	van Breemen and Piasecki, 1983
77	Corrieyairack	434 ± 9	Rb/Sr WRMI	0.7056	Clayburn, 1981
78	Allt Crom	c. 405	Rb/Sr WR	0.7056	Clayburn, 1981
81	Crathes	420 ± 2	K/Ar Bi	—	Brown et al., 1965
(iii)	Cairngorm Suite				
87	Moy (Saddle Hill)	407 ± 5	Rb/Sr WRI	0.7081 ± 0.0002	Zaleski, 1983
90	Ben Rinnes	411 ± 3	Rb/Sr WRI	0.7072 ± 0.0020	Zaleski, 1985
93	Monadhliath	419 ± 5	Rb/Sr WRI	0.7058 ± 0.0008	Harrison and Hutchinson, 1987
94	Cairngorm	408 ± 3	Rb/Sr WRI	0.7062 ± 0.0004	Harmon et al., 1984
95	Lochnagar	415 ± 5	Rb/Sr WRI	0.7065 ± 0.0002	Halliday et al., 1979
96	Glen Gairn	404 ± 6	Rb/Sr WRI	0.7081 ± 0.0015	Harrison and Hutchinson, 1987
98	Mount Battock	416 ± 4	Rb/Sr WRI	0.7058 ± 0.0004	Harrison and Hutchinson, 1987
102	Hill of Fare	413 ± 3	Rb/Sr WRI	0.7057 ± 0.0003	Halliday et al., 1979
103	Bennachie	400 ± 4	Rb/Sr WRI	0.7062 ± 0.0004	Darbyshire and Beer, 1988
104	Middleton	398 ± 19	Rb/Sr WR	0.7092 ± 0.0181	Darbyshire and Beer, 1988
105	Peterhead	406 ± 13	Rb/Sr Mins	—	Bell, 1968
Lower Old Red Sandstone volcanic rocks					
108	Lorn Plateau	400 ± 5	Rb/Sr WRI	0.7047	Clayburn et al., 1983
		421 - 413	Ar degassing	—	Thirlwall, 1988

NB All Rb/Sr and K/Ar ages recalculated using the decay constants of Steiger and Jaeger, 1977. WR = Whole rock (no isochron) WRI = Whole rock isochron WRMI = Whole rock – mineral isochron Mins = Mineral age (no isochron) Bi = Biotite Zr = Zircon Mon = Monazite Sph = Sphene

1 Average for complex 2 Appinitic diorite 3 Appinite 4 Meall Odhar Granite 5 Central Starav Granite 6 Tonalite and granodiorite 7 Granite.

The *Glen Kyllachy* intrusion (*49*) is essentially formed of a coarse-grained foliated biotite-granodiorite, which is locally hybridised through the assimilation of metasedimentary material. It is cut by leucocratic variants and a weakly foliated microcline-granodiorite (van Breemen and Piasecki, 1983). Relationships with the Findhorn granodiorite are complex and unknown in detail.

The *Maol Chnoc* granite and vein-complex (*50*) lie to the north-east of the Foyers granite. They comprise a complex network of ramifying medium-grained intrusions ranging in composition from granodiorite to granite, with felsite stockworks (Highton, 1987). The biotite-granodiorite is strongly deformed, with prominent small-scale shear zones, mostly resulting from the forceful emplacement of the Foyers Granite.

The *Strathspey Complex* (*51*) consists predominantly of biotite-muscovite-granite which is locally garnetiferous (Smith, 1970); the granite is seen to grade into pegmatitic patches. Emplacement appears to have been by passive stoping, but with many metasedimentary enclaves, preserving a ghost stratigraphy and structure. The margin of the body is a sheeted vein complex several kilometres wide. The complex postdates the local D_4 folds but predates the NE-striking Loch Tay fault system.

The *Loch Laggan Granite* (*52*) is an intense vein-complex of granites and pegmatites but also includes some aplites and microgranites (Anderson, 1956; Key et al., in press). It is older than the main phases of the Corrieyairack and Strathspey complexes.

(iii)   North-eastern diorites to granites

The members of this suite range from diorite to granodiorite, are in some cases foliated, and are highly xenolithic, with the more acid members carrying xenoliths of the more basic phases.

The *Kennethmont Complex* (*53*; Sadashivaiah, 1954; Read and Haq, 1965; Busrewil et al., 1975) was emplaced at the western end of the Insch basic mass and ranges from diorite to granite in composition. A Rb/Sr whole rock isochron of 453 ± 3 Ma and initial Sr isotope ratio of 0.7145 (Pankhurst, 1974) were obtained from the pink granite of the complex, which Busrewil et al. (1975) have suggested is a later, unrelated post-tectonic intrusion cutting the diorite and the xenolithic grey granite. The Kennethmont diorites were originally considered (Sadashivaiah, 1954) to be hybrids resulting from contamination of granitic magma by assimilation of gabbroic material from the Insch mass. However, Busrewil et al. (1975) regard the diorite as representing the primary Kennethmont magma, with the grey granite being a differentiate of it.

The *Syllavethy* intrusion (*54*; Gould, in press) varies from a foliated, xenolithic quartz-diorite to an unfoliated grey granodiorite. The *Tillyfourie Tonalite* (*55*; Harrison, 1987) consists of tonalite with minor granodiorite showing a marked foliation. Both are associated with an extensive tonalite–granodiorite vein complex in the area between them. No age data are available for either intrusion or for the vein complex, but the Tillyfourie Intrusion and the vein complex are cut by the late-tectonic Corrennie granite and the post-tectonic Crathes, Balblair and Bennachie granites, and the group is provisionally assigned to the late-tectonic suite. The small

Corrennie Granite (*84*), which forms an elongate body of pink, leucocratic granite with a foliation defined by the streaking out of quartz crystals, cuts the Tillyfourie intrusion.

POST-TECTONIC GRANITIC INTRUSIONS

The main phase of igneous activity in the Grampian Highlands is represented by the emplacement of large volumes of calc-alkaline magma during a period of post-orogenic uplift (Pitcher, 1982). These post-tectonic intrusions, the Newer Granites of Read (1961), are a diverse group in terms of mode of emplacement, petrology and geochemistry. Petrographically, they have been divided into three groups (Stevens and Halliday, 1984; Plant, 1986):

(i)   South Grampians Suite: diorite to granite plutons, mostly relatively small in size, in the south and south-east Grampians

(ii)   Argyll Suite: tonalite–granodiorite–granite complexes which show ring structure or compositional zoning, located up to 50 km south-east of the Great Glen Fault

(iii) Cairngorm Suite: intrusions of biotite-granite, some very large in size, concentrated in the northern part of the Grampians.

The three suites have all yielded emplacement ages in the range 420 to 395 Ma (Table 2). The three groups share essentially I-type isotopic and geo-chemical characteristics, with low initial $^{87}Sr/^{86}Sr$ ratios (0.704–0.708), although group (iii) is transitional to A-type. They are interpreted (Harmon and Halliday, 1980; Stevens and Halliday, 1984) as having formed from an admixture of a primary, mantle-derived component with a proportion of lower crustal material. Stevens and Halliday ascribed the geochemical differences between the suites to differences in lower crustal lithology and water content in different parts of the Grampians. Little or no Dalradian material appears to have been involved in the origin of the granites.

The tonalite–granodiorite–granite complexes of group (ii) commonly show marked enrichment in Sr and Ba, whereas the pink biotite-granites of group (iii) have low Sr and Ba and high Rb, Li, U and Th, due to their more advanced stage of magmatic differentiation (Plant et al., 1990). The larger complexes of the Argyll and Cairngorm suites are associated with closed negative Bouguer gravity anomalies, albeit superimposed on a composite regional negative anomaly; this has been interpreted as indicating that the plutons from Monadhliath to Mount Battock rise from an underlying batholith. Strong differences in magnetic susceptibility between phases have been recorded in several of the granites and this is reflected in annular magnetic anomalies, locally of significant (> 400 nT) amplitude, although the magnetic anomaly pattern in and around the granites might also be interpreted as due to ring-shaped bodies of basic igneous rocks or hornfelsed pelitic rocks at depth.

South Grampians Suite

This suite of largely granodioritic complexes is developed in the South-west and Southern Grampians; some are associated with NE-trending faults. The suite is characterised by a significant dioritic component, together with smaller amounts of more mafic rocks (peridotite, kentallenite (olivine-monzonite) and pyroxenite).

The *Garabal Hill–Glen Fyne Complex* (56) is divided into two parts by the Kinglas–Garabal fault. To the north-west of the fault lies the main part of the body, which comprises mostly granodiorite with K-feldspar megacrysts, with a subsidiary nonporphyritic granodiorite (Nockolds, 1941). Along its south-east margin, adjacent to the fault, the granodiorite has a contact with a small body of hypersthene-gabbro containing a minor ultramafic phase (augite-peridotite). To the south-east of the fault, contaminated pyroxene-mica-diorite with subsidiary pyroxenite and other ultramafic rocks crop out. Each successive phase carries xenoliths of one or more of the earlier phases. The more acid members define a liquid line of descent believed to be a product of differentiation from a dioritic magma. Nockolds (1941) and Nockolds and Mitchell (1948) concluded that the parental magma was of pyroxene-mica-diorite composition, and close to that of a pyroxene-andesite lava from Glencoe. The absence of an Fe-enrichment trend in the diorite complex suggests that the more basic rocks represent a substantial cumulus component.

The *Arrochar Complex* (57) forms an elongate, composite intrusion, predominantly pyroxene-mica-diorite, with subsidiary components ranging from peridotite/pyroxenite to tonalite and biotite-granodiorite. Internal contacts are sharp, with the successively younger phases both veining and bearing xenoliths of the older ones. The Arrochar body differs slightly from other members of the group in having enhanced V, Y, FeO, MgO and Co values. The parental magma of the intrusion is thought to be dioritic, with the various phases representing products of continuous differentiation. Again the ultramafic rocks are thought to represent cumulus phases.

The *Doune Farm Complex* (58) consists of seven separate diorite bodies with associated breccias, appinites and minor granodiorites. Early breccias and coarse-grained appinites are succeeded by several varieties of diorite. The *Inversnaid* body (59) is similar.

The *Glen Tilt Complex* (60) bears some similarity to the Garabal Hill–Glen Fyne Complex, comprising a main granitic intrusion, with a marginal basic to intermediate complex to the south-east (Deer, 1938; 1950; 1953). The marginal phases of the complex range in composition from augite-diorite through tonalite to granodiorite, with small enclaves of appinite. Contacts between the main phases are gradational, with evidence of hybridisation (Deer, 1938; 1950). The main intrusion, and a small satellitic body, consist of pink, coarse-grained biotite-granite. A related suite of NE-trending lamprophyre, microdiorite, felsite and quartz-porphyry dykes is well developed in country rocks to the south-west of the complex.

The *Glen Doll Diorite* (61) is the largest of a number of dominantly intermediate plutonic intrusions (61–64) that occur between Glen Shee and the south-eastern margin of the Cairngorm Granite (Barrow and Craig, 1912; Jarvis, 1987). It consists mainly of hornblende- and pyroxene-diorite, with small areas of gabbro, monzonite and granite. Some rocks are appinitic, and some gabbros and diorites show layering and cumulus textures. Partly assimilated metasedimentary xenoliths are common in the diorites, where enhanced zinc levels have been attributed to this assimilation. Contamination is particularly apparent in a 20 to 350 m-wide zone at the eastern margin, where diorite grades into monzodiorite and granite adjacent to hornfelsed rafts and xenoliths.

The *Comrie Complex* (*65*) consists of a peripheral xenolithic diorite, locally pyroxene-bearing, which is cross-cut by a highly leucocratic aplitic granite (Tilley, 1924). An extensive thermal aureole is developed in the low-grade country rocks, with pyroxene-, cordierite-, corundum- and spinel-bearing hornfelses.

Argyll Suite

This group of tonalite–granodiorite–granite intrusions comprises large multi-phase high-level complexes in the Lochaber and Lorn areas and smaller, mostly zoned plutons forming a well-defined NE-trending belt adjacent to the Great Glen Fault (Figure 26). Certain phases of the Skene Complex and some other small intrusions of the north-east Grampians have been included here because of petrographical similarity, without implying a genetic relationship to the Lochaber intrusions. The large plutonic complexes of Lochaber and Lorn postdate Lower Old Red Sandstone lavas of the Lorn Plateau. A close association between volcanic and plutonic rocks is seen at Glen Coe and Ben Nevis. The ring-faulting and rim-synclines suggest that there was only a thin cover (less than 3 km) above the Lochaber and Lorn plutons when they were emplaced (Roberts, 1966).

The *Moor of Rannoch Granite* (*66*) is the earliest intrusion of the Lochaber group (Hinxman et al., 1923). Its original roughly circular outline has been modified by displacement along the Ericht–Laidon Fault. An early phase of quartz-monzodiorite to monzogranite is intruded by a phase of monzodiorite and quartz-monzodiorite, and both phases are intruded by pink syenogranite. However, no chilled contacts have been detected between the monzodiorites and the syenogranite. The roughly circular shape of the pluton hides a complex sheeted geometry of individual phases, and shows evidence of forceful intrusion; a metamorphic aureole up to 700 m wide is developed.

Unlike the other members of this group, which are roughly circular or show a north-easterly elongation, the *Strath Ossian Complex* (*67*) shows a pronounced north-westerly elongation. It consists of three main lithologies. An older melanocratic granodiorite to quartz-diorite is intruded by the areally dominant medium- to coarse-grained, xenolithic hornblende-biotite-granodiorite. The third lithology is a feldspar-phyric granite, which forms a marginal facies transitional to the main granodiorite. The igneous complex was emplaced, with local disruption of country rock structures, into cold country rock (Key et al., 1993).

The *Ballachulish* pluton (*68*) is a composite intrusion comprising a zoned envelope of dioritic composition intruded by a younger core of porphyritic granite (Weiss and Troll, 1989). The inner part of the diorite consists of two-pyroxene-monzodiorite with cumulus texture, which exhibits a flow foliation marked by alignment of pyroxene and plagioclase crystals, and contains inter-cumulus opaque minerals, poikilitic biotite, and interstitial alkali feldspar. This grades outwards into grey quartz-diorite of variable grain size, which in many places is abundantly xenolithic. The main porphyritic granite forming the central part of the complex contains 5 to 15 per cent of perthite pheno-crysts set in a granodiorite matrix. A marginal hybrid facies of the granite contains partially resorbed mafic enclaves. A small stock of nonporphyritic fine-grained leucogranite, showing strong hydrothermal alteration, occupies the centre of the pluton. Satellitic intrusions of basic to intermediate, volatile-rich magma formed bodies of appinite and lamprophyre; pipes of explosion breccia are associated with these intrusions. They closely predate the main phase of diorite intrusion and show restricted hybridisation with quartz-diorites near the south-east margin of the Ballachulish pluton. Fluid pressures in the thermal aureole of the complex are estimated at 3.0 ± 0.5 kb (Pattison, 1989), with contact metamorphic temperatures ranging from 450°C to 800°C. Two zones of anatectic migmatite are genetically related to the emplacement and crystallisation of the quartz-diorite and granite respectively (Pattison and Harte, 1988).

The *Ben Nevis Complex* (*69*) consists of a sequence of concentric intrusions with a central block of Dalradian metasedimentary rocks and Lower Old Red Sandstone lavas faulted down in a cauldron subsidence structure (Anderson, 1935; Bailey, 1960; Haslam, 1968). The Outer Granite phase has three components. The earliest member is confined to the north-west, and consists of fine- to medium-grained quartz-diorite, bearing augite and rare hypersthene; in places, it can be seen to become more acid and more leucocratic upwards. The second member is a coarse-grained augite-diorite to tonalite, which becomes more acid and leucocratic both inwards and upwards. It everywhere has a gradational contact with the third member, a megacrystic hornblende-biotite-granite, which is believed to have been intruded while the quartz-diorite was still partly molten. The plutonic rocks crystallised under low pressure, at which the alumina-rich hornblende, typical of the more deep-seated plutons, was unstable. As a result, pyroxene crystallised in all but the most acid rocks, to be replaced in the final stages of solidification by a pale green alumina-poor hornblende. All members of the Outer Granite are cut by the NE-trending andesite dyke swarm centred on the complex. The Inner Granite has a sharp contact against the Outer Granite and truncates almost all of the dykes. It is a fine-grained, non-porphyritic leucocratic white to pink biotite-granodiorite. The outer contact with the porphyritic Outer Granite shows no sign of chilling but at its inner contact with the downfaulted metasedimentary rocks and lavas it becomes markedly chilled, with a fine-grained matrix enclosing phenocrysts of plagioclase and biotite. The chilling at the inner contact of the granite is attributed to subsidence of the central block of volcanic rocks in the liquid granitic magma. The adjacent *Mullach nan Coirean Granite* (*70*; Bailey, 1960) is a leucocratic biotite-granite which predates the Ben Nevis dyke-swarm.

The *Glencoe* cauldron subsidence (*71*; Figure 28) has a volcanic expression at the present level of erosion, except for the Fault Intrusion (Bailey, 1960; Roberts, 1966). This is a north-eastward extension of the Cruachan Granite of the Etive Complex, and is marked by a gradation in texture from coarse-grained tonalite or granodiorite to a microgranodiorite with phenocrysts of plagioclase, hornblende and biotite set in a fine-grained quartzo-feldspathic matrix. A thin band of flinty crush-rock occurs at the inner margin of the Fault Intrusion. The Fault Intrusion was regarded by earlier workers as being disrupted by the ring-faulting, but Roberts (1966) interpreted it as having been emplaced as a fluidised mass of crystals, rock fragments, liquid droplets and interstitial gas, and as having acted as a feeder to the Upper Group 2 and Group 5 ignimbrites of the volcanic succession.

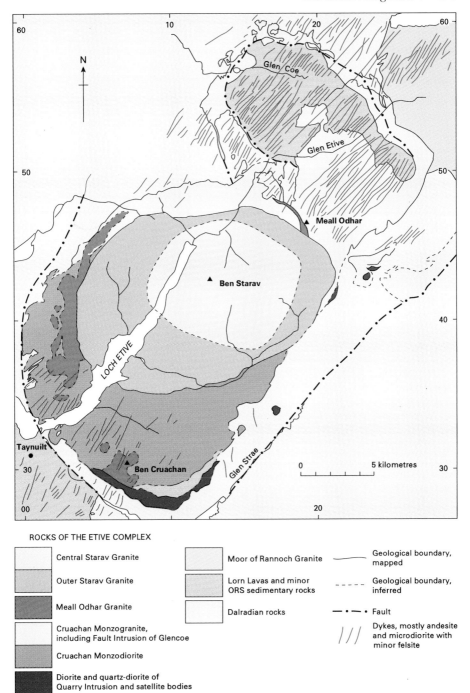

Figure 28   Simplified geological map of the Etive and Glencoe complexes (after Anderson, 1937 and Batchelor, 1987).

The *Etive Complex* (*72*; Figure 28) is an elliptical composite pluton in which three distinct intrusive phases are recognised (Anderson, 1937; Batchelor, 1987). It was emplaced at a high structural level, between 3 km and 6 km depth (Droop and Treloar, 1981), roughly contemporaneously with the Glen Coe cauldron subsidence. The first phase was the intrusion of several bosses and sheets of basic to intermediate material, ranging from olivine-monzonite (kentallenite) to quartz-diorite. The largest of these, the Quarry Intrusion, consists of medium-grained diorite and quartz-diorite, and forms an arcuate body along the south-east margin, separated from the main pluton by a screen of variably hornfelsed and foliated andesitic lavas with sedimentary intercalations. The second, Cruachan, phase of the complex ranges in composition from monzodiorite in the southern lobe to monzogranite in the northern lobe. The monzodiorite is heterogeneous, medium to fine grained, comprising subhedral, zoned plagioclase pheno- crysts with interstital quartz, feldspar, biotite and, more rarely, hornblende. The medium- to coarse-grained monzogranite of the northern lobe is more equigranular, though it locally contains megacrysts of K-feldspar. The facies becomes more silicic and finer grained northwards, grading into the Fault Intrusion of the Glencoe cauldron subsidence. The boundary between the two variants of the Cruachan phase is gradational (Batchelor, 1987). Emplacement of the Cruachan phase appears to have been through cauldron subsidence. The outcrop of the Cruachan phase is cut by irregular sheets and ring-dykes which dip inwards towards the Starav Granite. They are typically fine- to medium-grained pink monzogranites and, more rarely, syenogranites. Collectively they are referred to as the Meall Odhar Granite (Anderson, 1937), and show both sharp and transitional contacts with the Cruachan facies.

The Starav Granite, a body of medium-grained leucocratic monzogranite, was forcefully emplaced into the centre of the Etive Complex, with local devel- opment of a tectonic fabric in the earlier Cruachan phase and its contained microdiorite to felsite dykes. The megacrystic Outer Starav Granite was consid- ered to show a gradational contact with the inner nonporphyritic phase by Anderson (1937) and Bailey (1960). However, the absence of sphene from the inner phase granite contrasts markedly with the outer phase, suggesting that each phase represents a separate pulse of magma. Three petrochemically distinct zones are recognised in the Outer Starav Granite, becoming progres- sively more acid towards the centre. The Meall Odhar Granite has an arcuate outcrop, and is interpreted as a ring-dyke. Its composition lies close to the eutectic composition in the quartz–albite–orthoclase–water system. It is con- sanguineous with the Central Starav Granite, and the two bodies are probably contemporaneous (Batchelor, 1987).

The *Kilmelford Complex* (*73*) varies from diorite to granodiorite, with horn- blende-tonalite dominant (Peach et al., 1909, pp.67–71; Hill, 1905, pp. 98–100). There are three closely adjacent areas of outcrop, probably joined to each other at a shallow depth. The complex has a pronounced metamor- phic aureole.

The *Foyers Granite* (*74*) consists of two phases: an early fine- to medium- grained tonalite grading into locally porphyritic granodiorite, and a later pink medium-grained nonporphyritic monzogranite (Mould, 1946; Marston, 1971). The intrusion is funnel shaped overall, and the individual phases are irregular and, in places, vein-like in form. The foliation of the surrounding country rocks has been rotated into parallelism with the contacts of the complex pointing to a forceful mode of emplacement. A broad metamorphic aureole, containing sillimanite-bearing hornfelses, and with cordierite-K-feldspar migmatites in the innermost zone, is developed.

The *Findhorn Complex* (*75*) consists principally of of an elongate NE-trending mass of grey, medium- to coarse-grained, foliated biotite-granodiorite containing mafic schlieren and flattened, platy metasedimentary xenoliths. A number of small bodies of texturally heterogeneous pyroxene-bearing diorite and quartz-diorite, associated with a relatively melanocratic hornblende-granodiorite, cut the main granodiorite.

The *Boat of Garten* intrusion (*76*), lying in Strathspey, is largely obscured by drift. It consists of coarse-grained hornblende-biotite-granodiorite with abundant sphene and allanite. Locally it is megacrystic, with pink subhedral alkali feldspar phenocrysts up to 2 cm long.

The major portion of the *Corrieyairack Complex* (*77*) consists of a uniform pink-grey, medium-grained hornblende-biotite-granodiorite containing very few xenoliths (Key et al., in press.). However, the south-western portion is quite different, consisting of medium- to coarse-grained leucocratic granite choked with country-rock xenoliths. The two lithologies are separated by a narrow band of speckled black and white granodiorite. The south-western granite is roughly coeval with the Loch Laggan vein complex. The main granodiorite has not been directly dated but has been suggested to be coeval with the granodiorite phase of the adjacent Strath Ossian Complex (Anderson, 1956).

The *Allt Crom Granite* (*78*) is an elongate body lying between the Corrieyairack and Findhorn granodiorites. Recent mapping has shown that the southern part consists mainly of grey, equigranular biotite-granodiorite carrying large rafts of Grampian Group country rocks; a pink leucocratic monzogranite is present locally as a marginal facies. Further north the intrusion is mainly biotite monzogranite occurring for the most part as a vein complex but with some larger, mappable bodies.

In central Aberdeenshire a group of juxtaposed dioritic to granitic plutons was described by Bisset (1934) as the *Skene Complex*. The group is now believed to comprise intrusions of several different ages and affinities (Harrison, 1987). The Kemnay (*36*), Tillyfourie (*55*) and Corrennie (*84*), intrusions are probably late-tectonic; they have been described above. The Hill of Fare Granite (*102*) is a typical member of the Cairngorm Suite, and is described below. The remaining elements show a range from quartz-diorite to monzogranite, but the ring-structure typical of the Lorn and Lochaber plutons is absent.

The *Gask* (*79*) and *Torphins* (*80*) diorites (Harrison, 1987) consist largely of coarse-grained quartz-diorite and tonalite, but a fine-grained mafic quartz-diorite is present in places, often showing a complex relationship with the coarser-grained facies. The coarse-grained megacrystic *Crathes Granodiorite* (*81*) contains megacrysts of K-feldspar and locally becomes a monzogranite. It contains accessory hornblende and is locally rich in sphene. It is veined by the more melanocratic and finer-grained grey *Balblair Granodiorite* (*82*). The Balblair and Crathes granodiorites cut the Torphins Diorite and the Tillyfourie Tonalite, but are themselves cut by the Hill of Fare Granite. The roughly circular *Clinterty* mass (*83;* Munro, 1986b) consists of white coarse-grained granodiorite to granite.

Several additional granitic intrusions have been recognised during the course of recent mapping in the area to the west of the Skene Complex (Gould, in press). The *Logie Coldstone Tonalite* (*85*) and *Tomnaverie Granodiorite* (*86*; Read, 1927) are coarse-grained nonporphyritic grey granitoids. The *Kincardine O'Neil Granodiorite* (*87*) forms a group of small bodies of coarse-grained nonporphyritic granodiorite lying within a vein-complex. The *Lumphanan Granodiorite* (*88*) is finer grained and slightly more mafic, resembling the Balblair Granodiorite.

Cairngorm Suite

These plutonic bodies form a distinct suite within the northern Grampian Highlands, extending from the Monadhliath pluton in Strathspey to Peterhead (*89–107*); they are I-type granites with $^{87}Sr/^{86}Sr$ initial ratios of about 0.706 on the few examples measured. The intrusions comprise many textural varieties of biotite-granite; primary muscovite is rare (Stephens and Halliday, 1984; Plant et al., 1990). Most are pink coarse-grained rocks (the groundmass may reach 20 mm in some phases) and variants containing megacrystic K-feldspar are widespread. However, microgranites, mostly leucocratic and often porphyritic in places, form a sizeable component of some intrusions (e.g. Cairngorm and Mount Battock). Quartz is typically dark grey or brown and smoky and is called 'cairngorm(ite)' where it forms semi-precious crystals in drusy cavities. The main phases within each pluton are usually cross-cut by late pegmatite and aplite sheets and veins; some of these are locally silicified and/or brecciated, with faults in places developed along their contacts. Late hydrothermal alteration caused reddening of feldspars, associated with turbidity of quartz, over wide areas, and affects coarse-grained granites, microgranites and pegmatites. Epidotisation of feldspars, associated with quartz veining and brecciation, occurs in discrete areas. Some of the masses are bounded by faults, e.g. Bennachie. Contact metamorphic effects are small, but extend in places for up to 1000 m from the contact. These aureoles are often poorly developed in rocks previously affected by high-grade regional metamorphism or by the much more extensive contact metamorphism produced by the late-tectonic basic intrusions. Granites which show widespread alteration, the presence of aplitic and pegmatitic phases and abundant vuggy cavities are thought to have been emplaced at relatively high structural levels, possibly 5 to 8 km below surface (Harrison and Hutchinson, 1987). A large negative Bouguer gravity anomaly with an amplitude locally less than -60 mGal extends from the Monadhliath Granite to Mount Battock and Bennachie, indicating the presence at depth of a granite batholith. The exposed granites probably represent cupolas rising from this regional batholith. The thickness of granite underlying the Cairngorm pluton has been estimated at 12 km (Brown and Locke, 1979) but might be substantially thinner. Geochemical and geothermal data suggest that the deeper parts of the intrusions are slightly less acidic (e.g. granodiorite, tonalite) than the biotite-granites presently exposed (Webb and Brown, 1984). Significant, often annular, magnetic anomalies are associated with some of the intrusions (e.g. Monadhliath, Cairngorm, Lochnagar, Mount Battock, Hill of Fare). In some cases these can be ascribed to differences in the magnetite content of different phases of the granites, but in other cases basic rocks at depth or pelitic hornfelses around the margin of the pluton may be the source of the anomaly. The aplitic and pegmatitic phases of the granites show in many cases a marked enrichment in Rb, Li and other large-ion lithophile elements, and depletion in Sr and Ba. The granites of this group show relative enrichment in U and Th compared with the plutons of the South Grampians and Argyll suites.

Harrison and Hutchinson (1987) have made a controversial two-fold division of the post-tectonic granites of the eastern Grampians into an early approximately 415 Ma group (Table 2), showing forceful modes of emplacement or with strongly veined and magnetised aureoles, and a later approximately 408 Ma group which shows a more passive mode of emplacement, such as stoping or cauldron subsidence. However, this classification is decidedly speculative, as the mode of emplacement of several of the granites is still uncertain and the variation in trace element geochemistry is as great within the groups as between them. The inaccuracies in the ages quoted in Table 2 are too large for a reliable division on age alone. Both groups were emplaced at 5 to 8 km depth and show similar ranges of major and trace element and isotope geochemistry.

The Cairngorm (*94*), Ballater (*97*), Mount Battock (*98*) and Bennachie (*103*) granites have been investigated (Lee et al., 1984; Webb and Brown, 1984) as a possible source of 'hot dry rock' geothermal energy. Heat flow measurements were made down specially drilled boreholes, and samples from the boreholes and the outcropping granite were analysed for U and Th as well as for major and other trace elements. The results show that the granites have a high heat generation capacity at the levels sampled, but this is believed to decrease considerably in the deeper, unexposed levels of the granites. Combined with the low heat generation of the metasedimentary and basic igneous country rocks this gives rise to a much lower geothermal gradient than in the concealed Caledonian granites of northern England or the Hercynian granites of south-west England.

The *Saddle Hill Granite*, previously referred to as the Finglack alaskite, is a pink leucocratic coarse-grained granite, which occurs as three small intrusions cutting the Moy granite (*47*) to the north of Moy (Fletcher et al., 1995).

The *Auldearn Granite* (*89*) is a pink medium- to coarse-grained granite with orthoclase megacrysts. Its contact relations are unknown due to lack of exposure.

The composite *Ben Rinnes* pluton (*90*) crops out almost entirely to the south of the river Spey. The earliest phase, a reddened, foliated K-feldspar megacrystic granite, crops out along the southern and eastern margins, and is cut by a small ellipsoidal intrusion of foliated biotite-microgranite. The central part of the pluton comprises a grey, medium-grained porphyritic biotite-granite, which is locally xenolithic. These phases are cut by a chemically distinct medium- to coarse-grained leucocratic granite, relatively rich in opaque minerals, sphene and allanite, which forms an ellipsoidal mass in the west, and a sheet-like intrusion in the east. To the north of the River Spey, a few small exposures of granite, possibly forming a vein-complex or roof-zone, occur, together with a body of moderately coarse-grained augite-diorite, the Rothes or Netherly Diorite.

The *Glenlivet Granite* (*91*) is a coarse-grained leucocratic pink biotite-granite, with a porphyritic microgranite phase in its western part (Hinxman,

1896, pp.26–27). A small body of pink biotite-granite has been intruded and altered by the complex appinite–diorite *Dorback* intrusion (*92*; Zaleski, 1983).

The subcircular, stock-like *Monadhliath Granite* (*93*) comprises a main phase of medium- to coarse-grained, variably porphyritic, grey monzogranite and a younger, more evolved fine- to medium-grained biotite-monzogranite. Both phases are cut by sheets and irregular masses of porphyritic microgranite, aplite and pegmatite. Miarolitic cavities, with topaz and cairngorm, are locally abundant in the biotite-granite. The granitic components are highly reddened along NE-trending discrete zones, within which quartz-epidote veins, commonly associated with veins and masses of breccia, are confined (Highton, in press).

The *Cairngorm Granite* (*94*) is the largest of the Eastern Highlands granites, with an outcrop area of about 365 km². It is a steep-sided, stock-like body (Harry, 1965; Harrison, 1986), with an annular high magnetic anomaly pattern. The granite at the surface, however, is mostly nonmagnetic, probably due to alteration associated with the strong reddening of the rocks. The Main Granite is a coarse-grained, variably porphyritic, leucocratic red-pink biotite-granite. It intrudes the earlier coarse-grained, strongly porphyritic Glen Avon Granite, and is itself intruded by the finer-grained nonporphyritic Beinn Bhreac Granite. These three phases are intruded by flat-lying sheets of porphyritic microgranite and by the texturally similar Carn Ban Mor Granite. Pegmatite and aplite sheets and quartz veins cut all of the earlier phases. A layer up to 50 m thick of porphyritic microgranite, containing biotite, muscovite and garnet, is developed in places near the west contact of the Main Granite, possibly indicating local concentrations of volatiles near the contact. The Glen Avon Granite is geochemically less evolved, and poorer in radiogenic elements, than the main body of the intrusion. The contacts of the mass are sharp, subvertical, and cross cut the foliation of the country rock.

The Lochnagar and Glen Gairn granite plutons are chemically very similar and were emplaced into a zone of small precursor intrusions of diorite and granodiorite which extends from Glen Doll in the south to the eastern end of the Cairngorm Granite in the north. *Lochnagar (95;* Plate 11) was previously interpreted as having been emplaced by forceful intrusion and cauldron subsidence mechanisms (Oldershaw, 1974), but recent BGS work has cast doubt on this interpretation; intrusion by passive stoping seems more likely. The Lochnagar granites were emplaced as a zoned pluton of four main phases, each multicomponent. The earliest post-diorite phase was a large pluton of several coarse-grained and porphyritic granites, some containing a little hornblende. Subsequent phases comprise the central pluton of medium-grained inequigranular biotite-granite, which is cut by two bodies of microgranite, and several irregular bodies of very evolved pink and white leucogranites. On the basis of limited geochronological data, Lochnagar appears to be slightly older than the Cairngorm Granite (Table 2).

The earliest phase of the *Glen Gairn* pluton (*96*) consists of medium- to coarse-grained porphyritic hornblende-biotite-granite with an associated granite vein complex. A later phase, consisting of pink biotite-granite with an extensive aplitic, drusy and pegmatitic marginal facies, forms an arcuate body along the northern boundary of the pluton.

The *Coilacreich Granite* (*107*) consists of a medium-grained, white to pink lithium-mica-bearing granite with an aplitic-pegmatitic facies well-developed

Plate 11   Lochnagar (1155 m), south-west Aberdeenshire, from the north.
All the high ground in the distance is formed of various phases of the Lochnagar
Granite pluton; in the foreground is slightly older diorite. In the left distance are
spreads of glacial till and moraines below the corrie on Lochnagar (D 4249).

close to the margins and roof of the intrusion. Zinnwaldite occurs in a small
cupola exposed close to Gairnshiel Lodge (Hall and Walsh, 1972; Webb et al.,
1992). Greisen is developed in places, particularly in the zinnwaldite-bearing
granite. Associated aplite and pegmatite veins contain pyrite, sphalerite and
rare wolframite and beryl. A scheelite- and wolframite-bearing quartz vein
complex is also marked by marginal greisenisation of the country rock.

The western part of the *Ballater Granite* (*97*) consists of a pink, medium- to
coarse-grained, locally porphyritic granite, whereas the eastern part consists
of a very coarse-grained (2 cm groundmass) granite with K-feldspar
megacrysts up to 5 cm (Webb and Brown, 1984). The two granite variants
show a gradational contact. In the Pollagach Burn, the coarse-grained phase
intrudes the pelites, calc-silicate rocks and amphibolites of the Dalradian Tayvallich Subgroup. Wollastonite, diopside, idocrase and grossularite are
developed in the calcareous hornfelses, while andalusite, cordierite and,
more rarely, sillimanite and corundum are present in the pelitic hornfelses.
The granite is cut by a lead- and silver-bearing vein in the Pass of Ballater.

The large *Mount Battock* or *Kincardine Granite* (*98*) comprises several
irregular intrusions (Harrison and Hutchinson, 1987). The Main Granite is a
pink biotite-granite, moderately coarse-grained and nonporphyritic in the
western half of the outcrop, but coarser-grained and moderately porphyritic in
the eastern half. Bodies of coarse-grained, abundantly megacrystic granite
occur in the Birse area and in the vicinity of Clachnaben. A distinctive microgranite with scattered K-feldspar megacrysts up to 40 mm in size, the Water of
Feugh Granite, is exposed to the south of the Birse megacrystic granite. A
body of fine-grained lithium-mica-bearing granite occurs in the vicinity of

Bridge of Dye. Harrison and Hutchinson (1987) suggested that the major phases of the granite are probably fault-bounded, but recent work indicates that faulting is much more localised, although poor exposure precludes establishment of many of the age relationships. There are numerous sheets and veins of microgranite ranging from 1 m to 200 m in thickness and including a relatively biotite-rich variant. Hydrothermal alteration and reddening are widespread, especially to the north and south-west of Mount Keen. The principal phases of the Mount Battock pluton are more mafic and less geochemically evolved than those of the other East Grampians pink granites (Webb and Brown, 1984).

The *Cromar Granite* (*99*) is a roughly arcuate body of pink, nonporphyritic, moderately coarse-grained granite; the *Cushnie* (*100*) and *Ord Fundlie* (*101*) granites are small bodies of similar, but slightly finer-grained granite.

The medium- to coarse-grained pink granite of the *Hill of Fare* (*102*) is intruded into the Crathes (*81*) and Balblair (*82*) granodiorites along an arcuate contact (Harrison, 1987). There is some evidence of a fine-grained margin along the north-eastern contact. Irregular, often diffuse, patches of variably porphyritic leucocratic microgranite are present in the central parts of the pluton and it is cut, especially in the western part, by a suite of pink aplite sheets. At its south-western contact the Dalradian semipelites are hornfelsed for up to 300 m from the granite.

The *Bennachie Granite* (*103*; Plate 12; Webb and Brown, 1984) is bounded to the east and west by faults. Other contacts are steep, but some veining of the country rock can be seen in the south-west. The pluton comprises a coarse- to very coarse-grained (5–10 mm), variably porphyritic granite; a separate porphyritic microgranite phase occurs in the western part. The pluton is traversed by N–S-trending veins of reddened, brecciated and silicified aplite, the largest

Plate 12   Granite tors on Mither Tap, Bennachie, Aberdeenshire, with well-marked horizontal and subvertical joints (D 4523).

of which is coincident with the eastern contact. Beyond the eastern boundary fault, a number of small occurrences of fine-grained, reddened granite are interpreted as parts of a downfaulted eastern continuation of the Bennachie Granite. One of these, the *Middleton Granite* (*104*; Colman et al., 1989), is cut by quartz veins bearing molybdenum and tungsten mineralisation.

The *Peterhead Granite* (*105*) comprises a pink coarse-grained outer phase and a central finer-grained more-acid phase; both have been cut by microgranite dykes (Wilson, 1886; Buchan, 1934). The Peterhead Granite exposed in the coast section contains large xenoliths of the grey Forest of Deer Granite *(41)*. Although lithologically similar to the Cairngorm suite, the Peterhead Granite produces a relatively small Bouguer gravity anomaly (-8 Mgal).

The *St Fergus Granite* (*106*) is not exposed at surface, but has been proved in boreholes around the gas terminal site. Veins of pink and white granite crop out on the coast a few kilometres to the north (Peacock, 1983).

LATE- TO POST-TECTONIC MINOR INTRUSIONS

The Caledonian minor intrusions of the Grampian Highlands include plugs, sheets, and, more particularly, dykes. They can be divided into three geochemically distinct suites:

(i)   appinites and calc-alkaline lamprophyres—most abundant in Appin and Lorn, but extending through Lochaber to the East Grampians
(ii)   microdiorites—widespread throughout the Grampian Highlands
(iii)   felsites and quartz-feldspar-porphyries.

Appinites from Garabal Hill, Arrochar and Rubha Mor (Appin) have yielded U/Pb zircon and sphene ages of 422–429 Ma (Rogers and Dunning, 1991), indicating a restricted Middle Silurian range, possibly related to major transcurrent faulting. The other minor intrusions can only be dated by their field relations with the major plutonic phases; they cut some of the early phases of the postorogenic granitoid intrusions but predate some of the later phases.

Appinites and calc-alkaline lamprophyres

Many small intrusions of appinite and lamprophyre occur throughout the Grampian (and Northern) Highlands; they are respectively coarse-grained and fine-grained products of calc-alkaline magmatism. The presence of mafic hydrous phases, mainly hornblende and biotite, as euhedral phenocrysts indicates that the magma was volatile-rich (Rock, 1991). The lamprophyres occur as dykes and sheets, the appinites as plugs and irregular intrusions. The lamprophyres are mostly spessartites (hornblende phenocrysts with dominant plagioclase in the groundmass) with some vogesites (hornblende with orthoclase) and rare minettes (biotite with orthoclase).

The appinites typically comprise rocks with a fairly wide range of composition (the appinite suite); appinite itself is a kind of hornblende-melamonzonite. In the Ben Nevis district Bailey (1960) described appinite associated with augite-diorite, monzonite, kentallenite (olivine-monzonite) and cortlandtite (olivine-augite-hornblendite). Appinites in the Glen Roy area, where they are up to a kilometre across, exhibit a similar composition range (Key et al., in press). The rocks are characterised by extreme disequilibrium of mineral phases, with the mafic phases commonly showing reaction relationships, e.g. olivine mantled by pyroxene and pyroxene by amphibole or biotite.

Many of the appinite intrusions, notably those in the Appin and Glen Roy areas, are associated with pipes infilled with breccias composed of country rock fragments (Platten and Money, 1987). Several different lithologies, including many transitional types, may be present in a single mass of appinite, with the more acid members cutting the more basic; the lithological variation is ascribed to fractional crystallisation (Platten, 1991). A lamprophyre dyke associated with an appinite suite in the Loch Lomond area contains a varied suite of both cognate and upper crustal xenoliths (Dempster and Bluck, 1991).

Microdiorites

The microdiorites form sheets and dykes of intermediate composition, ranging from quartz-microdiorite to microgranodiorite. Typically, the dyke rocks have a fine-grained groundmass of andesine or oligoclase, hornblende, biotite, minor K-feldspar and quartz. The small plagioclase phenocrysts are generally andesine, often showing oscillatory zoning. Phenocrysts of hornblende and, more rarely, augite and/or biotite, may also be present. The textural difference between the microdiorites and the appinites is ascribed to the much lower volatile content of the former, the texture of which shows a much closer approach to equilibrium crystallisation. Alteration of the groundmass and ferromagnesian phenocrysts, which led in the past to some microdiorites being wrongly identified as lamprophyres (vogesites or minettes), is widespread.

Several episodes of calc-alkaline microdiorite intrusion occurred during the late-Caledonian magmatic episode. The earliest members are foliated and recrystallised sheets and dykes which have been affected by late-tectonic shearing; they are probably approximately coeval with the late-tectonic granites. Non-foliated and/or weakly foliated microdiorites, which cannot be related specifically to individual granitic complexes, are widespread throughout the Grampian Highlands and are believed to be roughly coeval with the main phase of post-tectonic granite intrusion.

All of the post-tectonic granitic complexes of the Lorn and Lochaber districts are spatially associated with extensive, dense swarms of NE-trending dykes of intermediate to acid composition, ranging from andesite and microdiorite to felsite. The Etive swarm cuts the Lorn Plateau Lavas, Glencoe volcanic rocks and the Moor of Rannoch Granite. Successive members of the Etive Granite Complex are cut by dykes associated with later phases of the complex, except the Central Starav Granite which is free of dykes. Similarly, the outer components of the Ben Nevis Complex are cut by a consanguineous suite of intermediate to acid dykes, which in turn are cut by the central granite. The vast majority of the dykes of this episode trend NE–SW, but there are a few earlier NW-trending dykes in the Glencoe cauldron subsidence. In the Glencoe area, the crustal extension represented by the emplacement of the dyke swarms has been estimated at 2.5 to 4 km over a distance of 9 km (Roberts, 1974). Suites of microdiorite and quartz-porphyry sheets and dykes are associated with the Glen Tilt Complex, particularly to the east of its main intrusive phases, and with the Foyers pluton.

Felsites and quartz-feldspar-porphyries

In the north-eastern Grampian Highlands, and extending as far west as Speyside and Glen Tilt, there is a prominent suite of felsite and quartz-feldspar-porphyry dykes, whose trends vary from north–south to NE–SW, but are frequently arcuate. These cut the earlier post-tectonic diorites and granites, but the later pink biotite-granites, with the exception of the Mount Battock Granite and the later phases of the Glen Gairn and Lochnagar granites, are not traversed by these dykes. A large sheet of particularly coarse-grained quartz-feldspar-porphyry, almost granitic in texture, occurs to the west of the Lochnagar Granite, and several irregular bodies of finer-grained felsitic rock are present nearby. In Strathnairn, stockworks of felsite sheets appear to be coeval with the Maol Chnoc vein complex, which predates the emplacement of the Foyers pluton.

LOWER OLD RED SANDSTONE VOLCANISM

Volcanic rocks in the Lower Old Red Sandstone occur principally in Lorn and Lochaber, with minor occurrences in the north-east Grampians, the Highland Border and Kintyre.

Lorn and Lochaber

Lower Old Red Sandstone volcanic rocks originally covered the whole area but are now preserved only in the Lorn Plateau and in downfaulted blocks within the Glencoe and Ben Nevis complexes (Bailey, 1960; Kynaston and Hill, 1908).

The *Lorn Plateau lavas* (*108*) have a present extent of 300 km^2 and a maximum preserved thickness of 800 m. Sedimentary rocks at the base of the succession on Kerrera contain Upper Lochkovian (Lower Devonian) fossils, implying, according to Morton (1979), a correlation with the Arbuthnott Group of the Midland Valley. The lavas are basalts and basaltic andesites, forming flows 5 to 30 m thick, with rare rhyolite flows up to 2 m thick. They were probably erupted from fissures fed by dykes and occasional small circular vents. Groome and Hall (1974) considered that appinites, rather than microdiorites, were the probable feeders to the lavas. Two intercalated flows of rhyolitic ignimbrite are thought to have originated from the Glencoe centre and were fed from the Fault Intrusion (Roberts, 1974).

The lavas are of potassic calc-alkaline type and show chemical similarities with the nearby lamprophyre intrusions; certain flows are anomalously rich in Mg, Ni and Cr. The lavas as a whole are richer in Sr, Ba, K, P and light rare-earth elements (LREE) than the equivalent age lavas of the Midland Valley and Southern Uplands, and the magmas are thought to be largely mantle-derived, with some contamination by crustal material, possibly of mafic granulite composition (Groome and Hall, 1974; Thirlwall, 1981; 1982). Clayburn et al. (1983) obtained a Rb/Sr isochron age of 400 ± 5 Ma (Early Devonian) for the Lorn Plateau lavas, very similar to the 401 ± 6 and 396 ± 12 obtained from the probably genetically related Etive granites. The initial $^{87}Sr/^{86}Sr$ ratio (0.7045–0.7050) of the lavas is also similar to that of the Etive granites (0.7055–0.7058). However, Thirlwall (1988) has suggested, on the basis of argon isotope work, that the Lorn Plateau lavas may be as old as 421 to 413 Ma, and that the Rb/Sr systematics of the lavas have been reset by the plutonic intrusions.

Glen Coe and Ben Nevis are two of the best-exposed examples of cauldron subsidence (Bailey, 1960). The volcanic sequence in *Glen Coe* (*109*; Figure 29; Plates 13 and 14) is preserved in a downfaulted block within the elliptical ring fracture. Roberts (1974) has postulated the following sequence of events and products in the Glencoe cauldron:

Initial stages of volcanic activity
Group 1—basalts and pyroxene-andesite lavas

First cycle of caldera formation
Lower Group 2—andesite and rhyolite lavas, interbedded with a thin
  ignimbrite layer near the top
Upper Group 2—rhyodacitic ignimbrite flow
Possible hiatus in volcanic activity
Group 3—breccias, grits and shales

Second cycle of caldera formation
Group 4—hornblende-andesite lavas
Group 5—rhyodacitic ignimbrite flow
Possible hiatus in volcanic activity
Group 6—grits and shales

Subsequent stages of volcanic activity
Group 7—andesite and rhyolite lavas, interbedded with a thin ignimbrite flow

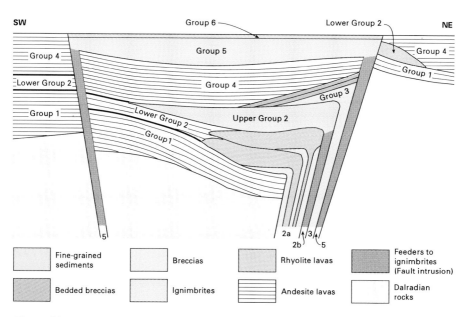

Figure 29   Volcanic evolution of the Glencoe cauldron (after Roberts, 1974).

Plate 13   Lower Old Red Sandstone lavas, Glen Coe, Lochaber.
Paler rhyolitic lavas and ignimbrites overlie darker basalts and andesites in the cliffs of
Aonach Dubh (left). The left slopes and summit of An t-Sron (right) are composed
mainly of granite of the Glencoe Fault Intrusion (D 1927).

The first stage of caldera formation was marked by a considerably larger
movement along the north-east part of the ring-fracture than along the south-
west part. However, during the second period of caldera formation the
amount of movement along the ring fracture was roughly uniform. The Fault
Intrusion follows the ring-fracture and shows flow structures; it was the feeder
for the Upper Group 2 and Group 5 ignimbrites, which appear to have completely filled the caldera and flowed possibly as far as the Lorn Plateau.

A 600 m-thick Lower Old Red Sandstone sequence unconformably
overlying Dalradian rocks is preserved by subsidence within the central ring-
structure of the *Ben Nevis* Complex (*110;* Bailey, 1960; Plate 15). It consists
principally of hornblende- and biotite-andesites with intercalated agglomer-
ates and sedimentary rocks, and is believed to be part of a regional lava cover,
otherwise completely removed by erosion from the Ben Nevis area.

North-east Grampians

The Tillybrachty Sandstone Formation of the largely sedimentary Old Red
Sandstone outlier at *Rhynie* (*111*) contains a single flow of highly vesicular
andesite lava (Trewin and Rice, 1992; Read, 1923, pp.180–182), which has
suffered potash metasomatism in the vicinity of the Rhynie hot spring system.
The Rhynie Chert, lying within the Dryden Flags Formation, has been interpreted as a siliceous sinter, related to the same hot spring system.

Plate 14   Glencoe Boundary Fault, Stob Mhic Mhartuin, Lochaber.

The fault separates the main Glencoe Fault Intrusion (top left), from bedded quartzites (bottom right). The Fault Intrusion, here a porphyritic microdiorite, is chilled against the fault zone which consists of brecciated quartzite in a streaky banded matrix of dark glassy material (pseudotachylite), well displayed below the hammer (D 1562).

Plate 15   North face of Ben Nevis, Lochaber.
The cliffs and summit area consist of a cylindrical mass of andesitic lavas which have
foundered into granite, exposed on the foreground ridge (D 2132).

A single outcrop of andesite in the *Cabrach* outlier (*112*) may be a lava and
the stratigraphical equivalent of that in the Tillybrachty Sandstone Formation
(Hinxman and Wilson, 1902, p.66). Andesite exposed in the *Gollachy Burn*
(*113*) near the Moray Firth coast has been variously interpreted as a lava flow
of Lower or Middle Devonian age, or as a sill intruding the underlying
Dalradian quartzites (Peacock et al., 1968, p.40).

Highland Border

A few outliers of Lower Old Red Sandstone rocks containing lavas overlie
Dalradian and Highland Border Complex rocks unconformably in the *Forest
of Alyth* area (*114*) and near *Monzie* (*115*). They occur as discontinuous
faulted exposures, but have been correlated with the main Strathmore
sequence of the Midland Valley by Armstrong and Paterson (1970). A distinc-

tive dacitic ignimbrite, the 'Lintrathen Porphyry', occuring from the Forest of Alyth to the North Esk, belongs to the Lochkovian Crawton Group, while the basaltic and andesitic lavas of the Forest of Alyth and Comrie are assigned to the immediately overlying Arbuthnott Group (also Lochkovian).

Kintyre

The New Orleans Conglomerate Formation (Friend and Macdonald, 1968) within the Lower Old Red Sandstone of Kintyre consists dominantly of coarse conglomerates of lava boulders, and has an estimated thickness of 890 m. Acid tuffs crop out at one locality. A nearby volcanic source is indicated. Precise correlation with the Lower Old Red Sandstone succession of the Stonehaven area is not possible, but the New Orleans Conglomerate Formation is believed to be equivalent to the Arbuthnott or Garvock groups (Lower Devonian).

9   Highland Border Complex

The Highland Border Complex comprises a group of highly varied lithologies: igneous and meta-igneous rocks, terrigenous mudstones, black shales, quartz arenites and minor limestones. The rocks occur as a series of generally fault-bounded slivers, ranging from a few metres to several kilometres in size, along the Highland Boundary Fault Zone between Stonehaven and Arran (Figure 30). They were originally referred to as the 'Highland Border Series', following the work of Barrow (1901). The term 'series' is now considered inappropriate for lithostratigraphical units and the terms

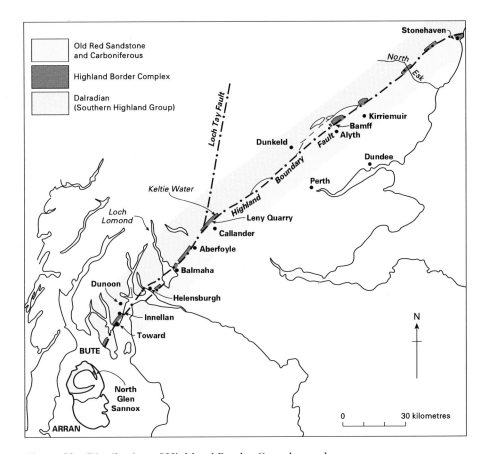

Figure 30   Distribution of Highland Border Complex rocks.

Highland Boundary Complex (Henderson and Robertson, 1982) or Highland Border Complex (Curry et al., 1984) have been proposed instead, the latter being the more commonly used. Subdivisions of this group of rocks have also in the past been called various 'series'; some of these are quoted in the following text, although most have now been renamed as 'formations' or informal 'assemblages'.

LITHOLOGY

The following account draws heavily on descriptions provided by Henderson and Robertson (1982), and Robertson and Henderson (1984). Where not otherwise referenced data have been taken from these two sources.

Igneous rocks

Ultramafic rocks, commonly with faulted contacts against other lithologies, occur at several localities along the Highland Border. They are invariably altered to serpentinites and carbonate-quartz rocks. The ultramafic rocks were derived mainly from harzburgites. Basic plutonic rocks or their metamorphic equivalents are present at a number of localities. Albitised gabbros occur at Aberfoyle in association with serpentinites (Jehu and Campbell, 1917) and amphibolites thought to be altered equivalents of gabbros are found in Arran. Hornblende-schists, of tholeiitic affinity and locally garnetiferous, are found at Aberfoyle and at Scalpsie Bay in Bute. At both localities the hornblende-schists are in contact with serpentinite. Immediately adjacent to the serpentinite at Aberfoyle the hornblende-schist belongs to the lower amphibolite metamorphic facies, the metamorphic grade falling to greenschist facies away from the contact (Henderson and Fortey, 1982). Basic volcanic rocks, commonly with intercalated sedimentary rocks, occur locally in the Highland Border Complex, notably at Stonehaven, the River North Esk and in Arran. The lavas are commonly altered to spilites and may be pillowed or massive. Brecciated lava has been described from Arran. Two magmatic types have been distinguished, one with mid-ocean ridge basalt (MORB) characteristics, the other of 'within plate' chemistry; the hornblende-schists are chemically identical to the MORB lavas.

Sedimentary rocks

The sedimentary rocks may be described in three groupings, namely, 'Margie unit', interlava sediments and 'Loch Lomond clastics'.

The 'Margie unit' is the predominant lithological group (and is not to be confused with the obsolete lithostratigraphical term 'Margie Series' (Figure 31)) and occurs extensively within the Highland Border Complex. It consists for the greater part of quartzose arenites with subordinate black shales and minor limestones (the most prominent being the Margie Limestone). Lithologically the group is similar to rocks of the Dalradian Southern Highland Group and distinguishing between the two has caused considerable confusion.

The interlava sedimentary rocks consist of siltstones, mudstones and cherts, the majority of which originated as terrigenous muds deposited as distal tur-

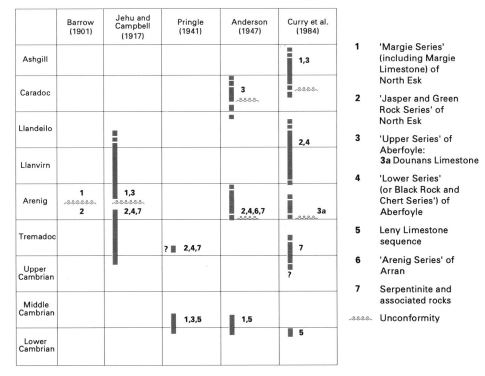

	Barrow (1901)	Jehu and Campbell (1917)	Pringle (1941)	Anderson (1947)	Curry et al. (1984)
Ashgill					1,3
Caradoc			3		
Llandeilo					2,4
Llanvirn					
Arenig	1 / 2	1,3 / 2,4,7		2,4,6,7	3a
Tremadoc			? 2,4,7		7
Upper Cambrian					?
Middle Cambrian			1,3,5	1,5	
Lower Cambrian					5

1 'Margie Series' (including Margie Limestone) of North Esk

2 'Jasper and Green Rock Series' of North Esk

3 'Upper Series' of Aberfoyle: 3a Dounans Limestone

4 'Lower Series' (or Black Rock and Chert Series') of Aberfoyle

5 Leny Limestone sequence

6 'Arenig Series' of Arran

7 Serpentinite and associated rocks

〜〜〜 Unconformity

Figure 31 Evolution of the stratigraphical interpretation of the Highland Border Complex rocks.

bidites with a subordinate amount of locally derived, largely volcanic, detritus. Certain iron-rich sedimentary rocks are regarded as hydrothermal precipitates.

The 'Loch Lomond clastics' is an informal term covering all rocks containing notable amounts of basic and ultramafic detritus. The grouping contains a number of small but stratigraphically significant units; these include the serpentinite rudites at Balmaha (Henderson and Fortey, 1982); the Loch Fad Conglomerate on Bute, which consists mainly of greywacke boulders with subordinate clasts of serpentinite; the basement breccias at Aberfoyle, which contain clasts up to 1 m in length of shale, limestone and volcanic material (Jehu and Campbell, 1917); the Dounans Conglomerate near Aberfoyle, which contains rare gabbro detritus (Henderson and Robertson, 1982); and the Green Conglomerate of Pringle (1941) in the North Esk, which contains clasts of lava, hornblende-gabbro and chert.

STRATIGRAPHY

Elucidation of the stratigraphy of the Highland Border Complex is greatly complicated both by relatively poor exposure and by the faulted contacts within the complex and the consequent tectonic repetition of units. This has led to a wide variety of interpretations and correlations over the years (see Figure 31 and reviews in Anderson, 1947b, Curry et al., 1984) and present understanding of the stratigraphy and relationships between the various components of the Highland Border Complex is still incomplete.

The first attempts to interpret the stratigraphy were made by Barrow (1901) and Jehu and Campbell (1917) working in the North Esk–Stonehaven area and the Aberfoyle area, respectively. They both recognised a group of lavas, black shales and cherts, apparently overlain by a group of coarse arenites and shales with subordinate but stratigraphically significant limestones. Pringle (1939), using the evidence of sedimentary structures, reversed the two groups placing the arenite-dominated sequence as the older. Pringle also determined a palaeontological age of late Lower Cambrian for the Leny Limestone at Callander (see also Cowie et al., 1972) which he believed to be the correlative of the limestones at Aberfoyle and the North Esk.

This view was accepted and expanded by Anderson (1947) in an extensive examination of the Highland Border. Anderson was impressed by the similarity between the Dalradian rocks and the adjacent arenite sequences of the Highland Border Complex and proposed that the latter (including the limestones) formed the uppermost part of the Dalradian sequence, which was therefore of Cambrian age. These Dalradian rocks he regarded as unconformably overlain by the black shale and lava sequences which he believed to be of Ordovician age.

In recent years, however, intensive palaeontological investigations of the Highland Border rocks have radically revised the stratigraphy. The results of this work are summarised in Curry et al. (1984), who suggest that the Highland Border Complex rocks rest unconformably on the Dalradian succession and range in age from Lower Cambrian to uppermost Ordovician. They divide the Highland Border Complex into four lithostratigraphical groupings (Figure 31). The first group comprises the serpentinites and associated ophiolitic rocks which are believed to be of pre-Arenig age. The black shale and limestone sequences of the Leny area, which are accepted as late Lower Cambrian age, may also be part of this grouping. The second group, of early Arenig age, comprises the serpentinite conglomerate at Balmaha, the Dounans (or Lime Craig) Conglomerate and the Dounans Limestone at Aberfoyle. The third group consists of the black shale and lava sequences which are ascribed a Llanvirn–Llandeilo age. This group is unconformably overlain by the youngest assemblage which comprises the arenite-dominated sequences (including the Margie Limestone). This last group is believed to be of Caradoc–Ashgill age; its base is an unconformity marked by the basal breccia of Jehu and Campbell (1917) at Aberfoyle and possibly the Green Conglomerate in the North Esk. Curry et al., (1984) point out that the upper group contains blackened and deformed fossils of Llanvirn–?Llandeilo age and unblackened fossils of Caradoc age. This is taken as evidence of a pre-Caradoc period of low-grade metamorphism and uplift.

The black shale and limestone sequences exposed around the Leny quarries are recognised to be of Cambrian age. However, their status within the Highland Border Complex stratigraphical sequences is complicated by their relationship to similar sequences in the nearby Keltie Water and the relationship between these sequences and the adjacent Dalradian rocks. Two factors are important in considering the problem. Firstly, detailed mapping by Pringle (unpublished manuscript, deposited with BGS) suggests that the Leny Limestone is the correlative of the limestones in the Keltie Water (cf. Harris, in discussion of Rogers et al., 1989); secondly, the black shales and limestones in the Keltie Water can be traced with apparent conformity into the underlying Dalradian grits (Anderson, 1947, p.495; Harris, 1969). However, Rogers et al. (1989) have published a U-Pb zircon age of 590 Ma as the date of intrusion of the Ben Vuirich Granite; this granite cuts early structures in the Dalradian rocks, implying that the Dalradian sequence and its earliest deformation phase(s) are Precambrian. This apparent contradiction in the available evidence is still unresolved.

STRUCTURE AND METAMORPHISM

Much debate has centred around the structural and metamorphic setting of the Highland Border Complex and its relationship with the Dalradian. Barrow (1901) originally separated the arenite sequences of the North Esk from the adjacent Dalradian rocks on the presence of clastic micas in the Highland Border Complex rocks and their absence from the Southern Highland Group rocks. This view, however, was not supported by subsequent workers. Jehu and Campbell (1917) suggested that the relatively high metamorphic grade of the hornblende-schists at Aberfoyle reflected locally greater cover and more intense shearing than in the adjacent rocks. Anderson (1947), in support of Clough (1897), noted that the Dalradian phyllites at Inellan, south of Dunoon, showed evidence of contact metamorphism at their junction with a serpentinite body. In general Anderson (1947) regarded the degree of metamorphism and 'intensity of cleavage development' as similar in both the Highland Border Complex and Dalradian rocks. A structural analysis of the North Esk section led Johnson and Harris (1976) to the conclusion that the earliest fold phase to affect the Dalradian was the same as the earliest phase recognised in the Highland Border Complex, thus in effect supporting Anderson's (1947) view. Harris (1969) argued that the Leny Limestone had a similar deformational history to the neighbouring Dalradian sequences.

Henderson and Robertson (1982) suggested that the carbonation of the serpentinites was an early event and had started before deposition of the serpentinite rudites at Balmaha. They also noted that the Dalradian rocks have been subjected to four major phases of regional deformation (see Harte et al., 1984) and considered the relationship of the Highland Border Complex to the Dalradian events, basing their arguments on a number of points. Firstly, they regarded the contact metamorphism of the Dalradian phyllites at Inellan as overgrowing the earliest Dalradian fabrics (D_1) with the porphyroblasts lineated and deformed by the second phase-structures (D_2), the degree of deformation increasing towards the serpentinite. They also recognised similar relationships at Toward, south of Dunoon, and at Scalpsie Bay on Bute. Secondly, they described high-strain zones, or mylonites, at a number of localities throughout the Highland Border Complex, including Arran, where folded mylonite zones and downward-facing structures were identified. These observations, along with regional considerations, led Henderson and Robertson (1982) to suggest that the Highland Border

Complex rocks had been brought into contact with the Dalradian on a series of thrusts, probably during the regional D_2 event. Frictional heat arising from the thrusting, allied to possible residual heat within the Highland Border Complex rocks, led to the development of the observed contact effects, analogous to effects on the soles of other obducted ophiolitic sequences (Henderson and Robertson, 1982; Curry et al., 1984). However, the presence of Caradocian fossils in parts of the Highland Border Complex, taken in association with the radiometric age (see above) which indicates that the early regional deformation in the Dalradian took place in the late Precambrian, makes the above structural interpretation untenable (Robertson and Henderson, 1984; Harris, 1991).

In an examination of the contradictary radiometric, structural and metamorphic evidence Harte et al. (1984) concluded that there was no good evidence that the Highland Border Complex rocks had a common structural history with the Dalradian rocks, at least before the regional D_4 event at about 460 to 440 Ma (see also Curry, 1986). Although the structural model of Henderson and Robertson (1982) has been largely abandoned, their interpretation that the contact metamorphic effects seen at Dunoon and elsewhere are the result of the tectonic emplacement of the Highland Border Complex rocks against the Dalradian is still regarded as generally valid (e.g. Curry et al., 1984).

TECTONIC SETTING

The early workers, including Anderson (1947), regarded the Highland Border Complex as essentially an upward extension of the Dalradian succession, the variable lithologies reflecting basin development with minor tectonism responsible for unconformable relationships.

The emergence of modern plate tectonic theories and the new evidence on the stratigraphy and structure of the complex have greatly changed modern interpretations of the tectonic setting. The Highland Border Complex was envisaged as developing initially within a marginal basin on the north side of the Midland Valley block (Longman et al., 1979). Robertson and Henderson (1984), basing their arguments largely on the chemistry of the rocks, suggested that the rocks of the complex developed in a small marginal basin next to a continental massif (analogous to the present Gulf of California) and were subsequently emplaced onto the Dalradian nappe pile. Curry et al. (1984) suggest that the Highland Border Complex formed in a basin on the landward side of a volcanic-arc massif, possibly in a faulted or fragmented basin, with subsequent obduction, faulting and deformation movements. Harte et al., (1984) concluded that the Highland Border Complex could not easily have been brought into thrust contact with the Dalradian and that it was probably emplaced against the Southern Highland Group rocks as part of a regional strike-slip regime allied to periodic uplift, the movements taking place in the late Ordovician and subsequent to the nappe-forming events in the Dalradian succession.

10   Devonian

Devonian rocks are largely confined to the peripheral areas of the Grampian Highlands, although scattered downfaulted outliers in the interior point to a more extensive original cover. The Caledonian Orogeny was complete by early Devonian times and the Grampian Highlands were then part of a very large landmass lying astride the equator. Under semi-arid conditions, vast thicknesses of continental sediment were deposited to form the Old Red Sandstone megafacies in two distinct basinal areas, the Argyll Basin and the Orcadian Basin. The Old Red Sandstone succession which borders the Grampian Highlands on the southern side of the Highland Boundary Fault, was deposited in the separate Midland Valley Basin (Cameron and Stephenson, 1985).

In northern Scotland, early Devonian extensional stresses gave rise to crustal fracturing which, in places, was accompanied by short-lived volcanic episodes. A series of half-graben developed in which a variety of alluvial, aeolian and fish-bearing lacustrine sediments accumulated. The largely extensional-tectonic non-marine sedimentary regime was maintained throughout the Devonian, but differential movements and intervals of heavy precipitation periodically led to interbasinal lacustrine conditions.

Traditionally, the Scottish Devonian succession has been subdivided into Lower, Middle and Upper Old Red Sandstone which, until recently, were regarded as separate units corresponding to the three Devonian subsystems. Fossil evidence, mainly based upon palynomorph (spore) assemblages, now demonstrates that the unit boundaries are diachronous and do not coincide with breaks in sequence at the Lower–Middle and Middle–Upper Devonian boundaries.

The largest area of preserved Devonian rocks occurs in Moray and Buchan, along the southern border of the former major Orcadian Basin. Three main subdivisions of the rocks, representing distinct environmental suites, are recognisable and they will be referred to as Lower, Middle and Upper Old Red Sandstone in this account. Correlations based upon fossil-fish assemblages are important in the lacustrine facies present, especially in the Middle Old Red Sandstone, where the Achanarras Assemblage is prominent. Most of the Orcadian Basin succession is represented in this region by Middle Old Red Sandstone sedimentary rocks. The Lower group is less extensively developed and is in places overstepped by later sequences. Due to erosion, relatively little of the Upper Old Red Sandstone is presently preserved on land.

Rocks of the Argyll Basin occur in the south-western part of the Grampian Highlands. The succession is dominated by a thick pile of andesitic lavas that overlie sedimentary rocks assigned to the Lower Old Red Sandstone. No rocks

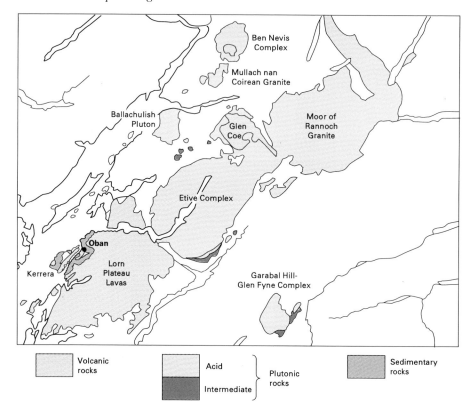

Figure 32   Sedimentary and volcanic rocks of the Lower Old Red Sandstone and their relationship to the Caledonian plutons (Argyll Suite) of the South-west Highlands.

of the Middle and Upper Old Red Sandstone are preserved in the onshore exposures of the Argyll Basin succession.

LOWER OLD RED SANDSTONE

The Lower Old Red Sandstone sedimentary rocks accumulated in small isolated fault-bounded basins and comprise conglomeratic fans, breccia screes and rock-falls of very locally derived rocks that pass basinwards into fluvial fine-grained sandstones and playa-lake carbonate-bearing mudstones.

In Argyll, the Lower Old Red Sandstone sedimentary rocks underlie volcanic rocks and are exposed on Kerrera, around Oban and in Glen Coe (Figure 32). They lens out eastwards beneath the Lorn Plateau lavas which form a 300 km² spread between Oban and Loch Awe; volcanic rocks are also preserved within Glen Coe and on top of Ben Nevis.

The Lower Old Red Sandstone rocks on Kerrera rest on an irregular Dalradian rock surface. They comprise a basal unit of breccia and conglomerate, which is succeeded by sandstones, shaly mudstones and up to 300 m of siltstones and thin limestones with, locally, intercalated lavas. The oldest flaggy shaly mudstones contain a varied fauna of cephalaspid and anaspid fish including *Cephalaspis lornensis* (Waterston *in* Craig, 1965), as well as the millipede *Kampecaris obanensis* and plant remains (Lee and Bailey, 1925). Such a fauna is considered to be Přídolí (uppermost Silurian) in age, similar to that of the Stonehaven Group in the Midland Valley succession, south of the Highland Boundary Fault. The basal beds at Oban are somewhat younger. Sections of sandstone with overlying thick grey shaly-mudstone-bearing conglomerates are well exposed. Andesite boulders within these conglomerates indicate there was volcanic activity in the Grampian region prior to sedimentation. At several localities, the mudstones have yielded fossil assemblages that include the eurypterid *Pterygotus anglicus* as well as *Kampecaris forfarensis, K. obanensis, C. lornensis, Mesocanthus mitchelli, Theolodus* sp., ostracods and plants. In Glen Coe, the oldest beds are breccias and conglomerates with green and blackish shaly mudstones; near the foot of Buchaille Etive Mòr, plant remains, including *Pachytheca fasciculata*, occur within the blackish mudstones.

At the southern end of Kintyre (Figure 35), there is a thick succession of Lower Old Red Sandstone rocks which Friend and Macdonald (1968) have subdivided into the following formations:

	Thickness m
BASTARD SANDSTONE FORMATION Purple sandstone with red siltstones; beds generally finely laminated and flat-bedded with some cross-bedding	100
NEW ORLEANS CONGLOMERATE FORMATION Generally coarse conglomerate with many large lava boulders; near the top interbedded with reddish purple sandstone and siltstone with some calcareous concretions; local pumice lapilli	890
GLENRAMSGILL FORMATION *Quartzite Conglomerate Member* Conglomerate with mainly quartzite clasts up to 1 m in diameter overlain by 200 m of purple sandstone and siltstone	300
Basal Breccia Member Poorly bedded breccia of Dalradian clasts; passing upwards into red sandstones and siltstones replete with fine-grained Dalradian detritus; lava fragments common near the top	150

In terms of the Midland Valley Old Red Sandstone succession, the Bastard Sandstone Formation is possibly the lateral equivalent of part of the Garvock Group and the lower formations parts of the Arbuthnott Group. Friend and Macdonald (1968) have suggested that the sediments were mainly derived from the north-west, in the area of Islay and Jura, and the volcanic fragments were transported from a Devonian lava field in northern Kintyre similar to that of the Lorn Plateau. On the southern coast of Kintyre, three vents cut through the junction of the Glenramsgill and New Orleans formations. Lava was intruded in only one of these vents, which largely comprise columns of gas-transported conglomerate and sandstone clasts. The Lower Old Red

Sandstone succession on Sanda Island, separated by faulting from that on the mainland of Kintyre, shows differences in detail, although the lithologies are generally similar.

Sedimentary and volcanic rocks of Devonian age occur on both sides of the Highland Boundary Fault in the Crieff and Blairgowrie areas; they form part of the Arbuthnott Group of the Midland Valley Old Red Sandstone succession described in the Regional Guide for the Midland Valley of Scotland (Cameron and Stephenson, 1985).

In the North-east Highlands, relatively small outliers at Tomintoul, Cabrach, Rhynie, Aberdeen and Turriff (Figure 33) are considered to be remnants of a more extensive development of the Lower Old Red Sandstone. Most represent irregular infillings of half-graben on the southern periphery of the Orcadian Basin. The Tomintoul Outlier is the sedimentary fill of an irregular NE-trending depression on the Dalradian basement. In the south-west, the outlier consists of a basal breccia overlain by a considerable, but unknown, thickness of coarse angular- and rounded-clast conglomerate. The sequence is well exposed in Ailnack Gorge (Plate 16), where the clasts are mainly of metasedimentary lithologies and up to 1 m in diameter. To the north-east, this conglomerate is overlain by red, immature medium-grained sandstones. The Cabrach Outlier is fault-bounded on its north-western side. Here a basal conglomerate is overlain by grey and red, friable, or red micaceous sandstone intercalated with coarse conglomerate layers; an outcrop of altered amygdaloidal andesite in the northern part of the outlier may represent a lava flow.

The Rhynie Outlier forms a 21 km-long NNE-trending outcrop some 12 km east of the Cabrach Outlier. It contains andesitic lavas and is fault-

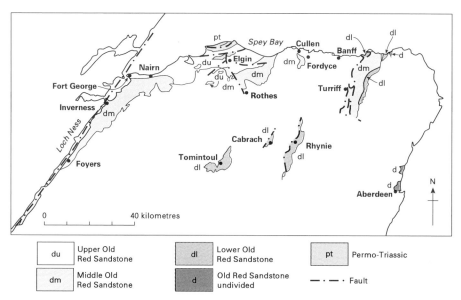

Figure 33   Old Red Sandstone and Permo-Triassic rocks of the northern Grampian Highlands.

Plate 16 Lower Old Red Sandstone unconformable on Dalradian, Ailnack Gorge, Tomintoul, Banffshire.

In the cliffs on the left, red-brown breccias and breccio-conglomerates dip north-west and overlie dark grey pelites, semipelites and limestones of the Appin Group (D 4010).

bounded on the western side. The rocks of this outlier may be the oldest in the Orcadian province; the sequence is (Gould, in press):

	Thickness m
DRYDEN FLAGS FORMATION Greenish grey flaggy siltstone with some mudstone and rare micaceous sandstone. Includes the *Rhynie Chert Member* near the base	260–800
QUARRY HILL SANDSTONE FORMATION Pale pink to grey massive sandstone with thin siltstone interbeds	0–400
TILLYBRACHTY SANDSTONE FORMATION Soft, whitish to deep purple sandstone with conglomerate lenses. Andesitic lava with minor tuff occur in the northern part of the outcrop	100–1400
CARLINDEN SHALE FORMATION Red and grey shaly mudstones with paler silty layers and calcareous sandstone beds	0–50
CORBIE'S TONGUE CONGLOMERATE FORMATION Compact conglomerate or breccia with some pebbly sandstone interbeds	0–20

Archer (1978) has shown that early sheet-flood conglomerates were succeeded by braided-stream deposits directed northwards and later by flood-plain sediments, the latter containing the Rhynie Chert sequence (Mackie,

1914). This chert incorporates remains of Devonian plants that grew on a peat bed close to a volcanic centre. Silica-rich volcanic waters rapidly flooded the local vegetation and prevented the microbial breakdown of plant tissues, thereby prefectly preserving their microscopic structures. Such remains include the psilophytes *Rhynia, Horneophyton* and *Asteroxylon* (Kidston and Lang, 1921) as well as fungi and myxophacean blue-green algae. In addition to the spectacular botanical remains, other fossils in the chert include the crustacean *Lepidocaris rhyniensis,* the arachnid *Paleocharinus rhyniensis* and the collembolids *Rhyniella praecursor* and *Rhyniognatha hirsti.* Trace-fossil burrows of annelids and possible lungfish (Archer, 1978) also occur along with spores of probable Siegenian age (Richardson, 1967). The Rhynie Chert is enriched in gold (0.18 ppm) and arsenic (79 ppm) (Rice and Trewin, 1988). In a fault zone of silicified brecciated chert and tuffaceous sandstone, values rise to 1.72 ppm and 89 ppm respectively and silicified boxworks are developed after pyrite. The whole cherty sequence is interpreted as a fully preserved precious-metal-bearing, hot-spring system.

The Turriff Outlier is the largest in the North-east Highlands. Here the Lower Old Red Sandstone is known as the Crovie Group (Read, 1923) and has been considered to be of Siegenian–Emsian age (Westoll, 1977). It rests unconformably upon the Dalradian Macduff Slate Formation and is unconformably overlain by Middle Old Red Sandstone.

Good sections of the Lower Old Red Sandstone can be examined at Crovie, New Aberdour (Sweet, 1985) and Gardenstown. Basal alluvial-fan sediments are overlain by a coarsening-upward sequence of deposits representing environmental changes from low-sinuousity ephemeral streams via floodplain meandering rivers to high-velocity braided rivers on an alluvial fan. Palaeocurrent data suggest a north-westerly dispersion. The unconformable junction with the Middle Old Red Sandstone is well exposed at Pennan. Scattered throughout the North-east Highlands are relatively small undated outliers of red cobbly to sandy deposits assigned to the Devonian. Outliers such as those at Aberdeen, near Towie, south-west of Keith and beneath the viaduct at Dufftown, may be remnants of the Lower Old Red Sandstone. At Buckie, a hornblende-andesite exposed in Gollachy Burn occurs also as clasts in the overlying Middle Old Red Sandstone conglomerate and may therefore belong to the Lower Old Red Sandstone. It is similar to an andesite cutting metamorphic rocks 5 km south-west of Cullen, in Banffshire, which may be of the same age.

MIDDLE OLD RED SANDSTONE

Rocks of the Middle Old Red Sandstone represent sediments laid down in numerous interconnected half-graben basins into which far-travelled clasts were deposited. This period of deposition was characterised by the development of ephemeral playa lakes with peripheral sandy wind-affected alluvial systems alternating with periods of extensive lake deepening when carbonate-mudstone sedimentation prevailed. Such lacustrine sediments are usually marked by particular fossil-fish assemblages, among which the oldest Achanarras Assemblage is the most widely recognised. In many places, the extension of the Orcadian Basin at this time is reflected by overstepping of the Lower Old Red Sandstone on to a variety of pre-Devonian rocks. Continuing movements on some of the half-graben faults resulted in conspicuous unconformities between the Lower and Middle Old Red Sandstone but, away from the fault zones, sedimentation was continuous without any break in sequence (Rogers, 1987). Palynological evidence indicates that the Middle Old Red Sandstone spans the Lower–Middle Devonian boundary, ranging in age from late Emsian to late Givetian, with the Achanarras Assemblage probably lying within the early Eifelian (Marshall *in* Rogers, 1987). There are no Middle Old Red Sandstone rocks in the Grampian Highlands south of Foyers. The main outcrops lie along the Great Glen between Foyers and Nairn, between Buckie and the Rothes Fault, and in the Turriff Outlier (Figure 33).

The sequence along the south-eastern side of the Great Glen rests upon a rolling surface of metamorphic and granitic rocks and the basal breccio-conglomerate forms a discontinuous spread incorporating a number of lenticular fans and possible canyon-fills. It is 75 m thick near Daviot and about 150 m thick near Cawdor, but over certain pre-Devonian knolls it is absent. The succession in the south-west, around Inverfarigaig, consists of thick lenticular granite-scree breccias, arkosic gritty sandstones, Moine-clast breccias and conglomerates, intercalated with and overlain by fine-grained sandstone, siltstone and rare shaly mudstone (Stephenson, 1972; Mykura, 1982). These sediments were involved in contemporaneous landslips and localised thrusting and to the north-east pass laterally into thick granite-clast-rich breccio-conglomerates derived from the south-west. North-east of Inverfarigaig, towards Inverness and Nairn, the conglomerates pass by intercalation into a succession of sandstones, flaggy siltstones and shaly mudstones commonly containing fish-bearing nodular limestone. In the area between Inverness and Nairn, the following succession has been established (Horne and Hinxman, 1914; Horne, 1923; Fletcher et al., 1995):

	Thickness m
HILLHEAD SANDSTONE FORMATION Sandstones, flags and fish-bearing shaly mudstones	900
INSHES FLAGSTONE FORMATION Grey and purple flaggy micaceous sandstones, some dark calcareous flags and laminated shaly mudstones with limestone nodules	300
LEANACH SANDSTONE FORMATION Red sandstones, some flaggy with interbedded shaly mudstones	500
NAIRNSIDE SANDSTONE FORMATION Grey and brown graded flaggy sandstones, siltstones and fish-bearing calcareous mudstones with limestone nodules; includes the *Easter Town Siltstone* and *Clava Mudstone* members	120
DAVIOT CONGLOMERATE FORMATION Bouldery to pebbly breccio-conglomerates with thin sandstones and a limestone bed	0–150

Exposures of the lacustrine mudstoneswss and siltstones of the Nairnside Sandstone Formation occur in the River Nairn at Easter Altlugie, in two sections at Clava, in the southern tributary of the River Nairn at Easter Town, at

Knockloan 5 km south of Nairn and at several localities near Lethan Bar and Easter Clune. The fauna is typical of the Achanarras Assemblage and includes *Cheirolepis* sp., *Coccosteus cuspidatus*, *Dipterus valenciennesis*, *Mesacanthus* sp. and *Osteolepis* sp. The lowest fish-bed appears to be that in Easter Town Burn and *Asmussia ['Estheria']* sp. is abundant beneath the fish-bearing Clava Mudstone. The best exposure of the Leanach Sandstone was at the Leanach Quarry on the southern flank of Culloden Moor. The fauna recorded by Taylor (*in* Horne, 1923, p.69) includes '*Coccosteus decipiens*', *Glyptolepis* sp., *Homosteus* sp. and *Pterichthyodes milleri*. From the evidence seen in the building stones hewn from this quarry, for the construction of the Clava Railway Viaduct, the Leanach sequence includes aeolian sandstones. Dark calcareous and slightly bituminous flags with limestone concretions within the Inshes Flagstone Formation exposed near Raigmore (Inverness) have yielded '*Coccosteus*' and *Osteolepis* species as well as plant remains. A higher fauna, comparable with that of the topmost 'Middle Old Red Sandstone' formations in north-eastern Caithness and Orkney, occurs in Hillhead Quarry, 12 km east of Inverness, where *Homosteus milleri* and *Millerosteus minor* have been identified in the Hillhead Sandstone.

Farther east, the Middle Old Red Sandstone sequence to the south-west of Buckie consists of a basal conglomerate overlain by a varied stack of thin conglomerates, sandstones and shaly mudstones containing a typical Achanarras Assemblage of *Cheiracanthus* sp., *Dipterus* sp., *Osteolepis* sp. and *Pterichthyodes* sp. (Peacock et al., 1969). A higher fish bed at Dipple, near Fochabers Bridge, has yielded *Dickosteus threiplandi*, a species typical of strata that overlie beds of Achanarras age elsewhere. In the north-east, the Middle Old Red Sandstone of the Turriff Outlier is represented by the 'Findon Group' (Read, 1923). This group comprises a basal slate-clast breccia up to 60 m thick overlain by the Findon Fish Bed and a sequence of breccias and conglomerates. The fish bed also contains abundant plant remains and occurs as a 2 m-thick section of grey and red shaly mudstone with limestone nodules in the Den of Findon containing a typical Achanarras Assemblage fauna (Read, 1923). In addition to the main outcrops, numerous patches of coarse-grained pebbly and cobbly red beds are scattered about the Moray Firth coast and its hinterland. All appear to be remnants of a former Middle Old Red Sandstone cover now preserved in hollows and depressions in the pre-Devonian basement. Notable sections occur on the coast in Sandend Bay and Cullen Bay, and inland in the Burn of Deskford.

UPPER OLD RED SANDSTONE

The Upper Old Red Sandstone in the main part of the Orcadian Basin is characterised by coarser-grained fluvial sediments than the underlying sequence and by the development of sabkha deposits, indicative of marginal marine conditions. The only Upper Old Red Sandstone in the Grampian Highlands lies on the southern side of the Moray Firth, between Fort George in the west and Spey Bay in the east, where they are peripheral to the basinal succession (Figure 33).

Although the relationship with the Middle Old Red Sandstone is generally unconformable, in the southern part of the Orcadian Basin, there is evidence to indicate that the junction largely represents a facies change without any break in sequence. In the Nairn–Elgin district, the basal beds appear to be of late Givetian age and the sedimentary sequence spans the Middle–Upper Devonian boundary. To date, no early Famennian sedimentary rocks have been recognised and there is the possibility of a sequence-break between late Frasnian and late Famennian–?Tournaisian beds around Elgin (Rogers, 1987). Because of lack of exposure, the nature of the boundary on the southern side of the Moray Firth is not resolved, although it is clear that the Upper Old Red Sandstone is diachronous from west to east and that the oldest beds are restricted to the western region (Figure 34). The overall shallow-water and rare dry-bed conditions are reflected by relatively rapid lateral facies changes and by some overstepping of units along the margins of fault-bounded sub-basins. The Upper Old Red Sandstone is made up essentially of sandstone and the recognised subdivisions are based mainly upon six distinctive fossil-fish assemblages (Traquair, 1896; 1897; 1905; Westoll, 1951; Tarlo, 1961; Miles, 1968) that have correlatives in Baltic, Belgium, Spitzbergen and Greenland successions.

The oldest unit is the Nairn Sandstone Formation, which comprises an irregular basal reddish conglomerate overlain by red, grey and yellow calcareous cross-bedded and flaggy sandstones containing thin beds of conglomer-

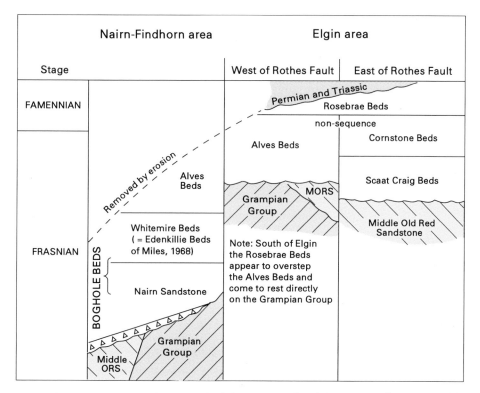

Figure 34 Probable correlation of the lithostratigraphical divisions of the Upper Old Red Sandstone in the Nairn–Elgin area (based on Peacock *et al.*, 1968).

ate and soft or shaly limestone-bearing mudstone. Good exposures occur on the beach north-west of Nairn, in Muckle Burn, at Glenshiel and in the Findhorn area. In the Findhorn area, the sandstone is faulted against Pre-cambrian gneisses (Black and Mackenzie, 1957) and some mudstone beds have been desiccated to clay galls. High in this section, a 3 m-thick calcrete bed named the Cothall Limestone (Parnell, 1983) is succeeded by 2.6 m of red and violet marl containing fossiliferous limestone concretions veined with calcite, cherty dolomite and pyrite; there are also patches of mamillate chal-cedony. The fish fauna of the Nairn Sandstone comprises two main faunules. The lower is characterised by *Asterolepis maxima, Psammolepis tesselata* and *P. undulata*, whilst the upper one, known as the Boghole faunule, is distin-guished by *A. alta* and *Eusthenopteron traquairi*; species common to both faunules include *Coccosteus magnus, Polyplocodus leptognathus* and *Holoptychius decoratus*. The first appearance of *Bothriolepis* has been taken to mark the base of another unit that bears the name Whitemire Beds (Figure 34). The Whitemire fauna is first encountered about 7 m above the Boghole faunule (Westoll, 1951) and is transitional in type between the Boghole and the overlying Alves faunas. The diagnostic species is *Bothriolepis taylori* and the assemblage includes *C.* ex. gr. *magnus* and *H. nobilissimus*, which are common in the older faunule, and *Psammosteus taylori, Cosmacanthus, Conchodus* and *H. giganteus* present in the Alves faunule.

In the Nairn–Findhorn area, a younger fauna occurs in a sequence of grey to reddish siliceous pebbly sandstones named the Alves Beds, which, in the Elgin area, rests unconformably on metamorphic rocks near Burgie (Figure 34). The diagnostic species are *B. alvesiensis* and *B. gigantea*, with *H. nobilis-simus* extending upwards into these beds. East of the Rothes Fault, the Alves succession is represented by two lithological units. The lower Scaat Craig Beds, comprising red and yellow sandstone and fine conglomerate, contain taxa such as *Cosmacanthus malcolmsoni, Conchodus ostreiformis* and coccosteomorph arthrodires, which are closely linked to the Whitemire fauna below. The Alves fauna, however, is distinguished by *P.* cf. *falcatus, Traquairosteus pustulatus* and *B. paradoxa*. The influence of an active Rothes Fault on Alves sedimentation is possibly reflected by the deposition east of the fault of a more carbonate-rich succession of pale grey and reddish brown marly sandstones, which overlies the Scaat Craig Beds. This sequence is characterised by sandy cherty calcrete beds representing palaeosols and is named the Cornstone Beds.

The highest strata in the Upper Old Red Sandstone are preserved in the Elgin area and form the Rosebrae Beds (Figure 34). They are well exposed in Quarry Wood, west of Elgin, and comprise brownish grey, yellow and reddish sandstone with only scarce pebbles. The fish fauna includes *Phyllolepis* cf. *woodwardi, B. cristata, B. laverocklochensis, Phaneropleuron* cf. *andersoni, Rhy-chodipterus elginensis* and *Glyptopomus elginensis*. Associated taxa include *B. alvesiensis, Conchodus, Eusthenopteron* and *H. nobilissimus*, all present in the underlying units. Palynological evidence (Marshall *in* Rogers, 1987) suggests that the Rosebrae Beds are late Famennian in age at their base and that the highest strata may transgress the Devonian–Carboniferous boundary; a non-sequence may separate the Alves and Cornstone beds from the Rosebrae Beds.

The bulk of the Upper Old Red Sandstone sediments are coarse grained and were laid down as small alluvial fans and by braided rivers marginal to alluvial plains and shallow lakes. An overall northerly flow is indicated by palaeocurrent structures (Westoll, 1977). The finer-grained sediments represent distal floodplain deposits, ponded lags in abandoned channels or lacustrine carbonate muds. Wet periods are indicated by fish-bearing limestones and dry periods by beds of wind-etched pebbles and palaeosol profiles.

Studies of heavy-mineral suites in the Elgin sequence (Mackie, 1897–1923) have shown contrasts between the Middle and Upper Old Red Sandstone sequences. Grains in the former are larger and more angular, having much garnet associated with subordinate iron oxide, rutile, monazite and, more locally, staurolite and epidote. In the Upper Old Red Sandstone, small rounded zircons predominate, with tourmaline, rutile, anatase and monazite also abundant.

Upper Devonian rocks overlying Dalradian and Cambro-Ordovician rocks along the Highland Border between Balmaha, east of Loch Lomond, and Kilcreggan, west of Helensburgh, are part of the Stratheden Group of the Midland Valley succession. They are described in the Regional Guide for the Midland Valley of Scotland (Cameron and Stephenson, 1985, p.31).

The Upper Devonian rocks of Kintyre are described along with the Carboniferous rocks of that area in Chapter 11.

11 Carboniferous

Small exposures of Carboniferous rocks occur along the River Awe, west of the Pass of Brander. In southern Kintyre, Carboniferous rocks of the Machrihanish Coalfield include both lavas and sedimentary rocks with several thick coals. Along the Highland Boundary Fault, in the Loch Lomond–Helensburgh area, Lower Carboniferous sedimentary rocks rest on the Upper Devonian and extend some distance into the Grampian Highlands (Figure 35).

River Awe

A few exposures on the banks of the River Awe, mainly upstream of Bridge of Awe, reveal a small outlier of Carboniferous rocks, less than 0.1 km^2 in area, resting on Old Red Sandstone lavas. The outlier is probably fault-bounded on the north-east (Pringle and MacGregor, 1940). About 18 m of Carboniferous sedimentary rocks are exposed; basal conglomerates are overlain by reddish gritty sandstones, reddish mottled mudstones, purplish shales and some paler, fine-grained sandstones, the finer beds containing poorly preserved plant remains. The best plant remains have been recorded from an outcrop 90 m upstream of Bridge of Awe and include a calamitid stem, *Asterocalamites,* and a fern rachis very like *Rhacopteris petiolata* (Göppert). Kidston (1899) considered these plants to be a Lower Carboniferous assemblage. Lithologically the Carboniferous rocks at Bridge of Awe have a strong resemblance to the lower strata in the outlier at Inninmore about 30 km to the north-west.

Kintyre

In southern Kintyre, the Lower Old Red Sandstone is overlain by a thin sequence of Upper Old Red Sandstone sedimentary rocks (McCallien, 1927). The lower part of the Upper Old Red succession is assigned to the *Stratheden Group* (Upper Devonian) and the upper, cornstone-bearing part to the *Kinnesswood Formation* in the *Inverclyde Group* of late Devonian to early Carboniferous age. These sedimentary rocks are unconformably overlain by up to 400 m of olivine-basalts, mugearites, trachyandesites and trachytes belonging to the *Clyde Plateau Volcanic Formation* (Strathclyde Group). At Skerry Fell Fad, a dome-like mass of trachyte probably represents a highly viscous lava capping its feeder pipe. There are also a number of plugs, sills and dykes ranging in

Figure 35   A   Carboniferous and Permian rocks of the South-west Highlands.
       B   Carboniferous rocks of southern Kintyre.

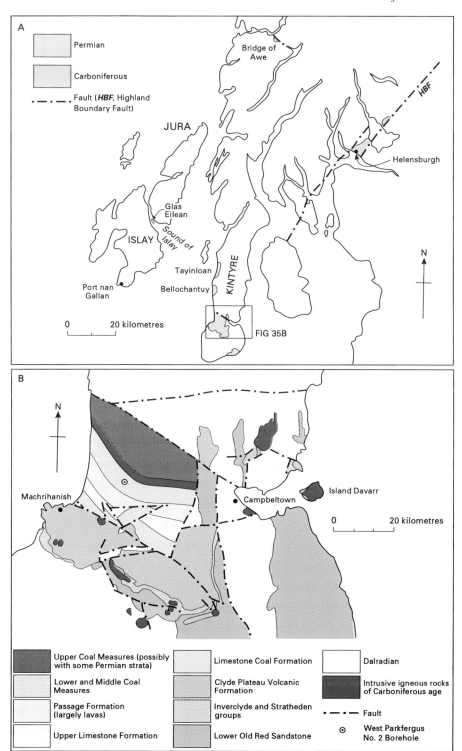

composition from olivine-dolerite to syenite and dacite (Macdonald, 1975). The lavas are overlain in places by reddish bauxitic clays, interpreted as detrital lateritic deposits formed by contemporaneous decomposition of the volcanic rocks (McCallien and Anderson, 1930).

The overlying Carboniferous sedimentary rocks form the Machrihanish Coalfield, which occupies the flat ground north and east of Machrihanish (McCallien and Anderson, 1930). The sequence resembles that of the Midland Valley and the same stratigraphical nomenclature can be applied (Figure 36). Sandstones containing a thin bed of limestone exposed on the shore at Machrihanish are doubtfully attributed to the *Lower Limestone Formation*. The *Limestone Coal Formation* is well developed in the south-western part of the coalfield, where it is about 100 m thick, and contains a number of coals, of which the Main Coal and to a lesser extent the Kilkivan Coal have been intermittently worked from the end of the eighteenth century. The Main Coal is 3 to 3.6 m thick, but the upper 1 m contains siltstone partings and is generally of inferior quality. The coal was worked from a number of collieries, principally the Argyll Colliery, which was abandoned because of major fire in 1925, and the Machrihanish Colliery, opened in 1944 and also closed because of fire in 1967. The Kilkivan Coal, some 35 m above the Main Coal, is up to 2.15 m thick. Other seams reach a considerable thickness in places, but are of limited lateral extent and are not sufficiently continuous to mine economically. The thick sandstone which forms the roof of the Main Coal was also mined as a source of moulding sand. In the eastern part of the Machrihanish Coalfield, borehole evidence shows that the lower part of the Limestone Coal Formation, including the economic coal seams, is absent and the sedimentary rocks lap on to an irregular surface of weathered lava and volcanic detritus *(Kirkwood Formation)*.

The *Upper Limestone Formation* is locally about 90 m thick, but is generally poorly developed. The formation includes limestones that have been correlated with the Index, Lyoncross and Orchard limestones of the Midland Valley sequence. South-west of the Drumlemble Fault, which bounds the Machrihanish Coalfield, rocks of the Upper Limestone Formation unconformably overlie the Clyde Plateau Volcanic Formation. A maximum of 20 m of limestone and limy shale rests on laterite in Tirfergus Glen and rests on bauxite in Torchoillean Burn (McCallien and Anderson, 1930). The unconformably overlying *Passage Formation* consists of sandstones and thin siltstones, interbedded with thick flows of basic lava and bands of reddish lateritic mudstone; the total thickness is approximately 150 m. In the West Parkfergus No. 2 Diamond Borehole, an intercalation of sedimentary rock above the lowest flow includes a marine band with remains of gastropods and ribbed brachiopods. The uppermost lavas have been weathered to bauxitic clay, indicating another break in the sequence. A 460 m-thick succession overlying the Passage Formation is assigned to the Communis, Modiolaris and Lower Similis-Pulchra chronozones of the *Lower* and *Middle Coal Measures*. The Vanderbeckei (Queenslie) Marine Band, which marks the boundary between these is present in the borehole and has yielded *Lingula* and one specimen of *Spirifer* (Manson, 1957; Brand, 1977). The coals in the Coal Measures appear to vary in both thickness and extent but locally seams up to 1.7 m thick are

Figure 36   Generalised sequence of the Carboniferous of southern Kintyre.

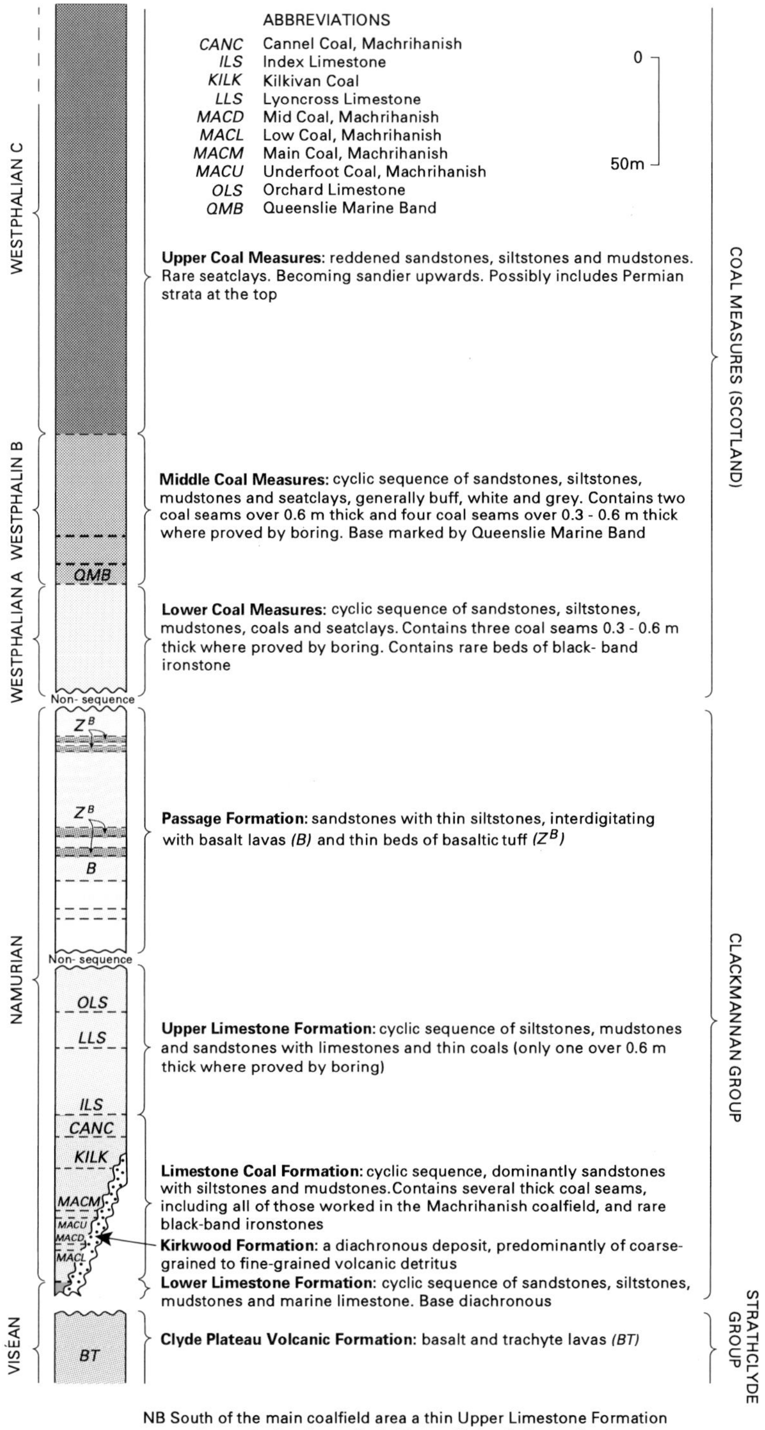

ABBREVIATIONS

CANC	Cannel Coal, Machrihanish
ILS	Index Limestone
KILK	Kilkivan Coal
LLS	Lyoncross Limestone
MACD	Mid Coal, Machrihanish
MACL	Low Coal, Machrihanish
MACM	Main Coal, Machrihanish
MACU	Underfoot Coal, Machrihanish
OLS	Orchard Limestone
QMB	Queenslie Marine Band

0 ⌐
50m ⌐ (scale bar)

WESTPHALIAN C
WESTPHALIN B
WESTPHALIAN A
NAMURIAN
VISÉAN

COAL MEASURES (SCOTLAND)
CLACKMANNAN GROUP
STRATHCLYDE GROUP

Upper Coal Measures: reddened sandstones, siltstones and mudstones. Rare seatclays. Becoming sandier upwards. Possibly includes Permian strata at the top

Middle Coal Measures: cyclic sequence of sandstones, siltstones, mudstones and seatclays, generally buff, white and grey. Contains two coal seams over 0.6 m thick and four coal seams over 0.3 - 0.6 m thick where proved by boring. Base marked by Queenslie Marine Band

QMB

Lower Coal Measures: cyclic sequence of sandstones, siltstones, mudstones, coals and seatclays. Contains three coal seams 0.3 - 0.6 m thick where proved by boring. Contains rare beds of black- band ironstone

Non- sequence

Z^B

Z^B

B

Passage Formation: sandstones with thin siltstones, interdigitating with basalt lavas *(B)* and thin beds of basaltic tuff *(Z^B)*

Non- sequence

OLS
LLS

Upper Limestone Formation: cyclic sequence of siltstones, mudstones and sandstones with limestones and thin coals (only one over 0.6 m thick where proved by boring)

ILS
CANC
KILK

Limestone Coal Formation: cyclic sequence, dominantly sandstones with siltstones and mudstones. Contains several thick coal seams, including all of those worked in the Machrihanish coalfield, and rare black-band ironstones

MACM
MACU
MACD
MACL

Kirkwood Formation: a diachronous deposit, predominantly of coarse-grained to fine-grained volcanic detritus

Lower Limestone Formation: cyclic sequence of sandstones, siltstones, mudstones and marine limestone. Base diachronous

BT

Clyde Plateau Volcanic Formation: basalt and trachyte lavas *(BT)*

NB South of the main coalfield area a thin Upper Limestone Formation succession rests directly on Clyde Plateau Volcanic Formation lavas and/or Kirkwood Formation

developed. Red sandstones exposed near the northern margin of the coalfield are probably reddened *Upper Coal Measures.*

Recent evidence from offshore surveys suggests that the Machrihanish Coalfield extends at least 15 km westwards into the Rathlin Trough. To the south, the succession in the Ballycastle Coalfield of County Antrim shows many similarities to the Limestone Coal Formation of Machrihanish (Wilson and Robbie, 1966).

Helensburgh–Loch Lomond

In the Helensburgh–Loch Lomond area, white, pink, and red-purple sandstones, with sporadic quartz pebbles and cornstone beds, overlie conglomerates of the Upper Devonian Stratheden Group (Paterson et al., 1990). The cornstone-bearing sandstones have been assigned to the Inverclyde Group (mainly Lower Carboniferous) because of their lithological similarity to beds of that age in the Midland Valley (Paterson and Hall, 1986). Exposure is confined to a few stream sections, but the junction with the underlying Devonian conglomerate is not exposed. The cornstones are impure irregular concretionary limestones up to 0.5 m thick which probably developed in soil profiles in Lower Carboniferous times. They usually occur at the top of upward-fining channel-fill fluvial sandstones developed in an evaporating environment after the channels were abandoned. A small fault-bounded area of grey mudstones with cementstones occurs in the Fruin Water, 2 km north-east of Helensburgh, where 5 m of grey mudstone with several thin dolomitic limestones (cementstones) and veins of gypsum overlie grey micaceous sandstone. On Ben Bowie, 2 km east of Helensburgh, the Inverclyde Group, about 170 m thick, is overlain by a 60 m-thick volcanic sequence consisting of basal tuffs overlain by feldspar-phyric basic lavas and mugearites. The volcanic rocks are correlated with the Clyde Plateau Volcanic Formation (Strathclyde Group).

East of Loch Lomond, the basal Carboniferous Kinnesswood Formation crops out on the hills north-east of Balmaha and overlies Upper Devonian conglomerates with slight angular unconformity. The sequence is about 50 m thick and includes well-developed cornstones up to 1.20 m thick which were formerly quarried and burnt for lime.

12   Permian, Triassic and Jurassic

PERMIAN AND TRIASSIC

Islay and Kintyre

Offshore-shelf studies along the western seaboard of Scotland have shown that rocks of Permo-Triassic age occur in the Rathlin Trough, extending as far north as the southern end of Jura (McLean and Deegan, 1978; Evans et al., 1979), and in the Arran Basin, with possibly a 1400 m succession preserved onshore on the Isle of Arran (see Warrington et al., 1980 for summary of literature). Although the submarine outcrops are extensive, onshore exposures in the South-west Highlands are very limited.

A 120 m Permian succession is exposed on Glas Eilean and Black Rock, low-lying islands in the Sound of Islay, 5 km south of Port Askaig (Figure 35A; Pringle, 1944; Upton et al., 1987). The succession dips WSW and is bounded to the west by a fault. The basal 6 m consist of conglomerate, made up of rounded to subangular clasts of Dalradian Jura Quartzite up to 30 cm in size, set in a gritty ferruginous matrix. The conglomerate is succeeded by 1 m of reddish brown fine-grained sandstone. The overlying lava succession is broken by two thin beds of flaggy sandstone and a bed of sandy limestone 0.3 m thick. The individual lava flows are each less than 2 m thick, blocky and with easily eroded slaggy tops. The flows are amygdaloidal with calcite filling the original vesicles. They were deposited subaerially, interrupting a period of shallow-water sedimentation. The lavas are alkali olivine-basalts with phenocrysts, mainly of olivine, but with minor plagioclase and augite. A few of the lower flows have microphenocrysts of Al-Cr spinel. A K/Ar age determination on one of the freshest lavas yielded an age of 285 ± 5 Ma (early Permian), virtually identical to the age of the Mauchline lavas in Ayrshire. At Port nan Gallan, one mile east of the Mull of Oa at the southern tip of Islay, there is a small outcrop of breccia composed of blocks of quartzite, limestone and schist, set in a matrix of bright red sandstone with well-rounded grains. It forms part of a sea stack and was first described by Peach (1907). It was recognised as Permo-Triassic by Pringle (1952) and forms a breccia infill within a vertical fissure cut in the Islay Limestone. A more extensive Permo-Triassic red sandstone cover, now eroded away, has left the underlying Dalradian quartzites on the Mull of Oa red-stained.

Along the western coast of Kintyre, soft, current-bedded, red sandstones and conglomerates form three large outcrops between Bellochantuy and Tayinloan (Figure 35A). They are well displayed in reefs between tide marks, and the basal beds are exposed in road sections at Bellochantuy and in stream sections east of the road, such as in Allt a Ghaoidh. These sections of the basal beds include coarse breccias with pebbles of vein-quartz, mica-schist and quartzite. Some pebbles show wind facetting and quartz grains in the sandstones are well rounded and polished. The exposures in the road sections at Bellochantuy show a steep and irregular junction with the Dalradian, probably indicating that the depositional basin was bounded by high land to the east (Pringle, 1952).

Moray Firth

Permian and Triassic sandstones are exposed north and west of Elgin and on the coast between Burghead and Lossiemouth (Figure 37); they are largely of aeolian origin and are the only onshore occurrences of the extensive Permo-Triassic basin deposits developed off the Moray Firth and Aberdeenshire coasts (Frostick et al., 1988; Andrews et al., 1990). The outcrop pattern in the Elgin area is controlled by two ENE-trending faults throwing down to the south. The sandstones have yielded fossil reptile remains, which have been studied by many palaeontologists (see A D Walker, 1961; 1964 and references therein). Watson (1904) and Watson and Hickling (1914) have shown that two distinct faunas are present. The lower fauna, in the Hopeman and Cutties Hillock sandstones is probably of latest Permian age and the upper fauna, in the Lossiemouth Sandstone, is late Triassic (Benton and Walker, 1985).

The succession in the Elgin area is (Peacock et al., 1968):

Stotfield Cherty Rock	
Sandstones of Lossiemouth, Spynie, and Findrassie	Upper Triassic
Burghead Beds	Triassic
Cutties Hillock and Hopeman sandstones	Lower Triassic and Upper Permian

CUTTIES HILLOCK AND HOPEMAN SANDSTONES

The Cutties Hillock Sandstone forms the highest part of the Quarry Wood ridge and occurs in isolated outcrops at York Tower and Carden Hill (Figure 37). It is a fine- to coarse-grained, yellow to brownish sandstone, commonly laminated, and varies from soft to hard and siliceous. The basal sandstone beds contain pebbles, some of which are thought to be wind etched (Mackie, 1901b), but Williams (1973) has shown that the sandstones more likely represent sheet-flood deposits. The overlying sandstones exhibit large-scale cross-bedding, have well-rounded quartz grains and are interpreted as having formed from wind-deposited barchan dunes. Fossil reptiles found just above the basal beds have been identified as a pareiasaur, *Elginia,* and two dicynodonts, *Geikia* and *Gordonia;* they were compared by Watson and Hickling (1914) to forms occurring at the Permian–Triassic boundary in Russia and in the Karoo of South Africa. More recently, Walker (1973) recognised the presence of a procolophonid in support of such a correlation. However, Benton and Walker (1985) have suggested that certain vertebrate tracks indicate a latest Permian age.

In the northern outcrop at Hopeman and Covesea, the basal beds are not exposed. Williams (1973) recognised four phases of aeolian deposition separated by periods of flooding, which are locally preserved as playa

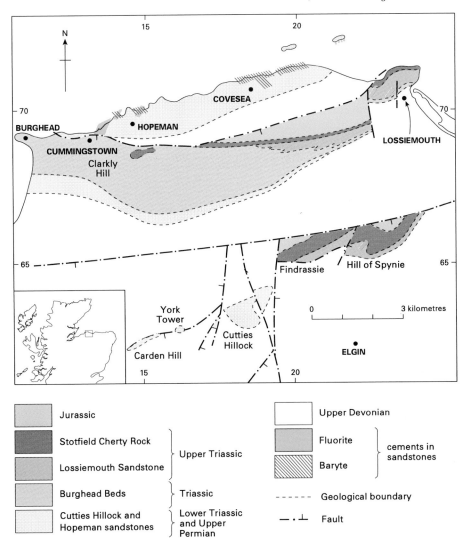

Figure 37   Permian and Triassic rocks of the Lossiemouth–Elgin area.

deposits. The oldest phase is made up of well-developed linear seif and star-shaped dunes (Clemensen, 1987); the succeeding phases are formed by crescentic barchan dunes. The seif deposits, exposed on the coast, were laid down by winds from the SSE on the edges of a sand sea with mountains to the south. The succeeding phases of barchan dune were deposited by winds blowing mainly from the north-east (Shotton, 1956). The sandstones interpreted as sheet-flood deposits and exposed in the Quarry Wood area are probably equivalent in age to the seif deposits.

The Hopeman Sandstone is up to 60 m thick in the Clarkly Borehole near Cummingstown, but thins eastwards, and is absent at Lossiemouth and Hill of Spynie. Contorted bedding in the Hopeman Sandstone is displayed in several exposures along the coast (Peacock, 1966). Williams (1973) has attributed the contortions to moistening or saturation of the sand during deposition of playa deposits, which were presumably subsequently removed by contemporaneous wind erosion. Glennie and Buller (1983), however, have suggested that the dune sands were moistened by a later Zechstein transgression, which caused partial homogenisation of the sand and created large-scale soft-sediment deformation structures. They also suggest that large pockets of air originally trapped within the dunes were later replaced by water, thus causing liquification of the original dune laminae and the disruption of structures.

The Hopeman Sandstone is in places cemented by fluorite, and patchy baryte mineralisation is well exposed in coastal sections north of Hopeman and Covesea (Mackie, 1901a).

BURGHEAD BEDS

The Burghead Beds are exposed around Burghead, where they reach a thickness of 60 m. They can be traced eastwards for some 6 km before passing laterally into the contemporaneous Lossiemouth Sandstone. They are unfossiliferous, and are interpreted as river-deposited, point-bar and floodplain deposits. The point-bar deposits were laid down in rivers of low sinuosity and high gradient and consist of laminated sandy conglomerates overlain by trough cross-bedded sandstones interbedded with siltstones exhibiting both ripple marks and horizontal lamination. Good sections occur at the Burghead coastguard station and in a quarry at Clarkly Hill. The foresets in the cross-bedded sandstones show that the direction of current movement was from the WSW. Sandstones exposed farther east along the coast, and at Lossiemouth, were formed as floodplain deposits.

LOSSIEMOUTH SANDSTONE

Aeolian sandstones of Upper Triassic age and partly equivalent in age to the Burghead Beds are exposed around Lossiemouth, and at Hill of Spynie and Findrassie. Their total thickness is less than 30 m. They are white, yellow and pink, fine- to coarse-grained sandstones that are siliceous at the top and calcareous towards the base. Joints within the sandstone commonly contain fluorite and baryte. Galena and fluorite are sporadic constituents of the sandstone matrix. The aeolian cross-bedding is generally obscured by silicification at the top of the sandstone, but is more pronounced in the lower calcareous part.

The Lossiemouth Sandstone in all three localities has yielded reptile fossils, including *Brachyrhinodon taylori, Erpetosuchus granti, Hyperodapedon gordoni, Leptopleuron lacertinum, Ornithosuchus longidens, Saltopus elginensis, Scleromochlus taylori* and *Stagonolepis robertsoni,* indicative of late Triassic age (Walker, 1961; 1964). Walker noted the complete articulation of the fossil remains, and concluded that they had been preserved in sediments laid down in their life environment. A more recent assessment of the Lossiemouth Sandstone fauna (Benton and Walker, 1985) has indicated the presence of a mixed assemblage of herbivores, carnivores and small omnivores.

The Lossiemouth Sandstone grades upwards into a coarse, silicified and calcareous sandstone named the *Sago Pudding Sandstone,* which reaches a thickness of 8 m and is characterised by large isolated quartz grains in a fine-grained matrix.

STOTFIELD CHERTY ROCK

At Lossiemouth and Hill of Spynie, the Lossiemouth Sandstone is overlain by a caliche horizon (Naylor et al., 1989), which consists of sandstone, veined and partly replaced by chert and limestone. The chert contains vugs filled with quartz, calcite and disseminated galena. The cherty rock was correlated by Judd (1873) with marly limestone and chert cropping out below the Lias at Golspie, thus establishing a connection with the sequence north of the Moray Firth.

JURASSIC

The only known onshore rocks of undoubted Jurassic age in the Grampian Highlands region were proved in a Geological Survey borehole sunk in 1964 just south-west of Lossiemouth (Berridge and Ivimey-Cook, 1967). Approximately 70 m of the borehole section consists of basal calcareous mudstones and marls, overlain by a rhythmic sequence of sandstones, siltstones and fissile mudstones, succeeded by coarse-granular, kaolinitic sandstones. The overall lithofacies is similar to that of the Lias. Ammonites diagnostic of the *Echioceras raricostatum* Zone (uppermost Sinemurian) occurred about 28 m above the base of the hole together with bivalves. The underlying 17 m of mudstones, siltstones and sandstones contain *Euestheria* and are probably equivalent to the White Sandstone of Dunrobin, Sutherland (Neves and Selley, 1975). The environments of deposition varied from marine to non-marine and the oldest sediments (marl and cementstone) were probably deposited in a lagoonal environment of low salinity. This passed via an estuarine sequence (with *Euestheria*) into a fully marine environment (with ammonites) and then into a freshwater sequence when the sandstones with kaolinite were formed. The ammonites from the borehole indicate a close correlation with those from the Lady's Walk Shale of the Dunrobin section, north of the Moray Firth.

13   Post-Caledonian minor intrusions

LATE CARBONIFEROUS QUARTZ-DOLERITE DYKE SUITE

Numerous east-trending quartz-dolerites occur in several parts of the Grampian Highlands (Figure 38). These are locally seen to cut the NNE- to

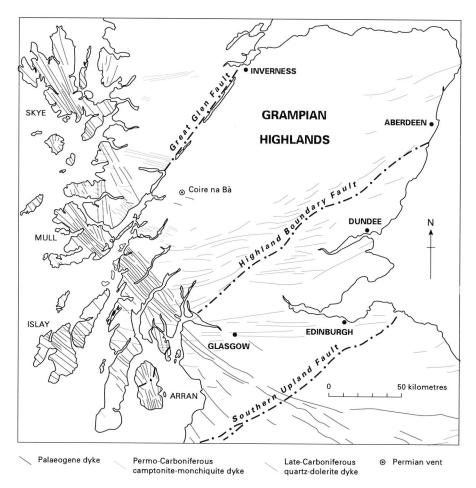

Figure 38   Distribution of post-Caledonian minor intrusions in the Grampian Highlands and adjacent regions.

NE-trending Late Caledonian dykes (Chapter 8) and are themselves cut by Permo-Carboniferous and Palaeogene dykes. They are most numerous in Argyll, good examples being seen at Carrick, Lochgoilhead and Restil (Cowal), and are quite common in Perthshire, with a few dykes also extending into Deeside and Buchan. Many of the dykes can be traced for considerable distances and can be readily located by aeromagnetic surveys due to their high magnetic susceptibilities. The Lochgoilhead dyke, for example, probably continues eastwards as far as Perth, a distance of over 100 km, while another, possibly discontinuous, dyke extends 65 km from Boddam, near Peterhead, to near Rhynie (Buchan, 1932; Gould, in press). The quartz-dolerite dykes are usually thick; Read (1923) recorded a thickness of 13 m at Auchinbradie, near Insch, and they are commonly over 10 m.

The dykes in the Grampian Highlands represent the northern part of the more widespread quartz-dolerite suite of the Scottish Midland Valley (Cameron and Stephenson, 1985) and of northern England (Dunham and Strasser-King, 1982). The trend of the main dyke swarm swings from ESE at Loch Awe in the South-west Highlands, through east–west to north-east in Aberdeenshire. The trend of the intrusions may have been influenced by the Highland Boundary Fault but Russell and Smythe (1983) interpret the arcuate trend of the swarm as the site of a nascent ocean rift formed at the beginning of the separation of Greenland from north-west Europe. The dykes form part of a much wider magmatic province extending into Scandinavia (Macdonald et al., 1981)

In the Midland Valley and northern England, there are good stratigraphical controls on the intrusion age. K-Ar ages of 302 to 297 Ma have been obtained on Midland Valley quartz-dolerite dykes and sills (Fitch et al., 1970; de Souza, 1979) and by analogy, the dykes within the Grampian Highlands were probably intruded during a narrow time interval—perhaps about 5 Ma—in latest Carboniferous to earliest Permian times (about 300 Ma).

Petrographically, the dykes are fairly uniform, comprising laths of basic plagioclase, ophitic augite, opaque phases and a glassy or micropegmatitic mesostasis, with or without pseudomorphed olivine, hypersthene or pigeonite. Amphibole and biotite may fringe augite and the opaque phases. Chemically, they divide into dominant tholeiites, with subordinate olivine tholeiites and tholeiitic andesites, through variations in normative quartz, differentiation index and Mg/Mg + Fe ratio (Macdonald et al., 1981). The dykes are relatively rich in Ti and Fe, and are closely comparable to certain basalts erupted in Iceland and Hawaii at the present time. Chemical variations along dykes are slight but some thicker dykes show more significant variations across their width.

PERMO-CARBONIFEROUS CAMPTONITE–MONCHIQUITE (ALKALINE LAMPROPHYRE) DYKE SUITE

The alkaline lamprophyres of the Grampian Highlands are chemically similar to suites in Norway, USA, Alaska and New Zealand (Rock, 1983). They were mostly intruded as dykes, though there are a few vents and plugs. The dominant rock types are camptonite, monchiquite and alkali basalt. Petrographically, most of the dykes consist of titanium-rich augite, biotite and amphibole (kaersutite), generally occurring as phenocrysts, in a groundmass of the same minerals plus feldspars (camptonite) and often abundant feldspathoids (monchiquite). Chemically, the dykes range from mildly silica-undersaturated alkali basalts through basanites to very strongly undersaturated nephelinitic compositions. Collectively, they are the most silica-poor and alkali-rich of any igneous rocks in the British Isles, and resemble certain recently erupted lavas and intrusive rocks of the Honolulu area of Hawaii and the East African Rift. The dykes are coeval with the Permian Glas Eilean lavas and probably with a 60 m-thick alkali olivine-dolerite sill which intrudes the Coal Measures of the Machrihanish outlier. The parent magmas of these rocks are probably closely related, though the more undersaturated differentiates are known only from the dyke swarm. Baxter (1987) postulates that they were formed by 0.4 to 2 per cent partial melting of a garnet-lherzolite source under-lying the spinel-bearing lithospheric mantle found as xenoliths in some monchiquites.

Some of the dykes retain fragments (xenoliths and xenocrysts) of the lower crust and mantle through which they have passed. The Colonsay, Machrihan-ish and Ardmucknish dykes have all yielded rich and highly significant mantle xenolith assemblages (Rock, 1983; Upton et al., 1983), as has the small volcanic agglomerate-nephelinite vent at Coire na Bà near Kinlochleven (Bailey, 1960). The xenoliths include a wide range of pyroxenites and perid-otites (olivine-pyroxene-rich rocks), many of them rich in biotite, tangible evidence that a heterogeneous mantle underlies this part of Scotland.

PALAEOGENE DYKES

In the South-west Highlands, an immense number of mostly basic Palaeogene dykes occur and trend uniformly NW–SE. There are several distinct major concentrations (Figure 38; Speight et al., 1982). That traced through Appin and Cowal represents the south-western extension of the regional linear swarm emanating from the Mull Central Complex; the regional linear swarm has locally resulted in 10 per cent crustal extension (about 60 to 70 dykes per kilometre traverse). The dyke swarms through Islay and Jura/Colonsay (F Walker, 1961), causing some 5 per cent crustal extension (23 dykes per km), have not been positively linked to any intrusive centre but aeromagnetic data suggest that they may emanate from the submarine Blackstones Centre west of Mull. Smaller swarms of dykes in northern Kintyre (McCallien, 1932) and Craignish (Allison, 1936) represent northern bifurcations of the Arran swarm and result in a maximum crustal extension of 5 per cent (about 12 dykes per km).

Petrologically, the dykes include both tholeiitic and alkali basalts and dolerites, mugearites, tholeiitic andesites, andesitic pitchstones and trachytes, with numerous composite and multiple examples. Little modern work has been carried out on the Palaeogene dykes in the Grampian Highlands but summaries of information for dykes of other areas are given by Thompson (1982a; b).

Two small Palaeogene bosses are noteworthy. An elongated boss on Maiden Island, Oban (Walker, 1939), shows an unusual occurrence of picrite marginal to olivine-dolerite. A second boss at Cnoc Rhaonastil, Islay (Walker and

Patterson, 1959; Hole and Morrison, 1992) is one of a very rare group of Palaeogene alkali dolerite intrusions (similar rocks form the much better-known Shiant Isles Sill-complex of the Minch), which carry nests of feldspathoidal syenite differentiate.

All these dykes and bosses can be dated, by analogy with adjacent areas, to the period 60 to 52 Ma (Macintyre et al., 1975).

14 Neogene

Following the Palaeogene volcanism in western Scotland the Neogene history of the Grampian Highlands was chiefly one of erosion. However, in the North-east Highlands there are deposits of flint and quartzite gravel of probable Neogene age classified as the *Buchan Gravels Group* (Hall, 1984). They occur in central Buchan as isolated spreads that underlie the highest ground in the area at elevations between 161 m and 120 m above OD, well above the present valleys of the rivers Ythan and Deveron. The Buchan Gravels are divided into the westerly quartzite-rich *Windyhills Formation* and the easterly flint-rich *Buchan Ridge Formation*. The type locality of the Windyhills Formation is Windyhills, 4 km north-east of Fyvie. The Buchan Ridge Formation principally underlies the watershed referred to in the literature as the 'Buchan Ridge', 11 km south-west of Peterhead (Figure 39). Both formations consist of bedded gravels with subsidiary sandy and silty units. Thicknesses of at least 14 m have been recorded for the Windyhills Formation and 25 m for the Buchan Ridge Formation. The deposits are deeply weathered and clasts other than flint or quartzite have been decomposed in whole or part to a kaolinitic clayey silt. The silty and sandy beds contain detrital quartz, flint, muscovite and kaolinite derived from a pre-existing deeply weathered land surface. Cross-bedding and imbrication of the gravel in the Windyhills Formation at Windyhills suggests transport and deposition by a NE-flowing river. The origin of the poorly exposed Buchan Ridge Formation is less clear and its genesis has been ascribed to fluviatile, beach and even glacial processes. Current opinion favours the first of these, though the top of the gravel is locally overlain by till and may be glacially disturbed. A maximum age for the gravels is provided by the occurrence of Cretaceous fossils in the flints. There is a close spatial relationship between the gravels and the remnant areas of clayey gruss (intensely weathered rock) (Figure 39) which may have developed under a subtropical climate, probably during Miocene times. Parts of the North-east Highlands may thus have remained close to the base-level of erosion in the area since the late Cretaceous marine transgression (Hall, 1991).

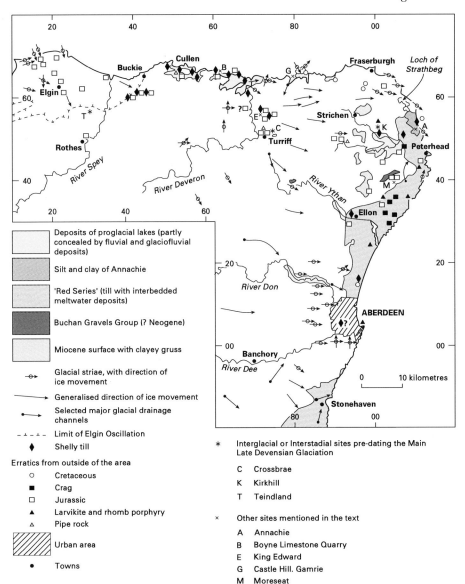

Figure 39 Selected aspects of the Neogene and Quaternary geology of the North-east Highlands.

15   Quaternary

Since the start of the Quaternary some 1.8 million years ago the Grampian Highlands have probably been glaciated many times. At first, glaciers were restricted to the mountains, but during the last 750 000 years there has been a rhythmic growth and decay of large ice sheets in the middle latitudes of Europe and these covered much of the Scottish mainland (Boulton et al., 1991). Periods of ice-sheet growth during the cold *stadials* were separated at intervals of about 100 000 years by relatively short *interglacials* during which climatic conditions were similar to those of the present day. The stadials included short warmer intervals (*interstadials*), which were rarely as warm as a typical interglacial.

As climatic change has had a dominant influence on sedimentation during the Quaternary and there has been little species evolution on which to base biostratigraphical subdivisions, the epoch is divided into a series of climato-stratigraphical stages. Although older deposits occur offshore (Andrews et al., 1990; Fyfe et al., 1993; Gatliff et al., 1994) glacial erosion has largely removed evidence of events in the Quaternary predating the last Main Late Devensian Glaciation and only the last five or so stages are presently known to be represented in the Grampian Highlands (Table 3). The repeated glaciation has resulted in the modification of the pre-Quaternary topography by the widening, straightening and deepening of pre-existing river valleys, the breaching of watersheds and the excavation of corries. Although most of the products of glacial erosion have been transported offshore, those pertaining to the latest glacial events remain, chiefly on low ground, in the form of hummocky moraines, till sheets and deposits of sand and gravel. Parts of the higher mountains are almost free of drift, excepting where mantled by block-fields or slope deposits, and till is thin or absent in parts of Argyll and Jura because higher precipitation in the western mountains resulted in more active ice streams that deposited the debris in what are now offshore areas. North-east Scotland was generally only weakly glaciated during the Quaternary and consequently it has yielded most evidence of glacial, interglacial and interstadial events predating the Late Devensian Glaciation, which culminated some 18 000 BP (radiocarbon years ago).

EVIDENCE FOR EVENTS PREDATING THE LATE DEVENSIAN

Pockets of deeply weathered igneous, metamorphic and sedimentary rocks have survived glaciation in several parts of the region, notably in the north-east (Figure 39). East of a line from Elgin to Dundee, the bedrock has been patchily but extensively decomposed to a gruss (granular sand), locally to depths of several tens of metres. It has been suggested that this weathering

Table 3 Subdivisions of the Quaternary of Scotland.

Isotope stage	Approx. age BP			Stage (chrono/climatostratigraphy)		Glaciations
1	0 – 10 000			Flandrian (Interglacial)	PG	—
2	10 000–26 000	DEVENSIAN	Late	Loch Lomond Stadial (10 000 –11 000 BP)	LG	Loch Lomond Readvance
				Windermere (Late-glacial) Interstadial (11 000–13 000 BP)		
				Dimlington Stadial (13 000 – 26 000 BP)		Main Late Devensian Glaciation
3	26 000–50 000			Middle		—
	50 000–122 000			Early		Early Devensian Glaciation?
4–5d						
5e	122 000–132 000			Ipswichian (Interglacial)		—
6–10	132 000–350 000			Wolstonian (Stadial)		Wolstonian (Saalian) Glaciation
11	430 000			Hoxnian (Interglacial)		—
12	480 000			Anglian (Stadial)		Anglian Glaciation

LG Late-glacial   PG Post-glacial

took place under the temperate conditions postdating the Miocene (Hall, 1986) and as such it is distinct from the clayey gruss associated with the Buchan Gravels Group. Pockets of deeply weathered rock also occur farther west, for example within the outcrop of the Foyers and Moy granites and in the Gaick Plateau area (Figure 40).

There are few documented sites where there is unequivocal evidence for depositional events predating the Main Late Devensian Glaciation and most occur in North-east Scotland (Gordon and Sutherland, 1993) (Figure 39). At Teindland, near Elgin, a fossil soil that has yielded both interglacial and interstadial pollen is overlain by deposits which have been classified by different workers as either solifluction deposits or till (Lowe, 1984). At Crossbrae, near Turriff, a thin layer of peat below soliflucted till has yielded radiocarbon ages in the range of 29 000 to 26 000 BP. A former quarry site at Kirkhill, near Strichen, revealed a sequence of tills and fossil soils interbedded with sand and gravel, including solifluction deposits (Hall and Connell, 1991). These deposits (Plate 17) occupy basins and channels between upstanding tor-like prominences of partly decomposed felsite. Many of the clasts in the gravels are formed of this rock type. The occurrence of erratics of biotite-granite and red sandstone in the lowermost beds is evidence for a glacial episode or episodes that predate the lower till. The tills themselves represent separate

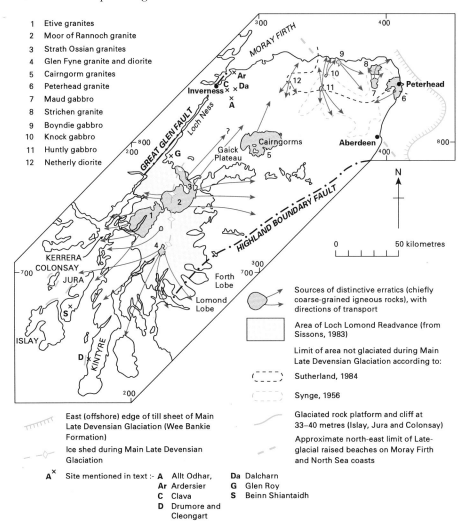

1 Etive granites
2 Moor of Rannoch granite
3 Strath Ossian granites
4 Glen Fyne granite and diorite
5 Cairngorm granites
6 Peterhead granite
7 Maud gabbro
8 Strichen granite
9 Boyndie gabbro
10 Knock gabbro
11 Huntly gabbro
12 Netherly diorite

Sources of distinctive erratics (chiefly coarse-grained igneous rocks), with directions of transport

Area of Loch Lomond Readvance (from Sissons, 1983)

Limit of area not glaciated during Main Late Devensian Glaciation according to:

Sutherland, 1984

Synge, 1956

Glaciated rock platform and cliff at 33–40 metres (Islay, Jura and Colonsay)

Approximate north-east limit of Late-glacial raised beaches on Moray Firth and North Sea coasts

East (offshore) edge of till sheet of Main Late Devensian Glaciation (Wee Bankie Formation)

Ice shed during Main Late Devensian Glaciation

Aˣ Site mentioned in text :- A Allt Odhar,   Da Dalcharn
                              Ar Ardersier   G Glen Roy
                              C Clava   S Beinn Shiantaidh
                              D Drumore and
                                 Cleongart

Figure 40   Selected aspects of the Quaternary geology of the Grampian Highlands.

glaciations and the fossil soils represent interglacial and/or interstadial periods that are older than the limit of radiocarbon dating (about 40 000 BP). It is not clear whether or not the upper till was deposited during the Main Late Devensian Glaciation or during some earlier glacial phase.

A fossil soil containing pollen of full interglacial aspect has been discovered at the base of a high river cliff at Dalcharn, near Cawdor (Figure 40; Walker et al., 1992). This deposit, which has been disturbed by frost action, over-ridden by glacier ice and tectonised, underlies a sequence of three tills with distinctive clast compositions and fabrics (Plate 18). At least one of the tills may predate the Main Late Devensian Glaciation. The palaeosol rests on, and incorporates, deeply weathered and whitened gravel, but the biogenic material cannot yet be ascribed certainly to a particular interglacial. Dalcharn provides the first

Plate 17 Interglacial deposits, Kirkhill Quarry, Aberdeenshire.
The upper till rests on head. The lower palaeosol (black at the top, mostly pale grey with brown iron pan at the base) is developed on lower sands and gravels (D 3716).

Plate 18   Glacial and interglacial deposits at Dalcharn, Nairnshire.

The hammer lies on an interglacial palaeosol containing several horizons of charcoal and compressed peat. This overlies deeply weathered whitened gravel. Both deposits are cut by thrust planes (brown) and have been disturbed by glacier ice which moved from right to left and lodged the overlying till (D 4275).

evidence that the northern Grampian Highlands were covered in pine forest during at least one interglacial stage of the middle to late Quaternary.

A few kilometres to the south of Dalcharn, a layer of compressed peat underlying unweathered lodgement till, but overlying weathered lodgement till, has been located in a river cliff of the Allt Odhar, Moy. This peat contains pollen, insect remains and plant debris indicating an appreciably cooler climate than that suggested for the Dalcharn site, and there is a convergence of evidence that it accumulated during an early-Devensian interstadial. It is the first site from the mainland of Scotland providing evidence of woodland during a Devensian interstadial.

Deposits of marine clay, either transported or in situ, have been found below till of the last Main Late Devensian Glaciation at several localities (Sutherland, 1981). Near Drumore and Cleongart in Kintyre (Figure 40), reddish brown till is underlain at about 55 m above OD by grey and brown pebbly and stone-free clays thought to have been deposited in a glaciomarine environment close to an ice front. The fossils in these beds are largely derived. On the north side of the Grampians at Clava, near Inverness, a raft of marine clay, at an altitude of about 150 m above OD and associated with shelly till, has yielded a fauna that is of a cooler-water aspect than the present Scottish marine fauna, but not high arctic. A combination of radiocarbon and amino-acid dating on shells from the clay suggests that the deposit formed during a mid-Devensian interstadial, prior to about 40 000 BP. The raft was probably carried from the Loch Ness basin during the expansion of the ice-sheet of the Main Late Devensian Glaciation (Merritt, 1992). Other localities where there are deposits of deformed and undeformed marine strata which could be either in situ or rafts are shown on Figure 39. An ice-transported raft of highly deformed marine clay is visible at the Boyne Limestone Quarry, east of Portsoy. Such clays are associated with shelly tills that contain fragments of temperate and cool-water molluscs which at one locality in Buchan have yielded a preliminary amino-acid age determination suggesting a Devensian age, that is post-125 000 BP.

Remains of rock platforms cut by the sea and backed by fossil cliff lines occur on parts of the west coast (Walker, Gray and Lowe, 1992). These, the 'pre-glacial beaches' of the Geological Survey memoirs, are to be found at heights ranging from a few metres above OD in Kintyre to 34 m above OD on Islay and Jura and over 40 m above OD on Colonsay (Figure 40). They are overlain by till in places and may have been glaciated more than once. It is believed that they were formed during periods of rapid combined marine and periglacial erosion. The high cliffs of parts of the Moray Firth and North Sea coasts were also shaped in part prior to the last glaciation. A more widespread rock platform, the Main Late-glacial Shoreline, is discussed below.

Although the distribution of erratics can be attributed largely to the Main Late Devensian Glaciation in the central and western part of the region, the varied transport directions of those derived from major igneous bodies in north-east Scotland is further evidence of a complex glacial history (Figure 40). Boulders of Norwegian larvikite and rhomb porphyry have been found at several localities, notably on the coast south of Aberdeen where some occur in gravels below the Main Late Devensian till. These may have been carried across the North Sea by Scandinavian ice at some period prior to the Ipswichian Interglacial. Other far-travelled erratics include 'Pipe rock', characteristic of the Cambrian of the North-west Highlands. Boulders of Jurassic and Cretaceous rocks, including the well-known Lower Greensand boulder at Moreseat, have probably been derived from the sea floor of the Moray Firth. When these erratics were transported is uncertain, but movement may have occurred during the last as well as earlier glaciations.

LATE DEVENSIAN

Events during the Late Devensian can be considered under three headings (Table 3). During the Dimlington Stadial, the *Main Late Devensian Glaciation* reached its maximum at about 18 000 BP when an ice sheet covered most, if not all, of the Grampians region. During the following period of ice wastage, sheets of gravel, sand and silt were laid down and the movement of any active ice was controlled by the local topography. This was followed by the almost complete disappearance of the ice sheet during the *Windermere (Late-glacial) Interstadial*. Glaciers of the *Loch Lomond Readvance* returned to the mountains during the cold period of the Loch Lomond Stadial between 11 000 and 10 000 BP, and possibly earlier. The Loch Lomond Stadial was followed by a rapid amelioration of climate at the beginning of the Flandrian, about 10 000 BP. The period of time between the end of the deglaciation of the Main Late Devensian ice sheet and the beginning of the Flandrian (Holocene) is known informally as Late-glacial, whereas the Flandrian is equivalent to Post-glacial (Gray and Lowe, 1977).

Main Late Devensian Glaciation

From the distribution of erratics, the ice-shed during this glaciation seems to have been over, or just west of, the Moor of Rannoch (Figure 40), with ice moving westwards across Islay and Jura and southwards and eastwards into the Midland Valley (Sutherland, 1991). A large ice stream from the Western and Northern Highlands flowed eastwards into the Moray Firth and another from the South-west Highlands extended north-eastwards along the southern flank of the Grampians. Between Fort Augustus and Glen Spean the distribution of erratics suggests that the major centres of glaciation migrated southwards from the Northern Highlands to the Moor of Rannoch. The highest summits are free of erratics and it is likely that they stood above the ice sheet for much of the Dimlington Stadial.

In north-east Scotland three separate ice streams appear to have coalesced during the Dimlington Stadial (Clapperton and Sugden, 1977). One powerful stream moved out of the Moray Firth into the North Sea; another originating in the South-west Highlands moved eastwards along Strathmore and then northwards along the North Sea coast towards Peterhead; and a third, more sluggish, possibly thinner ice stream moved eastwards towards the coast from the mountains and foothills of the eastern Grampian Highlands. Along the coast itself, tills formed by the ice moving from the interior commonly underlie tills formed by the Strathmore and Moray Firth ice streams, indicating that 'inland' ice expanded initially to the coast before flow weakened sufficiently to allow incursion of ice from the north and south. Further retreat of the inland ice early in the deglaciation, if not before, allowed ice-dammed lakes to form on the Moray and Banffshire coasts against the Moray Firth ice lobe and on the coast north of Aberdeen against the Strathmore lobe. Similar proglacial lakes occupied the lower reaches of the valleys of the Deveron, North and South Ugie, Ythan, Don and Dee (Brown, 1993) and spreads of interbedded silt and clay were deposited on the interfluves, where extensive tracts may have been inundated up to 80 m above OD (Figure 39).

In eastern Aberdeenshire, the deposits associated with the Strathmore ice stream, which are informally known as the 'Red Series', are the onshore equivalent of the offshore Wee Bankie Formation (Andrews et al., 1990; Gatliff et al., 1994). 'Red Series' tills are typically a vivid red-brown and contain rocks derived from Devonian strata in Strathmore and Permo-Triassic, Devonian and Cretaceous strata offshore. North of Aberdeen, the 'Red Series' comprises a complex sequence in which basal tills are overlain by interstratified flow tills, muds, silts, sands and gravels.

Tills deposited by the Moray Firth ice stream occur as far east as Peterhead. They are typically calcareous, fine-grained, dark grey and contain abundant fossils derived from Late Jurassic and Early Cretaceous strata, as well as rafts of these rock types ripped up from the floor of the Moray Firth. The tills formed by the 'inland' ice are typically sandy and their composition strongly reflects the character of the local bedrock, which is commonly deeply weathered.

The glacial landforms in the ground covered by the Strathmore and Moray Firth ice streams are sharp, whereas in much of the area overrun by 'inland' ice to the north and north-west of Ellon they are indistinct. It has, therefore, been suggested that parts of Buchan were ice-free during the Dimlington Stadial and that the last ice to cross the area was either Early Devensian or pre-dated the Ipswichian Interglacial. The difficulties of defining the boundaries of the 'unglaciated' area (if any) are illustrated by the two limits shown on Figure 40; other workers have placed the limit even farther west.

Although Buchan displays relatively little discernable evidence of glacial erosion, as shown by the survival of deeply weathered bedrock and patches of Neogene gravel, it does exhibit a network of glacial meltwater channels. It is known from Scandinavia and elsewhere that cold-based parts of ice sheets, such as those that may have covered Buchan, can cause minimal subglacial erosion and can leave delicate morphological features such as eskers and meltwater channels virtually unscathed (Hall and Sugden, 1987).

During retreat, the Main Late Devensian ice sheet probably remained active locally, taking the form of valley glaciers. The southern edge of such a glacier can be traced on the south side of Strathspey, near the Corrie Cas ski slopes, where it is defined by marginal channels and glaciofluvial deposits. Ice-marginal channels, terraces and benches are also displayed spectacularly in the gorge of the River Findhorn downstream of Tomatin, and in the valley of the River Nairn downstream of Daviot (Auton et al., 1990). The terraces of the River Nairn merge into Late-glacial beach deposits towards the coast. Local readvances such as the Elgin Oscillation in the east (Peacock et al., 1968), and a (probably later) stillstand or minor readvance (the 'Otter Ferry stage') in the west (Sutherland, 1984), interrupted the general retreat. Other readvances such as the Perth and Dinnet readvances have been reported in the literature, but the evidence is disputed and largely rejected. Another disputed readvance east of Inverness has recently been confirmed on the Ardersier Peninsula, although the timing of the event is unclear.

During deglaciation meltwaters laid down extensive spreads of sand and gravel, particularly on the low ground on the south side of the Moray Firth from the north end of the Great Glen to near Elgin, and also in Strathspey (Plate 19). The deposits include eskers, kames, kame-plateaux and kame-terraces, many having formed in ephemeral ice-marginal lakes as deltas or subaqueous fans (Fletcher et al., 1994). The Flemington Eskers, which extend over a distance of about 10 km to the south-west of Nairn, are probably the best example in Britain of a braided esker system that remains essentially unmodified by sand and gravel extraction (Auton, 1992). They are 5 to 10 m high, with intervening kettleholes. The Torvean Esker in Inverness, now largely quarried away, reached a height of some 60 m, making it one of the largest features of its kind in the country.

Plate 19   Glaciofluvial deposits, Findhorn valley at Quilichan, Nairnshire.
In the foreground are floodplain alluvium and low-lying fluvial and glaciofluvial
outwash terraces; in the middle ground irregular mounds of englacially and
subglacially deposited sand and gravel, and in the background stepped outwash fans
(D 4270).

In Scotland, the eustatic lowering of sea level caused by the abstraction of
water to form the continental and local ice sheets, was more than offset by
the isostatic depression of the land. Sea level initially rose quickly during
the deglaciation following the Main Late Devensian Glaciation and consequently the lobes of the ice sheet occupying the firths of eastern Scotland
retreated rapidly by the process of iceberg calving. Some of the associated
glaciomarine sediments are now raised above sea level. The oldest of these
deposits occur in the extreme north-east Grampians, where the glaciomarine St Fergus Silts at Annachie (Figure 39) are radiocarbon dated at
15 320 BP (Hall and Jarvis, 1989). Glaciomarine clays near Elgin, which
have yielded well-preserved remains of the brittle star *Ophiolepis gracilis*
Allman (Plate 20), can be compared with the *Errol Beds* of the Firth of Tay
which were laid down in the arctic conditions preceding the Windermere
Interstadial (Armstrong et al., 1985).

Plate 20   Quaternary fossils from Ardyne and Elgin.
A *Macoma calcarea*. Left valve × 3. Clyde Beds, Shell Pit, Ardyne. GSE 13140 MNS 4531.
B *Arctica islandica*. Left valve × 0.6. Clyde Beds, Elf Pit, Ardyne. GSE 14521 MNS 4532.
C *Ophiolepis gracilis*, natural size. Glaciomarine clays, Spynie, Elgin. Royal Museum of
Scotland RSM 1913.22.1 GN 035.
D *Chlamys islandica*. Lower (right) valve × 1.5. Clyde Beds, Elf Pit, Ardyne. GSE 14522
MNS 4530.
E *Bittium reticulatum*. × 16. Clyde Beds, Shell Pit, Ardyne. GSE 14520 MNS 4533.

GSE and RSM — registered specimen numbers. MNS and GN — registered photo numbers.

Windermere Interstadial and Loch Lomond Stadial

Deglaciation of the Firth of Clyde area was followed by the deposition of the *Clyde Beds*, which began to form about 13 000 to 12 800 BP, when glaciers finally retreated to within the mouths of the sea lochs in response to the sharp climatic amelioration marking the beginning of the Windermere Interstadial (Sutherland, 1984).

Pollen analysis of lake sediments and peat bogs shows that the climate warmed rapidly about 13 000 BP. The freshly deglaciated areas of the mainland were colonised by a pioneer vegetation of open habitat taxa which was followed by the immigration of crowberry heath, juniper and tree birch. There was a short-lived reversion to open habitat species in some areas at about 12 000 BP and during the succeeding Loch Lomond Stadial there was a more general return to open habitat vegetation and tundra. On the west coast, the raised marine shelly clays of the Clyde Beds include both cool-water interstadial faunas similar to those found today on the north-west Norwegian coast and an arctic fauna which can be referred to the Loch Lomond Stadial (Plate 20; Peacock, 1993).

The extent of deglaciation during the Windermere Interstadial is unknown because the evidence has largely been destroyed by glaciers during the subsequent *Loch Lomond Readvance*, but it is likely that some ice would have survived at high levels in the western mountains.

There is abundant evidence in the form of terminal moraines and the presence of hummocky gravel, till and fluted drift for renewed glaciation in the Loch Lomond Stadial, during which glaciers once more became extensive in the Central and South-west Highlands (Figure 40; Plate 21) (Gray and Coxon, 1991; Thorp, 1991). Glaciers in the upper Forth valley and the basins of Loch Lomond and Loch Creran ploughed through marine deposits of the Windermere Interstadial. An ice cap centred west of Rannoch Moor extended to 700 m above OD and ice at this time was some 600 m thick in the Great Glen. A smaller ice cap was present on the Gaick Plateau.

Ice crossing the Great Glen from the west and south dammed up lakes which gave rise to the shorelines known as the Parallel Roads of Glen Roy, Glen Gloy and Glen Spean (Figure 41A; Peacock and Cornish, 1989). Successive lakes in Glen Roy were controlled by outlets at present levels of 261 m, 325 m and 350 m above OD as the ice advanced and at 325 m and 261 m as the ice retreated (2 to 4 on Figure 41B). The maximum position of the readvance ice associated with the highest lakes is clear in Glen Roy and Glen Spean, where it is marked by terminal moraines and/or thick lake sediments, but less so in Glen Gloy. Here two positions have been suggested, one being east of the col (355 m) leading to upper Glen Roy. The latter position, however, may predate the Loch Lomond Stadial. The retreat of the ice from the maximum in these glens is thought to have been accompanied by catastrophic lake drainage, possibly accompanied by earthquakes.

During the Loch Lomond Stadial and during the earlier part of the retreat phase from the Late Devensian maximum, the arctic climate led to intense frost action which gave rise to blockfields and stone lobes on high ground and to patterned ground and solifluction sheets at all levels. Ice-wedge pseudomorphs are said to occur outside the limit of the Loch

Plate 21   Recessional moraine ridges composed mostly of sand and gravel that formed during the 'active retreat' (right to left) of a Loch Lomond Readvance glacier. Strath Fillan, near Tyndrum, Perthshire (D 2656).

Lomond Readvance glaciers; in present-day arctic climate environments such pseudomorphs indicate permafrost conditions and average annual temperatures several degrees below 0° C (Ballantyne and Harris, 1994). Fossil rock glaciers and protalus ramparts (ridges of debris formed at the base of snow slopes) were formed at several localities, mainly in the Cairngorms. A particularly striking example of a rock glacier can be seen on Beinn Shiantaidh in Jura (Figure 40). Landslips are common in the western Grampians, particularly within the limits of the Loch Lomond Readvance. They were probably triggered by a combination of factors such as unloading of rock faces as the ice melted, changes in water table and the presence of suitably oriented joints and bedding surfaces. It is likely that severe frost action at the shoreline played a part in the formation of the Parallel Roads of Glen Roy, which are partly incised in bedrock.

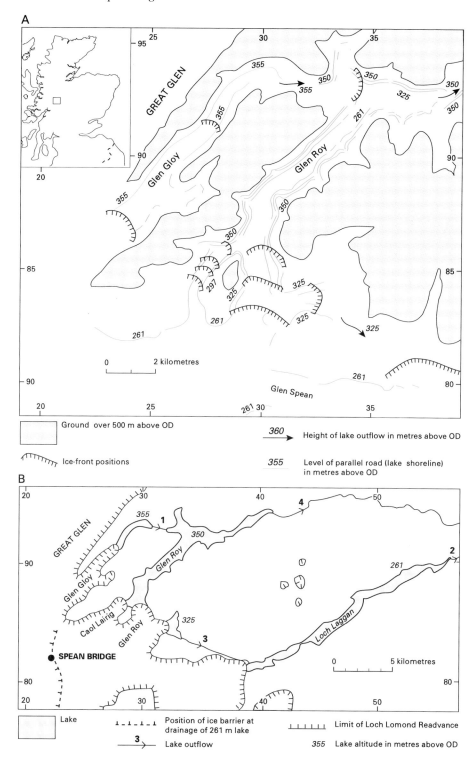

LATE-GLACIAL RAISED BEACHES AND ASSOCIATED MARINE DEPOSITS AND CLIFFLINES

Isostatic depression of the lithosphere by the Main Late Devensian ice sheet was sufficiently great to ensure that, although global eustatic sea levels were low during deglaciation, local sea levels around the Scottish ice sheet were high. In Late-glacial and Flandrian (Post-glacial) times, isostatic recovery sometimes failed to keep pace with the eustatic rise in sea level as the water was returned to the oceans. Consequently, raised beaches and associated marine deposits are now found well above OD on all coasts except part of Buchan (Smith and Dawson, 1983). Isostatic uplift was greatest where the ice load had been maximal, as a result of which the beaches tend to be tilted away from a centre of uplift in the South-west Highlands. Beaches of the same height in different places are therefore not necessarily of the same age. The raised beaches are divisible into a Late-glacial series (formed sometimes adjacent to retreating ice) and a Flandrian series (sometimes termed Post-glacial) which are separated in time, for the most part, by a period of relatively low sea level.

The widespread Late-glacial beaches formed partly during the retreat of the Main Late Devensian ice sheet and partly during the Windermere Interstadial and the Loch Lomond Stadial (Plate 22). They are usually in the form of accumulations of gravel which rest on platforms cut in drift. Heights of a little above 40 m above OD have been recorded both on the mainland adjacent to the Firth of Clyde and on Jura where a 'staircase' of 50 shingle ridges has been mapped. The regression of the sea from these high levels may have been briefly checked in the inner Firth of Clyde during the 'Otter Ferry stage'. In the east, raised beaches have been recorded up to 35 m above OD at Inverness and Stonehaven and at about 30 m above OD near Elgin.

To the east of Inverness and as far as Forres, some 40 km to the northeast, there are large tracts of moundy, ice-contact topography that lie below the elevation of the highest Late-glacial marine shoreline (the 'marine limit'), yet provide no evidence of any marine incursion. These low-lying areas are considered, therefore, to have remained covered either by ice or by sediment containing buried ice masses, until relative sea level had fallen. The absence of marine or littoral deposits in several large kettle-holes occurring within these tracts, indicates that ice remained inland of those coasts until relative sea level had fallen to less than 13 m above OD (Firth, 1989).

A well-marked platform and cliff occurs along much of the west coast, the base of the cliff sloping gently south and west from more that 10 m above OD in the inner sea lochs and firths to below high-tide level in Islay and southern Kintyre (Plate 21). This feature, which has been termed the Main Late-glacial Shoreline, has also been recorded near Inverness, where it is at about 2 m above OD. It is believed to have been formed by marine erosion accelerated

Figure 41   Parallel Roads of Glen Roy.

A   Ice front positions and lake shorelines during the Loch Lomond Stadial.

B   Lake outflows at the maximum of the Loch Lomond Readvance.

Plate 22   Rock platforms, near Port a'Chotain, north Islay.
The 30 m rock platform produced by marine erosion is overlain by Late-glacial beach gravels. In the foreground is the grass-covered platform of the 7 m Post-glacial raised beach (B 722).

by severe frost action during the Loch Lomond Stadial and part of the preceding interstadial.

FLANDRIAN

The abrupt increase of temperature following the Loch Lomond Stadial led to the rapid establishment of crowberry and juniper scrub with subsequent expansion of mixed deciduous woodland of birch and hazel, followed by oak and elm, or pine, and alder (Walker, 1984). The forests so formed declined after about 5000 BP, partly as the result of the cooler, wetter climate which favoured the growth of blanket peat and partly as a result of human activities.

Relative sea level fell following the final disappearance of the glaciers, to be followed by a marine transgression which culminated between about 6000 and 6800 BP (Smith and Dawson, 1983). This transgression formed the Flandrian raised beaches (the '25 ft' and '40 ft' beaches of older Geological Survey publications), which usually take the form of gravel storm beaches, but in sheltered situations estuarine silt and clay predominate. Lower beaches occur in places. The beaches tilt away from the centre of isostatic uplift, the Moor of Rannoch. The highest beaches occur 12–14 m above OD near Stirling and 10 m above OD at Inverness, but are at sea level on the Buchan coast at Fraserburgh; the average tilt is about 0.06–0.07 m/km. A gradient of 0.05 m/km has been calculated for the equivalent beach in the Firth of Lorn. Along the east coast a prominent fine-grained sand horizon, present in both estuarine and nearshore terrestrial sediments, was deposited at around 7000 BP in response to either a major storm surge in the North Sea basin or a tsunami resulting from a major submarine slide on the Norwegian continental slope (Long, Smith and Dawson, 1989).

Sandy beaches and blown sand are extensively developed on parts of the Moray Firth and North Sea coasts, but are less important in the west. Inland, the blanket peat referred to above is widespread in parts of the Grampians, but the extensive tracts formerly present on low ground have been greatly reduced to create agricultural land. The glacial deposits have been partly eroded and redistributed by streams to form the alluvial haughs and terraces seen in many Highland glens (Ballantyne, 1991).

16   Faulting and seismicity

FAULTING

The fault pattern of the Grampian Highlands consists of, in order of decreasing age:

Early deduced faults now represented by lineaments of various kinds, which exerted some controls on Dalradian sedimentation.

Ductile shear zones developed at elevated temperatures during the Caledonian Orogeny; represented by zones of mylonites and/or platy shear zones, and including the major slides of the Dalradian (ductile thrusts or lags).

Brittle faults developed towards the end and after the Caledonian Orogeny; the most important are a series of strike-slip faults.

Early deduced faults

The best-documented member of this group, the *Cruachan Lineament* (Figure 42) was first described by Graham (1986). The trans-Caledonoid trend of the lineament is strikingly picked out by a steep NW–SE gradient on the Bouguer gravity anomaly map and was interpreted by Graham as marking a change in the nature of the sub-Dalradian basement. The marked decrease in the thickness of synsedimentary basaltic rocks where they cross the lineament emphasises its influence on the sedimentation and igneous activity of the Dalradian.

Fettes et al. (1986), using stratigraphical, structural, geophysical and igneous criteria, subdivided the Dalradian into a number of fault-controlled basins, and showed that some of the faults continued to operate during the deformation and uplift phases of the Caledonian Orogeny. The largest structure, the *Deeside Lineament,* which crosses the Grampian Highlands for a distance of about 100 km, has no surface expression as a fault but is believed to have controlled the intrusion of the East Grampian Batholith.

On a smaller scale, the *Ossian Lineament,* which extends north-westwards from the Ericht–Laidon Fault to the Great Glen, is believed to have acted as a channel for the NW-elongate Strath Ossian Complex and for a suite of appinite plugs and intrusion breccias (Key et al., in press). It also appears to coincide with lithological facies changes in the Leven Schist suggesting, perhaps, that it existed during Lower Dalradian sedimentation.

Evidence for the renewed movement on some of the early deduced lineaments is provided by the Rothes Fault, (13 on Figure 42) along which there are marked changes in the thickness of the local Upper Old Red Sandstone

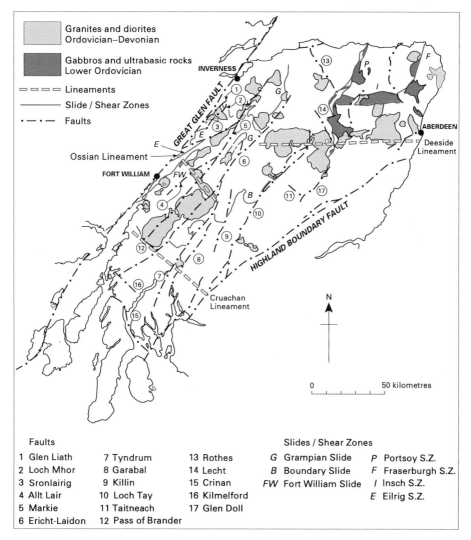

Figure 42 Distribution of lineaments, shear zones and major brittle faults in the Grampian Highlands.

succession, as well as of the Dalradian across it, indicating synsedimentary activity. The termination of the fault against the Permo-Triassic rocks provides an upper age limit to fault movements.

Ductile shear zones

As has been described in earlier chapters, the Dalradian of much of the Grampian Highlands is dominated by polyphase nappe folding and thrusting, with the development of a number of important and extensive slide zones, among which the *Boundary Slide, Grampian Slide* and *Fort William Slide* have been regarded as the most important. The recently recognised *Eilrig Shear Zone* of the Monadhliath Mountains (Phillips et al., 1993) is another. It has, however, long been recognised that the North-east Highlands is somewhat different. The area is characterised by the occurrence of large sheet-like syn- to late-tectonic basic and ultramafic masses (formerly known as the 'Younger Basics') which were partly deformed by the later stages of the Caledonian Orogeny. A detailed study of the area, augmented by magnetic surveys, has revealed the presence of a regional system of steeply inclined shear zones, the major ones, Portsoy, Fraserburgh and Insch (Figure 42), being located at the margins of these Younger Basics (Ashcroft et al., 1984). In places the shear zones have disrupted the gabbros and rotated originally flat-lying igneous layering into a vertical altitude. Basic mylonites have been formed locally and display steep down-dip extension lineations. Ashcroft et al. (1984) suggest that the main shear zones were in existence prior to the intrusion of the Younger Basics and acted as pathways along which the basic magma was emplaced. They recognised an even older history for the *Portsoy Shear Zone*, with the structure acting as a syndepositional fault during the sedimentation of Argyll Group and Southern Highland Group Dalradian rocks. Recent field work and detailed ground magnetic surveys over the region where the Portsoy and Insch lineaments converge has confirmed the main findings of Ashcroft et al. (1984) and produced firm evidence for the longevity of the shear zones (Fettes et al., 1991). The lack of continuity in Dalradian stratigraphy across the Portsoy Shear Zone, and the restriction of contemporaneous metavolcanic rocks to its immediate vicinity, is taken to indicate the influence of the shear during basin formation, thus influencing the depositional pattern and the subsequent metamorphic and tectonic histories. Some 10 to 15 km to the north-west of the Portsoy Shear Zone, the *Keith Shear Zone* has acted as a focus for the emplacement of granite sheets.

Brittle faults

The Grampian Highlands form a block sharply delineated to the north-west and south-east by the two largest and most important fractures in Scotland– the Great Glen Fault and the Highland Boundary Fault respectively, and contain a number of other major faults with a subparallel trend (Figure 42).

Research into the timing, sense of movement and amount of displacement of the *Great Glen Fault* has been almost continuous since Kennedy's conclusion that it is a major transcurrent fault with a sinistral displacement of about 100 km (Kennedy, 1946). A full account of the history of research is given in Johnstone and Mykura (1989). Numerous solutions, some speculative, have been advanced but the current consensus is that a main phase of sinistral transcurrent movement occurred at the end of the Caledonian Orogeny (Silurian) with reactivation, mainly as a normal fault, during and after deposition of the Old Red Sandstone (Smith and Watson, 1983; Rogers et al., 1989). Minor normal adjustments in the Permo-Triassic were followed by a period of limited dextral shift during late Triassic–early Cretaceous times (McQuillin et al., 1982; Andrews et al., 1990). That the Palaeogene Mull dyke swarm crosses the fault without deviation (Speight et al., 1982) suggests that, in the area of Mull at least, significant movement had ceased prior to dyke intrusion at 52 Ma. A much longer history for the Great Glen Fault was suggested by

Bentley (1988) who proposed that it had earlier acted as a strike-slip terrane boundary along which the 1800 Ma Rhinns terrane of Islay became juxtaposed against the Lewisian terrane before the deposition of the Dalradian.

The *Highland Boundary Fault,* which defines the contact of the Grampian Highlands with the Midland Valley graben to the south-east, also appears to have had a long and complex history. It is marked by the presence of discontinuous, fault-bounded slivers of Cambro-Ordovician ophiolitic rocks known as the Highland Border Complex (see Chapter 9). The absence from these rocks of clasts from the adjacent Dalradian rocks to the north-west, which were being uplifted and eroded during the Ordovician, suggests that the emplacement of the Highland Border Complex and the juxtaposing of the Grampian Highlands and Midland Valley took place in Silurian times (Bluck, 1985). There is little convincing structural evidence as to the mechanism of emplacement, but most recent authors favour large-scale strike-slip movements (e.g. Harte et al., 1984).

Subsequent movements on the Highland Boundary Fault during the Devonian and Lower Carboniferous were mainly normal, with downthrow to the south-east into the Midland Valley graben (Cameron and Stephenson, 1985; Paterson et al., 1990). East–west late Carboniferous quartz-dolerite dykes cross the fault zone without displacement, indicating that movement had effectively as ceased by that time, about 300 Ma (Anderson, 1947b; George, 1960).

Within the Grampian region the most prominent group of faults are those with a north-easterly trend, subparallel to the Great Glen and extending eastwards as far as the Glen Doll Fault (17 on Figure 42). On the basis of their trend, and the fact that several of them show a sinistral sense of movement, they have been genetically linked to the Great Glen Fault and accorded a similar age and history of movement. Watson (1984) emphasised the longevity of this fault set and stressed how reactivation might obscure evidence for early movements. The Glen Liath (1), Loch Mhor (2), Ericht–Laidon (6) and Tyndrum (7) faults all display a phase of movement older than the granites which they cut and Watson proposed that many of the main-phase Caledonian granites in the western Grampians used these fault fissures as pathways for intrusion.

Recently Treagus (1991), in a detailed study of the Ericht–Laidon (6), Tyndrum (7), Garabal (8), Killin (9) and Loch Tay (10) faults, has proposed that some of them were generated on the limbs of late ductile folds during the final stages of the Caledonian Orogeny. Thereafter, the main movements on the faults were dip-slip with a cumulative downthrow to the east of about 7 km, followed by a sinistral strike-slip phase with a cumulative displacement of 23 km.

The last significant movements on the Great Glen set of faults within the Grampian block are generally thought to have taken place at the end of the Caledonian Orogeny when the compressional ductile phase gave way to a period of vigorous brittle uplift and erosion. Regeneration of the Glen Liath Fault during the deposition of Middle Old Red Sandstone sediments (Mykura, 1982), and the presence of a mini-graben at the north-west end of the Markie Fault *(5),* attest to active faulting during Old Red Sandstone times.

By the end of the Devonian the eroded remnants of the Grampian block formed a topographic high on which little subsequent sedimentation took place. In Kintyre, the Devonian and Carboniferous rocks have been affected by faulting along various trends. There is evidence from the Carboniferous strata that some movement occurred during the Dinantian and Namurian. Only on the northern coast, where the southern margin of the Mesozoic Moray Firth Basin encroaches on to the land, is there evidence for substantial displacement of Permo-Triassic and Jurassic sedimentary rocks, by east–west normal faults. The most recent faults to affect the Grampian block are in the south-west, of which the Crinan *(15)* and Kilmelford *(16)* faults are the most prominent. They are contemporaneous with the intrusion of the Palaeogene dyke swarm which crosses the region without deviation and testify to the absence of any significant post-Palaeogene fault movements.

SEISMICITY

While from an examination of the structural geology one might expect the seismicity of the region to be reasonably uniform, such is far from the case. The west coast, particularly the area from Oban to Dunoon, is one of the most seismically active parts of Scotland (or Great Britain for that matter), while the east coast and Central Grampian Highlands are virtually free from earthquakes, only experiencing the vibration from earthquakes occurring elsewhere in Scotland or in the North Sea. The earthquake of 28 November 1880, with epicentre between Oban and Inveraray, is the largest known Scottish earthquake, with magnitude 5.2 on the Richter Scale. The largest recent earthquake in the Grampian Highlands area was the Oban earthquake of 29 September 1986, magnitude 4.1, which was widely felt along the west coast. An earthquake on 16 September 1985, magnitude 3.3 had an epicentre near Ardentinny and may have been associated with a NE-trending fault zone in the Loch Long area.

The reason for the localisation of earthquake activity is unclear, but there is a strong probability that it relates to old faults having been reactivated at the conclusion of the Loch Lomond Readvance phase of the last glaciation, and having remained active in the present regional stress regime.

The two major faults bounding the Grampian region, the Great Glen Fault and the Highland Boundary Fault, have long been assumed to be seismically active. This assumption has not withstood scrutiny. Many earthquakes formerly attributed to the Great Glen Fault have now conclusively been shown to have had their epicentres well away from it. This leaves the Inverness earthquakes of 1816, 1890 and 1901 which may have been caused by movement on a fault splaying northwards from the Great Glen Fault in the Aird area south-west of Inverness.

The other notable site of seismic activity is at Comrie, Perthshire, which is remarkable for its earthquake swarms, at the height of which scores of small earthquakes might occur in a single day. Two such swarms are well documented, the first lasting from 1788 to 1801, the second from 1839 to 1846. There may have been a previous swarm in the early 17th century and possibly another in the early 13th century. Since 1846 activity has been limited to a few small events at sporadic intervals. The largest of the well-documented Comrie earthquakes was that of 23 October 1839 (magnitude 4.8).

Traditionally, the Comrie earthquakes have been attributed to the Highland Boundary Fault, but this does not explain why this fault should produce earthquakes only at Comrie and nowhere else along the rest of its length. It is equally possible, and perhaps more likely, that some other fault, just north of Comrie, is responsible.

The 1839–1846 earthquake swarm is particularly remarkable for the scientific interest it aroused, and a number of early seismological advances were made in Scotland as a result. These included the first inverted pendulum seismometer, the first use of the word 'seismometer', the first seismometer network, and the first purpose-built seismological observatory (now restored and a tourist attraction at Comrie).

17 Economic geology

METALLIFEROUS MINERALISATION

Many occurrences of precious and base metals and of baryte have been discovered in the Grampian Highlands over the last quarter of a century as geochemical and other modern exploration techniques have been applied to the region; these add to the old records of metalliferous mineralisation (Wilson and Flett, 1921). Publications arising from the BGS Mineral Reconnaissance and Geochemical Survey programmes, plus the reports of exploration companies now on open file (Colman, 1990), form a metalliferous database comparable with that of any other region of similar size in the world. Geochemical atlases covering the Grampians record well over 100 locations of significant mineralisation (e.g. Gallagher, 1990; 1991a; Gallagher and Young, 1993) and provide multi-element data on drainage samples from 47 000 sites at a mean density of one sample/1.5 km^2. Also available are comprehensive gravity and aeromagnetic maps; airborne electromagnetic and radiometric surveys of some areas have been placed on open file by exploration companies.

Eight types of mineralisation are described on the basis of the 28 selected occurrences, numbered with references, in Table 4 and located on Figure 43. Of principal economic significance are Dalradian stratabound deposits and post-Caledonian vein deposits, probably of late Silurian to Carboniferous age, cutting Dalradian rocks. Rocks of the Dalradian Supergroup are considered to have been the principal crustal reservoir of metals in the Grampians region (Simpson et al., 1989; Plant et al., 1991).

Mineral production

Mining for lead, with silver as a valuable by-product, and to a lesser extent for copper, nickel and manganese, was quite widespread in the Grampian Highlands in the past, extending into the present century only at Tyndrum (*21* on Figure 43) where about 10 000 tonnes of lead ore were produced from veins in the periods 1741 to 1862 and 1916 to 1925. Some 200 t of zinc ore was extracted from dumps in this final period. Wilson and Flett (1921) record an output of 1400 t lead and 0.5 t silver in the period 1862 to 1880 from Islay, principally from veins at Mulreesh (*28*). Smaller mines at Abergairn (*9*) and Corrie Buie (*19*) exploited lead veins and the Kilmartin veins (*26*) yielded copper, as did a diorite-granite intrusion containing disseminated sulphides at Tomnadashan (*17*). Some 400 t of ore containing nickeliferous pyrrhotite were raised from the Coille-braghad deposit (*24*) which, like the old copper mine of Abhainn Strathain at Meall Mor (*27*) and the lead-copper trial at

McPhun's Cairn *(25)*, is classified as Dalradian stratabound in type. Iron was extracted in the eighteenth century, and manganese in the nineteenth, from breccias in the Dalradian at the Lecht (*7*).

Bedded baryte and zinc-lead sulphide deposits in Dalradian rocks of the Aberfeldy district, found in the 1970s (Coats et al., 1984) are among the most important mineral discoveries to be made in Britain this century. Foss Mine (*13*), the largest baryte producer in Britain, and the Ben Eagach Quarry (*12*; Plate 23) have yielded more than 0.5 Mt of direct shipping-grade ore (defined by a minimum specific gravity of 4.2 gcm^{-3}) since 1984 for use in drilling fluid in North Sea hydrocarbons operations. Underground production of 0.2 Mt/annum over 30 years is planned for the adjacent Duntanlich deposit (*11*) (Butcher et al., 1991). A small amount of baryte was quarried at Balfriesh (*3*) around 1980. Economic evaluations have also taken place of gold-bearing structures at Cononish (*22*), presently regarded as Britain's premier gold-silver deposit, and at Calliachar Burn (*15*) where a little gold was extracted in 1991.

Dalradian stratabound mineralisation

Metasedimentary rocks of the Easdale and Crinan subgroups in the Argyll Group contain stratabound deposits of baryte, barium silicates, base metal sulphides and chromian minerals at locations over some 200 km of the regional strike (Figure 40). Mineralisation is recognisable in at least six horizons despite Caledonian deformation and amphibolite-grade metamorphism (Smith et al., 1984).

Deposits of baryte accompanied by sulphidic quartz-celsian rocks and barium-enriched mica-schists extend at intervals over 7 km of the strike-length of the Ben Eagach Schist (Easdale Subgroup) near Aberfeldy (*11–13*). The Duntanlich deposit is sited towards the base of the formation on the strike extension of the deposit being quarried (in 1993) at Ben Eagach, forming a mineralised zone tens of metres thick running for some 2 km. Foss Mine on the other hand, lies in a baryte bed close to the stratigraphical top of the Ben Eagach Schist, at a similar level to the subeconomic mineralised zone adjacent to the Duntanlich deposit (Gallagher, 1991b). Although of no economic significance, unusual barium minerals–barian muscovite containing up to 8% BaO and the barium feldspar celsian—are the major components of the mineralised zones. Another barium feldspar, hyalophane, and the hydrated barium silicate, cymrite, are also present. Overall the Aberfeldy section of the Ben Eagach Schist probably represents the highest concentration of barium known worldwide. The bedded baryte of the Aberfeldy deposits is composed of anhedral grains 0.1–2.0 mm across with accessory carbonate, magnetite, sulphides (mainly pyrite) and rare fuchsite (the chromian mica). Thin veins of coarse-grained crystalline baryte are ascribable to postmetamorphic remobilisation.

Company drilling at Duntanlich (*11*) defined a high-grade baryte bed 5–13 m thick (maximum 28 m), extending over 0.8 km of strike and to at least 550 m below surface. The bed is enclosed by quartz-celsian rocks in which sphalerite and galena can attain econonomic grades. These rocks exhibit fragmental textures interpreted as a consequence of growth fault activity during basin-floor ore formation. The upper mineralised zone at Duntanlich, which runs for 0.6 km, contains faulted units of baryte and quartz-celsian

Table 4   Principal metalliferous mineral occurrences in the Grampian Highlands, numbered from north to south as on Figure 43.

No.	Name	Type	Mineralogy	Reference
1	Stotfield	G	Gl	Naylor et al., 1989
2	Littlemill	C	Cp PGM Pn Po	Fletcher and Rice, 1989
3	Balfreish	F	Bt	Gallagher, 1984
4	Arthrath	C	Cp PGM Pn Po	Rice, 1975; Gallagher, 1991a
5	Kelman Hill	C	PGM Po	Gunn et al., 1990
6	Rhynie	E	(As Au Sb)	Rice and Trewin, 1988
7	Lecht	B	Mx	Smith, 1985; Nicholson and Anderton, 1989; Smith et al., 1991
8	Gairnshiel	D	Bm Ct Fl Mb Sp Wo	Webb et al., 1992
9	Abergairn	H	Fl Gl Sp	Dunham, 1952; Gallagher, 1991a
10	Coire Loch Kander	A	Bt Gl Sp	Gallagher et al., 1989; Fortey et al., 1991; Fortey et al., 1993
11	Duntanlich			Coats et al., 1981; Fortey and Beddoe-Stephens, 1982; Moles, 1982; Willan and Coleman, 1983
12	Ben Eagach	A	Bt Cn Sp Gl	
13	Foss			
14	Glen Lyon	A	Gl Sp	Coats et al., 1984
15	Calliachar Burn	H	Ap Cp Gl Go Sp	Mason et al., 1991
16	Loch Lyon	A	Bt Cn Gl Sp	Coats et al., 1984
17	Tomnadashan	D	Ap Bm Cp Mb St	Pattrick, 1984
18	Corrycharmaig	C	Ch	Harrison, 1985; Hawson and Hall, 1987
19	Corrie Buie	H	Bm Cp Gl Go Sp	Pattrick, 1984
20	Auchtertyre	A	Cp Gl Mb Sp	Fortey and Smith, 1986; Smith et al., 1988; Scott et al., 1988
21	Tyndrum	H	Bt Cp Gl Sp Tt Ur	Pattrick et al., 1988; Pattrick et al., 1991
22	Cononish	H	Cp Gl Go Sp Te	Parker et al., 1989; Earls et al., 1992
23	Lagalochan	D	Ap Cp Gl Go Sp St Te	Harris et al., 1988
24	Coille-braghad	A	Po Cp	Wilson and Flett, 1921
25	McPhun's Cairn	A	Cp Gl Sp	Smith et al., 1977; Willan and Hall, 1980
26	Kilmartin	H	Cp	Wilson and Flett, 1921
27	Meall Mor	A	Cp Gl Sp St	Smith et al., 1978; Mohammed, 1987
28	Mulreesh	H	Cp Gl Sp	Wilson and Flett, 1921; Barnett, 1959

Brackets indicate chemical elements, not minerals.

Types of metalliferous mineral occurrence
H   Vein
G   Hosted by Permo-Triassic sedimentary rock
F   Hosted by Middle Devonian sedimentary rock
E   Epithermal
D   Associated with granite and diorite
C   Associated with basic and ultramafic rocks
B   Manganese
A   Dalradian stratabound

Table 4 *(continued)*   **Minerals**

Ap	Arsenopyrite	Gl	Galena	Sp	Sphalerite
Bm	Bismuth minerals	Go	Gold	St	Stibnite
Bt	Baryte	Mb	Molybdenite	Te	Telluride species
Ch	Chromite	Mx	Manganese oxides	Tt	Tetrahedrite
Cn	Celsian	Pn	Pentlandite	Ur	Uraninite
Cp	Chalcopyrite	PGM	Platinum-group	Wo	Wolfamite
Ct	Cassiterite		minerals or elements		
Fl	Fluorite	Po	Pyrrhotite		

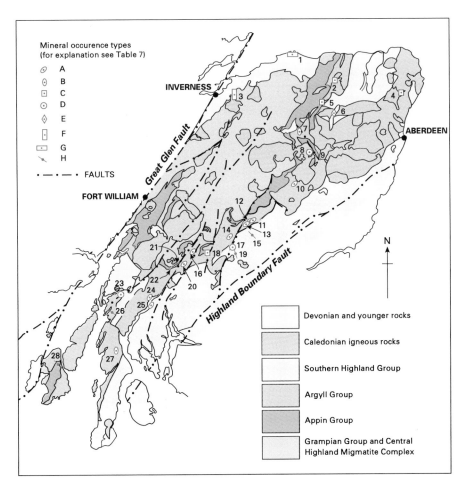

Figure 43   Principal metalliferous mineral occurrences in the Grampian Highlands.

rock. It has an outcrop width of 100 m made up largely of barium-enriched graphitic muscovite-schist, calcareous-schist and muscovite-schist. The Ben Eagach quarry (*12*) has operated since 1990 with an annual production of 10 000 t of baryte from a folded bed averaging 2.5 m in thickness over a strike-length of 250 m. The associated quartz-celsian rocks can be rich in

Plate 23   Stratabound baryte deposit, Ben Eagach Quarry, Aberfeldy, Perthshire.
The baryte bed (in the Ben Eagach Schist of the Easdale Subgroup) has been
prepared for extraction. In the distance are hummocky moraines around Loch
Deruelich, and Strathtay. (MNS 5591).

sphalerite and galena, which have been locally remobilised, and exhibit spots
of fuchsite. At this locality are present a 4.3 m-thick bed of manganoan calcite
containing fine-grained sphalerite and galena of ore grade, sulphidic
dolomitic quartz-rock, barium-enriched muscovite-schist and sphalerite-
galena veins in cherty quartz-celsian rock. Foss Mine (*13*) has produced about
50 000 t of baryte annually since 1984 from a bed averaging 4 m in thickness
in the underground and openpit workings where extraction has been success-
fully adapted to the pronounced folding. It forms part of mineralised zone
1.8 km in strike-length and 60 to 100 m thick which has been tested by
drilling through 250 m of vertical interval.
    Deposits closely similar to Foss in composition and lithostratigraphical po-
sition are preserved 45 km to the WSW at Loch Lyon (*16*) and 45 km north-

east at Coire Loch Kander (*10*) along the regional strike of the Ben Eagach Schist and its lateral equivalent in the north-east, the Glas Maol Schist. On Beinn Heasgarnich and in Allt Chall south of Loch Lyon, a thin (1–2 mm) layer of calcareous schist containing variable amounts of baryte, barium silicates (barian muscovite, celsian and hyalophane), sphalerite and galena is conspicuously exposed a few metres beneath the base of the Ben Lawers Schist. The horizon extends for 3 km on the upper limb of the recumbent Ben Lui fold and on the lower limb it is traceable for at least 1 km. At Allt an Loch near Coire Loch Kander, bedded baryte-quartz rock, 4.5 m thick in a drill intersection, can be followed for 0.7 km in thin graphitic schist lying directly beneath amphibolite of the Ben Lawers Schist (Forty et al., 1993). In the corrie itself sphalerite, galena and iron sulphides are common in a 15 m-thick band of barian quartzite which displays a mineralogy unique to Britain, including armenite [$BaCa_2Al_3(Al_3Si_9O_{30})2H_2O$], hyalophane, baryte, salitic pyroxene and tremolite-actinolite, attributable at least in part to the contact metamorphic effects of a Silurian diorite stock. This intrusion also postdates a thin baryte-galena vein which is therefore among the oldest recorded in Europe.

The stratabound baryte and related deposits in the Ben Eagach Schist are of synsedimentary-exhalative origin. The sulphur isotope value of Aberfeldy baryte ($\delta^{34}S + 33\%$) is close to that of late Neoproterozoic seawater and most of the associated sulphides are also isotopically heavy (about + 24%), indicating sulphur of hydrothermal origin. Metalliferous brines are believed to have been exhaled into small rifted basins floored by carbonaceous mud.

Sulphide concentrations unaccompanied by barium minerals also occur in the Ben Eagach Schist, notably near Dericambus in Glen Lyon (*14*) where sphalerite-galena-pyrrhotite-pyrite assemblages can be traced along 300 m of strike and across strike for 200 m in quartzites of the Ben Eagach Schist transitional with the underlying Carn Mairg Quartzite. Remobilisation of the sulphides into late metamorphic quartz segregations is unusually common.

In the upper part of the Ben Lawers Schist, which overlies the Ben Eagach Schist, a thick zone of weakly cupriferous pyrite enrichment running southwestwards from Glenshee to Tyndrum (Smith et al., 1984) is regarded as volcanogenic in origin (Scott et al., 1991). The Ardrishaig Phyllite, probably a lateral equivalent of the Ben Lawers Schist, hosts nickeliferous sulphide at Coille-braghad (*24*), originally described as a metasomatic replacement deposit (Wilson and Flett, 1921), and Cu-Pb-Zn sulphides at McPhun's Cairn (*25*). Lying above the Ben Lawers Schist in the Tyndrum district, the Ben Challum Quartzite hosts two developments of low-grade base metal sulphides near Auchtertyre (*20*). The lower one extends for 8 km and is about 80 m thick; the upper one is 10 to 20 m thick on Ben Challum and incorporates 1 m-thick units containing 3% Zn which are considered to be siliceous exhalites. To the south-west, on Creag Bhocan, chalcopyrite and pyrite occur at the same level in the Ben Challum Quartzite which is overlain by a carbonate-fuchsite horizon at the base of the Ben Lui Schist Formation in the Crinan Subgroup. At Meall Mor (*27*), the Erins Quartzite, which lies in the same subgroup, carries a thick pyritic zone containing chalcopyrite and other sulphides. Within this zone, copper mineralisation was worked in the past at

Abhainn Srathain from the margins of metabasaltic bodies, suggesting that the mineralisation was partly volcanogenic in orgin.

Vein deposits

Wilson and Flett (1921) document more than 50 mines and trials on metalliferous veins within the Grampian Highlands. With few exceptions, the deposits occur in the south-west of the region and in Dalradian rocks. Past metal production was principally from veins of the Tyndrum district (the Hard, Clay, Eas Anie, Crom Allt and Meall Odhar veins) and is estimated to total 6000 t of lead, 1 t of silver, 100 t of zinc and 100 t of copper. Modern exploration has located gold-bearing structures, notably at Cononish (*22*) near Tyndrum and at Calliachar Burn (*15*) near Aberfeldy, and some small baryte and base metal veins (Gallagher, 1991b).

The celebrated Tyndrum veins occupy fractures running subparallel to and on the north-west side of the Tyndrum Fault, which trends around 040°, (Patrick, 1985, fig. 1). Host rocks are mainly psammites of the Appin Group transitional with psammites of the underlying Grampian Group. Younger rocks of the Argyll Group lying south-east of the fault rarely contain metalliferous veins (Smith et al., 1984).

The Cononish gold-silver vein is 0.2 to 6 m thick and has been proved over 0.7 km of strike-length and up to 0.5 km below surface within a fault structure traceable for 2.5 km. Ore reserves of 0.75 Mt grading 10 g/t Au and 43 g/t Ag have been defined. Quartz veins containing visible gold were first located at the near-surface intersection of the fault structure with the Ben Eagach Schist, lying above the much older metasedimentary rocks at a slide junction (Gallagher, 1991b, fig. 16.18). The fault trends 050°, terminating north-eastwards against a barren quartz vein, the Mother Reef. Pyrite is the dominant sulphide of the Cononish structure, occurring with fine-grained galena in an early phase of white quartz and in a later phase of mottled quartz formed as a result of brecciation. Gold occurs in the pyrite and galena, usually as particles less than 20 μm in size. Chalcopyrite, sphalerite and minor amounts of haematite, covellite, tellurides and native silver are also present. Gold values are lower in a phase of grey pyritic quartz and absent from cross-cutting white quartz.

The Cononish gold vein is best developed where the fault intersects psammitic rocks, which display alteration up to 15 m from the vein contact. An outer chloritised zone is succeeded by a sericitised zone and, within 2 m of the vein, intensely altered and reddened psammite, which has been haematitised, silicified and pyritised, can carry economic gold values. The presence of red psammite clasts within the early phase of white quartz signifies alteration and brecciation of wallrock prior to vein formation. The gold-bearing structure is cut by a late Carboniferous basic dyke and by the Eas Anie Vein (Plate 24). This vein is characterised by coarse-grained galena, calcite and baryte with only minor pyrite; it is not gold-bearing, nor are the Hard and other veins mined for lead in the past.

At Tyndrum Mine *(21)* the Hard Vein dips steeply south-east to terminate against the Tyndrum Fault, occupied by the later Clay Vein. Levels 145 to 365 m in length were driven through 230 m of vertical interval along veins up to 6 m thick containing coarse-grained galena and sphalerite, together with chalcopyrite and a little pyrite. Massive quartz is the main gangue mineral, accom-

Plate 24   Eas Anie Vein, Cononish gold prospect, Tyndrum, Perthshire; a complex breccia vein (right) cut by shears (centre). Gold occurs with pyrite and galena scattered through the vein quartz and in altered country rock (MNS 5694).

panied by local concentrations of calcite and baryte. Ore textures are typical of growth into 'open space', namely vuggy breccias and banded veins. Silver- and cadmium-rich tetrahedrites occur in the massive galena. Minor amounts of uraninite have been recorded from a cross-course vein.

Lead was mined on Islay from at least 12 veins of diverse trend (035°–135°) cutting both the Islay and the older Ballygrant limestones of the Dalradian Appin Group (Gallagher and Young, 1993). The principal veins at Mulreesh (*28*) were up to 1 m thick and 250 m in strike-length, containing chalcopyrite, pyrite and sphalerite as well as galena, set in a gangue of calcite, dolomite and quartz. A vein at nearby Kilsleven was first worked for copper, and native silver is reported from Gartness. Modern drilling at Mulreesh identified narrow sub-vertical fault breccias containing up to 3.1% Zn, 0.1% Pb and 15 g/t Ag, suggesting that the mineralisation is zinc-dominant and silver-enriched. One vein is cut by a NW-trending basalt dyke, presumably of Palaeogene age.

East of Loch Tay, at Corrie Buie (*19*), an outlier of the Loch Tay Limestone on the inverted limb of the Tay Nappe is cut by narrow quartz-sulphide veins trending 160°. Silver-rich galena, minor sphalerite, chalcopyrite and iron sulphides, and rare native bismuth and gold occur in a gangue of quartz and lesser carbonate. Adits uncovered in recent gold exploration follow veins for at least 200 m. Precious metals are also reported from chalcopyrite-rich ore mined from a quartz-calcite vein trending north-west in a metabasaltic body in Argyll Group rocks at Kilmartin (*26*).

Gold-bearing structures at Calliachar Burn south-west of Aberfeldy (*15*) trend 150° in garnetiferous mica-schist, psammitic schist and hornblende-schist of the Southern Highland Group. Although thin (0.1–0.5 m), the structures can contain up to 350 g/t Au, including visible gold in quartz veinlets and in goethite. Mineralisation was accompanied by movement on the structures, resulting in deformation of the sulphides. Hornblende-schist wallrocks are bleached over a few centimetres with formation of ferroan dolomite. In the hypogene sulphide assemblage (Table 4), electrum (50–65 wt% Au) forms clusters of inclusions on fractures in pyrite and larger grains at pyrite–galena boundaries. Sphalerite is replaced in part by chalcopyrite. At surface, pyrite is oxidised to goethite, limonite and jarosite, and galena to anglesite and pyromorphite. Chalcopyrite breaks down to covellite and native copper, and arsenopyrite to scorodite. The nearby Urlar Burn veins (Wilson and Flett, 1921) are also NW-trending, at right angles to the regional Caledonide grain. Galena from the veins contains the tellurides altaite, hessite and coloradoite.

Other types of mineralisation

Vein and Dalradian stratabound deposits apart, at least six other types of mineralisation are recognisable in the Grampian Highlands (C to H on Table 4). Iron and manganese were formerly worked at the Lecht (*7*) from goethite and cryptomelane deposits in post-Dalradian explosive-intrusion breccias. Associated minerals are todorokite, cacoxenite and lithiophorite which contains up to 145 ppm Th. Drilling in the 1980s indicated a resource of 0.25 Mt grading 7% MnO and high levels of barium and of zinc (2.5% Zn over 11 m) in the breccias. Stratiform manganese-rich garnet intersected in Argyll Group pelitic metasedimentary rocks adjacent to the worked breccia deposits represents a potential source of manganese.

The basic and ultrabasic rocks of the region contain nickel-copper sulphides, elevated values of the platinum-group elements and ilmenite, but only at Corrycharmaig (*18*) has there been any exploitation, in this instance of chromite in altered serpentinite which contains considerable quantities of magnesite. The serpentinite at Corrycharmaig lies at the base of the Ben Lui Schist (Crinan Subgroup), in a similar position to the chromian mineral horizon near Auchtertyre (*20*). Complex platinum-bearing grains intergrown with nickel arsenide are reported from another serpentinite at Kelman Hill (*5*). Syn- to late-tectonic basic and ultramafic bodies in Aberdeenshire have been systematically explored and Ni-Cu mineralisation located in the contact zone of the Huntly–Knock mass at Littlemill (*2*) and in xenolithic norite of the Arthrath–Dudwick intrusion (*4*). Massive pyrrhotite, accompanied by pentlandite and chalcopyrite, is up to 20 m thick in the Littlemill ore zone and can contain up to 3% Ni, 6.5% Cu together with traces of platinum, palladium and gold. Sulphides are also concentrated in olivine-bearing cumulates and in graphitic, pyroxenic pegmatites within the Huntly–Knock mass.

Metalliferous minerals are known from many of the Caledonian granitic bodies in the Grampian Highlands but none of the occurrences are of economic significance. Some copper was extracted last century from a diorite mass at Tomnadashan on the east side of Loch Tay (*17*). Sulphides are concentrated at the edges of lenses of granodiorite and granite within the mass; calcite, quartz, siderite and baryte are associated minerals. A rich mineralogy has recently been described from an outcrop of zinnwaldite-bearing granite within the Coilacreich pluton at Gairnshiel, west of Ballater (*8*). Wolframite, cassiterite and scheelite are found in quartz veins and their silicified wallrocks. A second assemblage infilling cavities in the veins and dissolution zones in the granite comprises either molybdenite or sphalerite with pyrite, chalcopyrite and cassiterite, together with rare stannite and argentiferous cosalite.

The porphyry and breccia complex at Lagalochan near Kilmelford (*23*) also displays a wide range of mineralisation. An early hypogene stage of Cu-Au-(Mo) mineralisation is associated with breccias and granodiorite–diorite intrusives. Electrum (55–94 wt% Au) forms irregular inclusions up to 100 μm in size in pyrite and chalcopyrite. Subsequently, a Pb-Zn-Ag-Au-As-Sb suite developed as veins and disseminations in shear zones cutting porphyry breccia. Locally high silver values are related to inclusions of argentiferous tetrahedrite, lead-antimony sulphosalts and native silver in sphalerite and in galena which is itself non-argentian. A third phase of mineralisation is represented by Pb-Zn-Ag carbonate veins. Intense sericite-quartz-pyrite and carbonate alteration is associated with the mineralisation.

Small but distinctive amounts of Au, As and Sb occur in altered lavas and plant-bearing siliceous sinter at the faulted western margin of the Rhynie outlier of Lower Devonian rocks (*6*). Andesitic lavas and tuffs are intensely altered to K-feldspar, silica, mica, chlorite and pyrite. The alteration and metalliferous enrichment are ascribed to epithermal hot spring activity related to a nearby volcanic vent which developed in the final stages of Caledonian magmatism. The Rhynie deposit is the only example of epithermal gold concentration known in Britain.

Disseminated baryte has been worked from Middle Old Red Sandstone sedimentary rocks resting on a late Caledonian felsite body at Balfreish (*3*). The baryte is irregularly developed in the matrix of a breccio-conglomerate, most probably as a diagenetic constituent, and also forms small lenses and veins. The boundary of the Orcadian Basin in the north of the Grampian Highlands is an important mineralisation control. Younger arenaceous sediments in the Permo-Triassic sequence of the Elgin area commonly contain minor amounts of galena, haematite, fluorite and baryte. The Stotfield Cherty Rock, interpreted as a calcrete horizon, was worked in the past for galena at (*1*).

CONSTRUCTION AND INDUSTRIAL MINERALS

Brick clay

Superficial deposits have been worked for making bricks, tiles and pipes in widespread areas of coastal Aberdeenshire, Banffshire and Morayshire. The raw materials used include boulder clay, glaciolacustrine deposits, raised marine deposits and alluvium, but a combination of relatively high costs of firing, ration-alisation of the industry and remoteness from major markets in central Scotland have resulted in the closure of all the brickworks in the Grampian region.

Building stone

The demand for dimension stone for building virtually disappeared in the 1950s, but a general dissatisfaction with the aesthetics and lasting qualities of concrete, together with reawakening of architects to the visual appeal of natural stone, has led to an upsurge in the demand for granite and sandstone blocks. Until recently the main requirement has been for facing or cladding and for selected replacement of stone in restoration projects, but a market has re-emerged for load-bearing cube stone.

Good-quality freestone is worked from quarries in the Permian sandstone near Elgin. In Aberdeenshire, the Caledonian granites were the best and most highly exploited building stones but extraction of cube stone is very small and the Aberdeen stone-polishing industry now (1993) uses mainly imported materials.

Crushed rock aggregate

The overriding property determining the suitability of a rock for making aggregate is its crushing strength, which is mostly governed by the nature of the rock, but can be affected quite significantly by the state of weathering and alteration. The mechanical properties, other than crushing strength, that are critical in more specific applications are resistance to polishing and wear, which governs the suitability of rock chippings for road surfacing, and drying shrinkage, which governs the versatility of aggregate for concreting purposes.

The Grampian Highlands are well endowed with resources of rock suitable for producing aggregate (Smith, 1989a). Granites are potentially suitable for most road-making and concreting applications and some of those occurring along the western seaboard have good export potential. Quarries are concentrated in the Aberdeen area where the granites are crushed to produce aggregate and reconstituted 'Fyfestone' (Plate 25). Gabbro is worked for aggregate at Balmedie and Pitcaple to the north of Aberdeen. Most other quarries in the Grampians exploit Dalradian psammites.

Diatomite

Small deposits of diatomite occur at sites formerly occupied by lochs, but the only major potential source is the Muir of Dinnet, between Ballater and Aboyne in Aberdeenshire, where the diatomite was recognised in 1882 and was exploited spasmodically on a small scale until 1918 (Gould, in press). The fine-grained nature of diatomite, together with the open structure and shape of the diatom frustules, renders it a useful filtering medium, which is its main commercial use.

Fireclay

A reddish lateritic clay occurs at the base of the Coal Measures in the Carboniferous of the Machrihanish coalfield. Although sometimes referred to as

Plate 25 Kemnay Quarry, Aberdeenshire.
The Kemnay granite is worked for aggregate, dimension stone and cladding. In the
background is the granite mass of Bennachie (D 4337).

a bauxitic clay, it contains an excess of silica and is not truly bauxitic. Although the deposit is a potential resource of high-alumina fireclay, it is un-
likely to be extracted in the foreseeable future because present production
fully satisfies, and at times exceeds, demand and there are more readily acces-
sible deposits in the Midland Valley, closer to markets for the material.

Limestone and dolomite

The Dalradian Supergroup contains four major developments of metamorphosed carbonate rock, some of which extend over strike-lengths of about
100 km and are up to several tens of metres thick. They show considerable
variation in outcrop dimension, purity and dolomite content. The Blair
Atholl Subgroup limestones are, on average, more pure than the other Dalradian carbonate formations.

Limestone is currently produced at five quarries, mostly for agricultural use
but also, to a limited extent, for concrete aggregate (Grout and Smith,
1989a). Limestones of the Ballachulish Subgroup are worked at Parkmore,
Dufftown and at Torlundy, to the north of Fort William. A thick limestone in
the Blair Atholl Subgroup is worked at Shierglas Quarry near Blair Atholl and
the Boyne Castle Limestone of the Tayvallich Subgroup is worked on the
coast between Banff and Portsoy. The Loch Tay Limestone is exploited at Cal-
liburn, near Campbeltown.

Dolomitic limestone, with potential uses in the manufacture of rock wool, was formerly worked in the Appin Phyllite and Limestone Formation at Durar, Appin, but extraction ceased because of variation in quality. Dolomitic limestone also occurs at this stratigraphical level elsewhere in the Grampian Highlands.

Peat

Deposits of peat are both widespread and abundant throughout the region, especially the 'hill peat' type on the higher uplands, although there are also considerable areas of 'basin peat' on the plateau districts of Aberdeenshire.

Utilisation of the peat as fuel, generally by individual users, has not been important in the present century, but persists on a small scale in some areas and has a specialised application in the distillation of whisky. Commercial production of peat for fuel and horticultural usage has been undertaken in northern Aberdeenshire and at Moy, south-east of Inverness, but this kind of exploitation is generally limited by the inaccessibility of large-scale deposits that would respond to mechanical extraction, by the costs of transportation and by concerns regarding conservation of the environment.

Sand and gravel

Haulage costs make up a substantial proportion of the delivered price of sand and gravel. Consequently, despite rationalisation of the industry, sand and gravel is still exploited from many localities scattered widely across the area. It is mainly derived from 'dry' workings in Quaternary glaciofluvial, alluvial, beach and sand dune deposits for use in the manufacture of concrete and concrete products. It is traditionally preferred to crushed rock aggregate because of its lower cost and better workability. More than half of the sand and gravel production is used in the manufacture of concrete. About 26 per cent of production is used in road sub-bases and embankments, a further 11 per cent is used as 'mortaring' and 'sharp' sand in the building industry, and fine-grained sand for mixing with bitumen to make asphalt accounts for another 11 per cent or so of production (BGS, 1991).

The aggregates industry is very important to the local economy, but extraction of sand and gravel, in particular, is placing increasing pressure on planning authorities to balance necessary mineral exploitation with competing uses of land and with conservation of the environment (Merritt, 1992). Descriptive accounts of sand and gravel resources of the whole of the Grampian Highlands are contained in the Institute of Geological Sciences Reports 77/2, 77/6, 77/8, 77/9 and 78/8. In addition more detailed assessments of the sand and gravel resources for the area between Peterhead and Stonehaven are available in BGS Mineral Assessment Reports 58, 76, 146, 148 and 149 and for part of Lower Speyside (Report 41).

Serpentine

Serpentinite intrusions occur in association with the Dalradian succession in Banffshire, Aberdeenshire, Angus and Perthshire, and in the Cambro-Ordovician Highland Border Complex. Besides its traditional use as an ornamental stone, for which it was worked in the past, notably at Portsoy, this rock has some potential as a source material for refractory brick manufacture, but much of the Highland Border material is carbonated or otherwise impure.

Silica rock and silica sand

Some of the most pure Dalradian quartzites were thought to have potential for silica-brick manufacture in lump form, but only the Binnein Quartzite of the Loch Leven–Glencoe area has proved to be of sufficient purity (Smith, 1989b). The Appin Quartzite near Kentallen was formerly quarried for use in grinding-tubs in the pottery industry. Quartz veins occur in many parts of the Grampian Highlands, but are generally thin and impersistent. A thick vein of high-purity quartz occurs near Dalwhinnie and may have potential for use in the preparation of a decorative aggregate.

A siliceous sandstone of considerable purity, belonging to the Limestone Coal Formation, occurs in the Machrihanish Coalfield, where it was mined, mainly as a source of moulding sand.

Slate

Reserves of slate are abundant in the North-east and South-west Grampian Highlands and along the Highland Border, but workings have been discontinued as a result of competition from Welsh slates, which are of superior quality, and from cheaper manufactured roofing materials.

Talc

Steatite (massive, fine-grained talc), soapstone and potstone (impure talcose rock), are normally found as alteration products of serpentinised ultramafic igneous rocks that have been subjected to intense shearing or deformation. The main resources to have been exploited have been in crush zones of the Portsoy Serpentinite but small quantities have been recovered from numerous serpentinite bodies occurring between Portsoy and Blairgowrie. Talc is generally located along shear planes in these bodies, whereas part of a small intrusion at Corrycharmaig, near Killin has been altered to an aggregate of talc and breunnerite. In the Inellan and Toward areas of Argyllshire, on the western shore of the Firth of Clyde, lenticular veins of talc are present in a belt of serpentinite lying between two faults in the Highland Boundary Fault Zone. Overall, the resources of talc in the Grampian Highlands are small, scattered and restricted to relatively low-value industrial grades (Grout and Smith, 1989b).

References

Most of the references listed below are held in the Library of the British Geological Survey at Keyworth, Nottingham. Copies of the references can be purchased subject to the current copyright legislation.

AFTALION, M, VAN BREEMEN, O, and BOWES, D R. 1984. Age constraints on basement of the Midland Valley of Scotland. *Transactions of the Royal Society of Edinburgh: Earth Sciences,* Vol. 75, 53–64.

ALLAN, W C. 1970. The Morven–Cabrach basic intrusion. *Scottish Journal of Geology,* Vol. 6, 53–72.

ALLISON, A. 1936. The Tertiary dykes of the Craignish area, Argyll. *Geological Magazine,* Vol. 73, 73–87.

— 1941. Loch Awe succession and tectonics–Kilmartin–Tayvallich–Danna. *Quarterly Journal of the Geological Society of London,* Vol. 96, 423–449.

AMOS, B J. 1960. The geology of the Bowmore district, Islay. Unpublished PhD thesis, University of London.

ANDERSON, E M. 1923. The geology of the schists of the Schiehallion district. *Quarterly Journal of the Geological Society of London,* Vol. 79, 423–445.

ANDERSON, J G C. 1935. The marginal intrusions of Ben Nevis, the Coille Lianachain complex and the Ben Nevis dyke swarm. *Transactions of the Geological Society of Glasgow,* Vol. 19, 225–269.

— 1937. The Etive granite complex. *Quarterly Journal of the Geological Society of London,* Vol. 93, 487–532.

— 1942. The stratigraphical order of the Dalradian schists near the Highland Border in Angus and Kincardine. *Transactions of the Geological Society of Glasgow,* Vol. 20, 223–237.

— 1947a. The Kinlochlaggan Syncline, southern Inverness-shire. *Transactions of the Geological Society of Glasgow,* Vol. 21, 97–115.

— 1947b. The geology of the Highland Border, Stonehaven to Arran. *Transactions of the Royal Society of Edinburgh,* Vol. 61, 479–515.

— 1948. Stratigraphic nomenclature of Scottish metamorphic rocks. *Geological Magazine,* Vol. 85, 89–96.

— 1953. The stratigraphical succession and correlation of the late pre-Cambrian and Cambrian of Scotland and Ireland. *Report of the 19th International Geological Congress,* Vol. 1, 9–19.

— 1956. Moinian and Dalradian rocks between Glen Roy and the Monadhliath Mountains, Inverness-shire. *Transactions of the Royal Society of Edinburgh,* Vol. 63, 15–36.

ANDERTON, R. 1975. Tidal flat and shallow marine sediments from the Craignish Phyllites, Middle Dalradian, Argyll, Scotland. *Geological Magazine,* Vol. 112, 337–340.

— 1976.   Tidal shelf sedimentation: an example from the Scottish Dalradian. *Sedimentology*, Vol. 23, 429 – 458.

— 1979.   Slopes, submarine fans and syn-depositional faults: sedimentology of parts of the Middle and Upper Dalradian in the SW Highlands of Scotland.   483–488 *in* The Caledonides of the British Isles — reviewed. HARRIS, A L, HOLLAND, C H, and LEAKE, B E (editors). *Special Publication of the Geological Society of London*, No. 8.

— 1980.   Did Iapetus start to open during the Cambrian? *Nature, London*, Vol 286, 706–708.

— 1982.   Dalradian deposition and the late Precambrian–Cambrian history of the N Atlantic region: a review of the early evolution of the Iapetus Ocean.   *Journal of the Geological Society of London*, Vol. 139, 421–431.

— 1985.   Sedimentation and tectonics in the Scottish Dalradian.   *Scottish Journal of Geology*, Vol. 21, 407–436

— 1988.   Dalradian slides and basin development: a radical interpretation of stratigraphy and structure in the SW and Central Highlands of Scotland.   *Journal of the Geological Society of London*, Vol. 145, 669–678.

ANDREWS, I J, and six others, 1990.   *United Kingdom offshore regional report: the geology of the Moray Firth*.   (London: HMSO for the British Geological Survey.)

ARCHER, R. 1978.   The Old Red Sandstone outliers of Gamrie and Rhynie, Aberdeenshire.   Unpublished PhD thesis, University of Newcastle upon Tyne.

ARMSTRONG, M, and PATERSON, I B. 1970.   The Lower Old Red Sandstone of the Strathmore Region.   *Report of the Institute of Geological Sciences*, No. 70/12.

— — and BROWNE M A E. 1985.   Geology of the Perth and Dundee district. *Memoir of the British Geological Survey*, Sheets 48W, 48E, 49 (Scotland).

ASHCROFT, W A, and BOYD, R. 1976.   The Belhelvie mafic igneous intrusion, Aberdeenshire — a reinvestigation.   *Scottish Journal of Geology*, Vol. 12, 1–14.

— KNELLER, B C, LESLIE, A G, and MUNRO, M. 1984.   Major shear zones and autochthonous Dalradian in the northeast Scottish Caledonides.   *Nature, London*, Vol. 310, 760–762.

— and MUNRO, M. 1978.   The structure of the eastern part of the Insch Mafic Intrusion, Aberdeenshire.   *Scottish Journal of Geology*, Vol. 14, 55–79.

ASHWORTH, J R. 1975.   The sillimanite zones of the Huntly–Portsoy area in the north-east Dalradian, Scotland.   *Geological Magazine*, Vol. 112, 113–224.

— 1976.   Petrogenesis of migmatites in the Huntly–Portsoy area, north-east Scotland. *Mineralogical Magazine*, Vol. 40, 661–682.

— 1979.   Comparative petrography of deformed and undeformed migmatites from the Grampian Highlands of Scotland.   *Geological Magazine*, Vol. 116, 445–456.

— 1985.   Introduction. 1–35 in *Migmatites*. ASHWORTH, J R (editor).   (Glasgow and London: Blackie.)

— and McLELLAN, E L. 1985.   Textures.   180–203 in *Migmatites*. ASHWORTH, J R (editor).   (Glasgow and London: Blackie.)

ATHERTON, M P. 1977.   The metamorphism of the Dalradian rocks of Scotland. *Scottish Journal of Geology*, Vol. 13, 331–370.

AUTON, C A. 1992.   Scottish landform examples—6: The Flemington Eskers. *Scottish Geographical Magazine*, Vol. 108, 190–196.

— FIRTH, C R, and MERRITT, J W. 1990.   *Beauly to Nairn: field guide*.   (Cambridge: Quaternary Research Association.)

BAILEY, E B. 1910.   Recumbent folds in the schists of the Scottish Highlands. *Quarterly Journal of the Geological Society of London*, Vol. 66, 586–620.

— 1917.   The Islay Anticline (Inner Hebrides). *Quarterly Journal of the Geological Society of London*, Vol. 72, 132–164.

— 1922.   The structure of the South-west Highlands of Scotland. *Quarterly Journal of the Geological Society of London*, Vol. 78, 82–127.

— 1923.   The metamorphism of the south-western Highlands of Scotland. *Geological Magazine*, Vol. 60, 317–331.

— 1925.   Perthshire tectonics: Loch Tummel, Blair Atholl and Glen Shee. *Transactions of the Royal Society of Edinburgh*, Vol. 53, 671–698.

— 1934.   West Highland tectonics: Loch Leven to Glen Roy. *Quarterly Journal of the Geological Society of London*, Vol. 90, 462–523.

— 1938.   Eddies in mountain structure. *Quarterly Journal of the Geological Society of London*, Vol. 94, 607–625.

— 1960.   The geology of Ben Nevis and Glencoe and the surrounding country (2nd edition). *Memoir of the Geological Survey, Scotland,* Sheet 53 (Scotland).

— and McCALLIEN, W J. 1937.   Perthshire tectonics: Schiehallion to Glen Lyon. *Transactions of the Royal Society of Edinburgh*, Vol. 59, 79–118.

— and MACGREGOR, M. 1912.   The Glen Orchy Anticline, Argyllshire. *Quarterly Journal of the Geological Society of London*, Vol. 68, 164–179.

BAKER, A J. 1985.   Pressures and temperatures of metamorphism in the eastern Dalradian. *Journal of the Geological Society of London*, Vol. 142, 137–148.

— 1987.   Models for the tectonothermal evolution of the eastern Dalradian of Scotland. *Journal of Metamorphic Geology*, Vol. 5, 101–118.

BALDWIN, C T, and JOHNSON, H D. 1977.   The Dalradian rocks of Lunga, Luing and Shuna. *Scottish Journal of Geology*, Vol. 13, 143–154.

BALLANTYNE, C K. 1984.   The Late Devensian Periglaciation of Upland Scotland. *Quaternary Science Reviews*, Vol. 3, p. 311–344.

— 1991.   Holocene geomorphic activity in the Scottish Highlands. *Scottish Geographical Magazine*, Vol. 107, 84–98.

— and HARRIS, C. 1994.   *The periglaciation of Great Britain.* (Cambridge: Cambridge Press.)

BAMFORD, D. 1979.   Seismic constraints on the deep geology of the Caledonides of northern Britain.   93–96 *in* The Caledonides of the British Isles—reviewed. HARRIS, A L, HOLLAND, C G and LEAKE, B E. *Special Publication of the Geological Society of London*, No. 8.

— NUNN, K, PROEDEHL, C, and JACOB, B. 1978.   LISPB-IV. Crustal studies of northern Britain. *Geophysical Journal of the Royal Astronomical Society*, Vol. 54, 43–60.

BARNETT, G W T. 1959.   Lead in Islay.   65–76, 84 in *The future of non-ferrous mining in Great Britain and Ireland.* (London: Institution of Mining and Metallurgy.)

BARR, D. 1985.   Migmatites in the Moines.   225–264 in *Migmatites.* ASHWORTH, J R (editor). (Glasgow and London: Blackie.) .

BARROW, G. 1893.   On an intrusion of muscovite-biotite gneiss in the southeast Highlands of Scotland and its accompanying metamorphism. *Quarterly Journal of the Geological Society of London*, Vol. 19, 33–58.

— 1901.   On the occurrence of Silurian (?) rocks in Forfarshire and Kincardineshire along the Eastern Border of the Highlands. *Quarterly Journal of the Geological Society of London*, Vol. 57, 328–345.

— 1904. Moine gneisses of the east central Highlands and their position in the Highland sequence. *Quarterly Journal of the Geological Society of London*, Vol. 60, 400–444.

— 1912. On the geology of Lower Dee-side and the southern Highland Border. *Proceedings of the Geologists' Association*, Vol. 23, 275–290.

— and CRAIG, E H C. 1912. The geology of the districts of Braemar, Ballater and Glen Clova. *Memoir of the Geological Survey, Scotland*, Sheet 65 (Scotland).

BATCHELOR, R A. 1987. Geochemical and petrological characteristics of the Etive granitoid complex, Argyll. *Scottish Journal of Geology*, Vol. 23, 227–249.

BAXTER, A N. 1987. Petrochemistry of late Palaeozoic alkali lamprophyre dykes from N Scotland. *Transactions of the Royal Society of Edinburgh: Earth Sciences*, Vol. 77, 267–277.

— and MITCHELL, J G. 1984. Camptonite–monchiquite dyke swarms of Northern Scotland; age relationships and their implications. *Scottish Journal of Geology*, Vol. 20, 297–308.

BEDDOE-STEPHENS, B. 1990. Pressures and temperatures of Dalradian metamorphism and the andalusite–kyanite transformation in the north-east Grampians. *Scottish Journal of Geology*, Vol. 26, 3–14.

BELL, K. 1968. Age relations and provenance of the Dalradian Series of Scotland. *Bulletin of the Geological Society of America*, Vol. 79, 1167–1194.

BENTON, M. 1977. *The Elgin reptiles*. Moray Society—Elgin Museum. (Aberdeen: Aberdeen People's Press.)

BENTON, M J, and WALKER, A D. 1985. Palaeonology, taphonomy and dating of Permo-Triassic reptiles from Elgin, north-east Scotland. *Palaeontology*, Vol. 28, 207–234.

BENTLEY, M R. 1988. The Colonsay Group. 119–130 in *Later Proterozoic stratigraphy of the Northern Atlantic regions*. WINCHESTER, J A (editor). (Glasgow and London: Blackie.)

— MALTMAN, A J, and FITCHES, W R. 1988. Colonsay and Islay: a suspect terrane within the Scottish Caledonides. *Geology*, Vol. 16, 26–28.

BERRIDGE, N G, and IVEMEY-COOK, H C. 1967. The geology of a borehole at Lossiemouth, Morayshire. *Bulletin of the Geological Survey of Great Britain*, No. 27, 155–169.

BISSET, C B. 1934. A contribution to the study of some granites near Aberdeen. *Transactions of the Edinburgh Geological Society*, Vol. 13, 72–88.

BLACK, G P, and MACKENZIE, D H. 1957. Supposed unconformities in the Old Red Sandstone of western Moray. *Geological Magazine*, Vol. 94, 170.

BLUCK, B J. 1985. The Scottish paratectonic Caledonides. *Scottish Journal of Geology*, Vol. 21, 437–464.

BLYTH, F G H. 1969. Structures in the southern part of the Cabrach igneous area, Banffshire. *Proceedings of the Geologists' Association*, Vol. 80, 63–79.

BOOTH, J E. 1984. Structural, stratigraphic and metamorphic studies in the SE Dalradian Highlands. Unpublished PhD thesis, University of Edinburgh.

BORRADAILE, G J. 1973. Dalradian structure and stratigraphy of the northern Loch Awe district, Argyllshire. *Transactions of the Royal Society of Edinburgh*, Vol. 69, 1–21.

— 1979. Pre-tectonic reconstruction of the Islay anticline: implications for the depositional history of Dalradian rocks in the SW Highlands. 229–238 *in* The Caledonides of the British Isles–reviewed. HARRIS, A L, HOLLAND, C H, and LEAKE, B E (editors). *Special Publication of the Geological Society of London*, No. 8.

— and JOHNSON, H D. 1973. Finite strain estimates from the Dalradian Dolomitic Formation, Islay, Argyll, Scotland. *Tectonophysics*, Vol. 18, 249–259.

BOULENGER, G.A. 1903.   Some reptilian remains from the Trias of Elgin. *Philosophical Transactions of the Royal Society*, Vol. 196, 175.

BOULTON, G S, PEACOCK, J D, and SUTHERLAND, D G. 1991.   Quaternary.   503–543 in *Geology of Scotland* (3rd edition). G Y CRAIG (editor). (London: The Geological Society of London.)

BOWES, D R, and CONVERY, H J E. 1966.   The composition of some Ben Ledi grits and its bearing on the origin of albite schists in the south-west Highlands.   *Scottish Journal of Geology*, Vol. 2, 67–75.

— and WRIGHT, A E. 1967.   The explosion-breccia pipes near Kentallen, Scotland, and their geological setting.   *Transactions of the Royal Society of Edinburgh*, Vol. 67, 109–143.

— — 1973.   Early phases of Caledonian deformation in the Dalradian of the Ballachulish district, Argyll.   *Geological Journal*, Vol. 8, 333–344.

BOYD, R, and MUNRO, M. 1978.   Deformation of the Belhelvie mass, Aberdeenshire. *Scottish Journal of Geology*, Vol. 14, 29–44.

BRADBURY, H J. 1985.   The Caledonian metamorphic core: an Alpine model.   *Journal of the Geological Society of London*, Vol. 142, 129–136.

— HARRIS, A L, and SMITH, R A. 1979.   Geometry and emplacement of nappes in the Central Scottish Highlands.   213–220 *in* The Caledonides of the British Isles— reviewed. HARRIS, A L, HOLLAND, C H, and LEAKE, B E (editors). *Special Publication of the Geological Society of London*, No. 8.

— SMITH, R A, and HARRIS, A L. 1976.   "Older" granites as time-markers in Dalradian evolution.   *Journal of the Geological Society of London*, Vol. 132, 677–684.

BRAND, P J. 1977.   The fauna and distribution of the Queenslie Marine Band (Westphalian) in Scotland.   *Report of the Institute of Geological Sciences*, No. 77/18.

BRITISH GEOLOGICAL SURVEY. 1991.   *United Kingdom Minerals Yearbook 1990.* (Keyworth, Nottingham: British Geological Survey.)

BROWN, G C, and LOCKE, C A. 1979.   Space-time variations in British Caledonian granites: some geophysical correlations.   *Earth and Planetary Science Letters*, Vol. 45, 69–79.

BROWN, I M. 1993.   Pattern of deglaciation of the last (Late Devensian) Scottish ice sheet: evidence from ice-marginal deposits in the Dee valley, north-east Scotland. *Journal of Quaternary Science*, Vol. 8, 235–250.

BROWN, P E. 1991.   Caledonian and earlier magmatism.   229–295 in *Geology of Scotland* (3rd edition). CRAIG, G Y (editor). (London: The Geological Society of London.)

— MILLER, J A, GRASTY, R L, and FRASER, W E. 1965.   Potassium–argon ages of some Aberdeenshire granites and gabbros.   *Nature, London*, Vol. 207, 1287–1288.

BUCHAN, S. 1932.   On some dykes in East Aberdeenshire.   *Transactions of the Edinburgh Geological Society*, Vol. 12, 323–328.

— 1934.   The petrology of the Peterhead and Caringall granites.   *Report of the meeting of the British Association for the Advancement of Science, Aberdeen, 1934*, 303.

BUSREWIL, M T, PANKHURST, R J, and WADSWORTH, W J. 1973.   The igneous rocks of the Boganclogh area, NE Scotland.   *Scottish Journal of Geology*, Vol. 9, 165–176.

— — — 1975.   The origin of the Kennethmont granite-diorite series. *Mineralogical Magazine*, Vol. 40, 367–376.

BUTCHER, N J D, BURNS, A R, and GALLAGHER, M J. 1991.   The proposed Duntanlich baryte mine: from exploration to extraction in a glaciated national scenic area, the Grampian Highlands of Scotland.   In *Exploration and the environment*. LUMB, A J, BROWN, M J, and SMITH, C G (editors). (Edinburgh, and Keyworth, Nottingham: British Geological Survey.)

CAMERON, I B, and STEPHENSON, D. 1985. *British regional geology: the Midland Valley of Scotland* (3rd edition). (London: HMSO for British Geological Survey.)

CHINNER, G A. 1960. Pelitic gneisses with varying ferrous/ferric ratios from Glen Clova, Angus, Scotland. *Journal of Petrology*, Vol. 1, 178–217.

— 1966. The distribution of pressure and temperature during Dalradian metamorphism. *Quarterly Journal of the Geological Society of London*, Vol. 122, 159–186.

— 1978. Metamorphic zones and fault displacement in the Scottish Highlands. *Geological Magazine*, Vol. 115, 37–45.

— 1980. Kyanite isograds of Grampian metamorphism. *Journal of the Geological Society of London*, Vol. 137, 35–39.

— and HESELTINE, F J. 1979. The Grampian andalusite/kyanite isograd. *Scottish Journal of Geology*, Vol. 15, 117–127.

CLAPPERTON, C M, and SUGDEN, D E. 1977. The Late Devensian glaciation of north-east Scotland. in *Studies in the Scottish Lateglacial environment*. GRAY, J M, and LOWE, J J (editors). (Oxford: Pergamon Press.)

CLARKE, P D, and WADSWORTH, W J. 1970. The Insch layered intrusion. *Scottish Journal of Geology*, Vol. 6, 7–25.

CLAYBURN, J A P. 1981. Age and petrogenetic studies of some magmatic and metamorphic rocks in the Grampian Highlands. Unpublished PhD thesis, University of Oxford.

— HARMON, R S, PANKHURST, R J, and BROWN, J F. 1983. Sr, O and Pb isotope evidence for the origin and evolution of the Etive Igneous Complex, Scotland. *Nature, London*, Vol. 303, 492–497.

CLEMMENSEN, L B. 1987. Complex star dunes and associated aeolian bedforms, Hopeman Sandstone (Permo-Triassic), Moray Firth Basin, Scotland. 213–231 *in* Desert sediments: ancient and modern. FROSTICK, L E and REID, I (editors). *Special Publication of the Geological Society of London*, No. 35.

COATS, J S, FORTEY, N J, GALLAGHER, M J, and GROUT, A. 1984. Stratiform barium enrichment in the Dalradian of Scotland. *Economic Geology*, Vol. 79, 1585–1595.

— SMITH, C G, FORTEY, N J, GALLAGHER, M J, MAY, F, and McCOURT, W J. 1980. Stratabound barium-zinc mineralization in Dalradian schist near Aberfeldy. *Transactions of the Institution of Mining and Metallurgy (Section B: Applied Earth Science)*, Vol. 89, B110–122.

— and six others, 1981. Stratabound barium-zinc mineralisation in Dalradian Schist near Aberfeldy, *Mineral Reconnaissance Report, British Geological Survey*, No 40.

— PEASE, S F, and GALLAGHER, M J. 1984. Exploration of the Scottish Dalradian. 21–34 in *Prospecting in areas of glaciated terrain*. (London: Institution of Mining and Metallurgy.)

COLMAN, T B. 1990. *Exploration for metalliferous and related minerals in Britain : a guide* (Keyworth, Nottingham : British Geological Survey.)

— BEER, K E, CAMERON, D G, and KIMBELL, G S. 1989. Molybdenum mineralisation near Chapel of Garioch, Inverurie, Aberdeenshire. *British Geological Survey Technical Report*, No. WF/89/3 (= *Mineral Reconnaissance Report*, No. 100).

COWARD, M P. 1983. The thrust and shear zones of the Moine thrust zone and the NW Scottish Caledonides. *Journal of the Geological Society, London*, Vol. 140, 795–811.

COWIE, J W, RUSHTON, A W A, and STUBBLEFIELD, C J. 1972. A correlation of the Cambrian rocks in the British Isles. *Special Report of the Geological Society of London*, No. 2.

CRAIG, E H C, WRIGHT, W B, and BAILEY, E B. 1911.   The geology of Colonsay and Oronsay with part of the Ross of Mull.   *Memoir of the Geological Survey of Scotland,* Sheet 35 (Scotland).

CRAIG, G Y (editor). 1991.   *Geology of Scotland* (3rd edition).   (London: The Geological Society of London.)

CUMMINS, W A, and SHACKLETON, R M. 1955.   The Ben Lui recumbent syncline. *Geological Magazine,* Vol. 92, 353–363.

CURRY, G B. 1986.   Fossils and tectonics along the Highland Boundary Fault in Scotland.   *Journal of the Geological Society of London,* Vol. 143, 193–198.

— BLUCK, B J, BURTON, C J, INGHAM, J K, SIVETER, D J, and WILLIAMS, A. 1984.   Age, evolution and tectonic history of the Highland Border Complex, Scotland. *Transactions of the Royal Society of Edinburgh: Earth Sciences,* Vol. 75, 113–133.

DARBYSHIRE, D P F, and BEER, K E. 1988.   Rb-Sr age of the Bennachie and Middleton granites, Aberdeenshire.   *Scottish Journal of Geology,* Vol. 24, 189–193.

DEER, W A. 1938.   The diorites and associated rocks of the Glen Tilt Complex, Perthshire. I—The granitic and intermediate hybrid rocks.   *Geological Magazine,* Vol. 75, 174–184.

— 1950.   The diorites and associated rocks of the Glen Tilt Complex, Perthshire. II—Diorites and appinites.   *Geological Magazine,* Vol. 87, 181–195.

— 1953.   The diorites and associated rocks of the Glen Tilt Complex, Perthshire. III—Hornblende schist and hornblendite xenoliths in the granite and diorite. *Geological Magazine,* Vol. 90, 27–35.

DEMPSTER, T J. 1983.   Studies of orogenic evolution in the Scottish Dalradian. Unpublished PhD thesis, University of Edinburgh.

— 1985.   Uplift patterns and orogenic evolution in the Scottish Dalradian.   *Journal of the Geological Society of London,* Vol. 142, 111–128.

— and BLUCK, B J. 1991a.   The age and tectonic significance of the Bute amphibolite, Highland Border Complex, Scotland.   *Geological Magazine,* Vol. 128, 77–80.

— — 1991b.   Xenoliths in the lamprophyre dykes of Lomondside: constraints on the nature of the crust beneath the southern Dalradian.   *Scottish Journal of Geology,* Vol. 27, 157–165.

— and HARTE, B. 1986.   Polymetamorphism in the Dalradian of the Central Scottish Highlands.   *Geological Magazine,* Vol. 123, 95–104.

DEWEY, J F, and SHACKLETON, R M. 1984.   A model for the evolution of the Grampian tract in the early Caledonides and Appalachians.   *Nature, London,* Vol. 312, 115–121.

DE SOUZA, H A F. 1979.   The geochronology of Scottish Carboniferous volcanism. Unpublished PhD thesis, University of Edinburgh.

DICKIN, A P, and BOWES, D R. 1991.   Isotopic evidence for the extent of the early Proterozoic basement of Scotland and northwest Ireland.   *Geological Magazine,* Vol. 128, 385–388.

DIXON, J S. 1905.   *Final report of the Royal Commission of Coal Supplies.*   (London: HMSO.)

DOWNIE, C. 1975.   The Precambrian of the British Isles: Palaeontology.   113–115 *in* A correlation of Precambrian rocks in the British Isles, HARRIS, A L, and five others (editors). *Special Report of the Geological Society of London,* No. 6.

— LISTER, T R, HARRIS, A L, and FETTES, D J. 1971.   A palynological investigation of the Dalradian rocks of Scotland.   *Report of the Institute of Geological Sciences,* No. 71/9.

DROOP, G T R, and CHARNLEY, N. 1985. Comparative geobarometry of pelitic hornfelses associated with the newer gabbros: a preliminary study. *Journal of the Geological Society of London*, Vol. 142, 53–62.

— and TRELOAR, P J. 1981. Pressures of metamorphism in the aureole of the Etive Granite Complex. *Scottish Journal of Geology*, Vol. 17, 85–102.

DUNHAM, A C, and STRASSER-KING, V E H. 1982. Late Carboniferous intrusions of northern Britain. 277–283 in *Igneous rocks of the British Isles*. SUTHERLAND, D S (editor). (Chichester: John Wiley and Sons.)

DUNHAM, K C. 1952. Fluorspar (4th edition). *Special Reports on the Mineral Resources of Great Britain, Memoir of the Geological Survey of Great Britain*, Vol. 4.

DYMOKE, P L. 1989. Geochronological and petrological studies of the thermal evolution of the Dalradian, South-west Scottish Highlands. Unpublished PhD thesis, University of Edinburgh.

EARLS, G, PARKER, R T G, CLIFFORD, J A, and MELDRUM, A H. 1992. The geology of the Cononish gold-silver deposit, Grampian Highlands of Scotland. 89–102 in *The Irish minerals industry. 1980–1990*. BOWDEN, A A, EARLS, G, O'CONNOR, P G, and PYNE, J F (editors). (Dublin: Irish Association for Economic Geology).

ELLES, G L. 1926. The geological structure of Ben Lawers and Meall Corranaich (Perthshire). *Quarterly Journal of the Geological Society of London*, Vol. 82, 304–331.

— and TILLEY, C E. 1930. Metamorphism in relation to structure in the Scottish Highlands. *Transactions of the Royal Society of Edinburgh*, Vol. 56, 621–646.

— 1931. Notes on the Portsoy coastal district. *Geological Magazine*, Vol. 68, 24–34.

EVANS, D, KENOLTY, N, DOBSON, M R, and WHITTINGTON, R J. 1979. Geology of the Malin Sea. *Report of the Institute of Geological Sciences*, No. 79/15.

— WILKINSON, G C, and CRAIG, D L. 1979. The Tertiary sediments of the Canna Basin, Sea of the Hebrides. *Scottish Journal of Geology*, Vol. 15, 329–332.

— CHESHER, J A, DEEGAN, C E, and FANNIN, N G T. 1982. The offshore geology of Scotland in relation to the IGS Shallow Drilling Programme, 1970–1978. *Report of the Institute of Geological Sciences*, No. 81/12.

EYLES, C H, and EYLES, N. 1983. Glaciomarine model for upper Precambrian diamictites of the Port Askaig Formation, Scotland. *Geology*, Vol. 11, 692–696.

EYLES, N, and CLARK, B M. 1985. Gravity-induced soft sediment deformation in glaciomarine sequences of the Upper Proterozoic Port Askaig Formation, Scotland. *Sedimentology*, Vol. 32, 789–814.

FAIRCHILD, I J. 1980a. Sedimentation and origin of a late Precambrian "dolomite" from Scotland. *Journal of Sedimentary Petrology*, Vol. 50, 423–446.

— 1980b. Stages in a Precambrian dolomitisation, Scotland: cementing versus replacement textures. *Sedimentology*, Vol. 27, 631–650.

— 1985. Petrography and carbonate chemistry of some Dalradian dolomitic metasediments: preservation of diagenetic textures. *Journal of the Geological Society of London*, Vol. 142, 167–185.

— and HAMBREY, M J. 1984. The Vendian succession of northeastern Spitsbergen: petrogenesis of a dolomite-tillite association. *Precambrian Research*, Vol. 26, 111–167.

FETTES, D J. 1970. The structural and metamorphic state of the Dalradian rocks and their bearing on the age of emplacement of the basic sheet. *Scottish Journal of Geology*, Vol. 6, 108–118.

— 1971. Relation of cleavage and metamorphism in the Macduff Slates. *Scottish Journal of Geology*, Vol. 7, 248–253.

— 1983.   Metamorphism in the British Caledonides.   205–219 in *Regional trends in the geology of the Appalachian–Caledonian–Hercynian–Mauritanide Orogen.* SCHENK, P E (editor). (Dordrecht: Reidel.)

— GRAHAM, C M, HARTE, B, and PLANT, J A. 1986.   Lineaments and basement domains; an alternative view of Dalradian evolution.   *Journal of the Geological Society of London,* Vol. 143, 453–464.

— — SASSI, F P, and SCOLARI, A. 1976.   The lateral spacing of potassic white micas and facies series variations across the Caledonides.   *Scottish Journal of Geology,* Vol. 12, 227–236.

— HARRIS, A L, and HALL, L M. 1986.   The Caledonian geology of the Scottish Highlands.   303–334 in *Synthesis of the Caledonian rocks of Britain. Proceedings of the NATO Advanced Study Institute.* FETTES, D J, and HARRIS, A L (editors). (Dordrecht: Reidel.)

— LESLIE, A G, STEPHENSON, D, and KIMBELL, S F. 1991.   Disruption of Dalradian stratigraphy along the Portsoy Lineament from new geological and magnetic surveys.   *Scottish Journal of Geology,* Vol. 27, 57–73.

— and six others. 1985.   Grade and time of metamorphism in the Caledonide Orogen of Britain and Ireland.   41–53 *in* The nature and timing of orogenic activity in the Caledonian rocks of the British Isles. HARRIS A L (editor). *Memoir of the Geological Society of London,* No. 9.

— MENDUM, J R, and SMITH, D I. 1992.   Geology of the Outer Hebrides.   *Memoir of the British Geological Survey.*

— and MUNRO, M. 1989.   Age of the Blackwater mafic and ultramafic intrusion, Banffshire.   *Scottish Journal of Geology,* Vol. 25, 105–111.

FIRTH, C R. (1989).   Late Devensian raised shorelines and ice limits in the inner Moray Firth area, northern Scotland.   *Boreas,* Vol. 18, 5–21.

FITCH, F J, MILLER, J A, and WILLIAMS, S C. 1970.   Isotopic ages of British Carboniferous rocks.   *Comptes Rendues du Sixième Congrès International sur la Géologie Carbonifère* (Sheffield, 1967), Vol. 2, 771–789.

FITCHES, W R, and MALTMAN, A J. 1984.   Tectonic development and stratigraphy at the western margin of the Caledonides: Islay and Colonsay, Scotland.   *Transactions of the Royal Society of Edinburgh: Earth Sciences,* Vol. 75, 365–382.

FLETCHER, T A and RICE, C M. 1989.   Geology, mineralization (Ni-Cu) and precious metal geochemistry of Caledonian mafic and ultramafic intrusions near Huntly, northeast Scotland.   *Transactions of the Institution of Mining and Metallurgy (Section B: Applied Earth Science),* Vol. 98, B185–200.

FLETCHER, T P, AUTON, C A, HIGHTON, A J, MERRITT, J W, ROBERTSON, S, and ROLLIN, K E. 1995.   Geology of the Fortrose and eastern Inverness district.   *Memoir of the British Geological Survey,* Sheet 84W (Scotland).

FORTEY, N J, and BEDDOE-STEPHENS, B. 1982.   Barium silicates in stratabound Ba-Zn mineralisation in the Scottish Dalradian.   *Mineralogical Magazine,* 46, 63–72.

— and SMITH, C G. 1986.   Stratabound mineralisation in Dalradian rocks near Tyndrum, Perthshire.   *Scottish Journal of Geology,* Vol. 22, 377–393.

— NANCARROW, P M A, and GALLAGHER, M J. 1991.   Armenite from the Middle Dalradian of Scotland.   *Mineralogical Magazine,* 55, 135–138.

— COATS, J S, GALLAGHER, M J, GREENWOOD, P G, and SMITH, C G. 1993.   Dalradian stratabound baryte and base metals near Braemar, NE Scotland.   *Transactions of the Institution of Mining and Metallurgy (Section B: Applied Earth Science),* Vol. 102, B55–64.

FRANCIS, E H, FORSYTH, I H, READ, W A, and ARMSTRONG, M. 1970.   The geology of the Stirling district.   *Memoir of the Geological Survey of Great Britain,* Sheet 39 (Scotland).

FRIEND, P F, and MACDONALD, R. 1968. Volcanic sediments, stratigraphy and tectonic background of the Old Red Sandstone of Kintyre, W. Scotland. *Scottish Journal of Geology*, Vol. 4, 265–282.

FROSTICK, L, REID, I, JARVIS, J, and EARDLEY, H. 1988. Triassic sediments of the Inner Moray Firth, Scotland: early rift deposits. *Journal of the Geological Society of London*, Vol. 145, 235–248.

FYFE, J A, LONG, D, and EVANS, D. 1993. *United Kingdom offshore regional report: the geology of the Malin–Hebrides sea area.* (London: HMSO for the British Geological Survey.)

GALLAGHER, J W. 1983. The north-east Grampian Highlands: an investigation based on new geophysical and geological data. Unpublished PhD thesis, University of Aberdeen.

GALLAGHER, M J. 1984. Barite deposits and potential of Scotland. *Transactions of the Institution of Mining and Metallurgy (Section A: Mineral Industries)*, Vol. 93, A130–132.

— 1990. Metalliferous mineralisation. In *Regional geochemical atlas: Argyll* (Keyworth, Nottingham: British Geological Survey.)

— 1991a. Metalliferous mineralisation. 21–23 in *Regional geochemistry of the East Grampians area.* (Keyworth, Nottingham: British Geological Survey.)

— 1991b. Metalliferous minerals. 567–586 and 590–595 in *Geology of Scotland* (3rd edition). CRAIG, G Y (editor). (London: The Geological Society of London.)

— and YOUNG, B. 1993. Metalliferous mineralisation. In *Regional geochemical survey of southern Scotland.* (Keyworth, Nottingham: British Geological Survey.)

— and six others. 1989. Stratabound barium and base-metal mineralisation in Middle Dalradian metasediments near Braemar, Scotland. *Mineral Reconnaissance Report, British Geological Survey*, No. 104.

GARSON, M S, and PLANT, J. 1972. Possible dextral movements on the Great Glen and Minch faults, Scotland. *Nature, London, Physical Science*, Vol. 240, 31–35.

GATLIFF, R W, and eleven others. 1994. *United Kingdom offshore regional report: the geology of the central North Sea.* (London: HMSO for the British Geological Survey.)

GEORGE, T N. 1960. The stratigraphical evolution of the Midland Valley. *Transactions of the Geological Society of Glasgow*, Vol. 24, 32–107.

— 1966. Geomorphic evolution in Hebridean Scotland. *Scottish Journal of Geology*, Vol. 2, 1–34.

GIBBONS, W, and GAYER, R A. 1985. British Caledonian terranes. 3–16 in *The tectonic evolution of the Caledonian–Appalachian orogen.* GAYER, R A (editor). (Wiesbaden: Vieweg.)

GLENNIE, K W, and BULLER, A T. 1983. The Permian Weissliegend of NW Europe, the partial deformation of aeolian dune sands caused by Zechstein transgression. *Sedimentary Geology*, Vol. 35, 43–81.

GLOVER, B W. 1993. The sedimentology of the Neoproterozoic Grampian Group and the significance of the Fort William Slide between Spean Bridge and Rubha Cuil-cheanna, Inverness-shire. *Scottish Journal of Geology*, Vol. 29, 29–43.

— and WINCHESTER, J A. 1989. The Grampian Group: a major Late Proterozoic clastic sequence in the Central Highlands of Scotland. *Journal of the Geological Society of London*, Vol. 146, 85–97.

GOODMAN, S, and WINCHESTER, J A. 1993. Geochemical variations within metavolcanic rocks of the Dalradian Farragon Beds and adjacent formations. *Scottish Journal of Geology*, Vol. 29, 131–141.

GORDON, J E, and SUTHERLAND, D G (editors). 1993. *Quaternary of Scotland: Geological Conservation Review.* (London: Chapman and Hall.)

GOULD, D. In press. Geology of the country around Inverurie and Alford. *Memoir of the British Geological Survey*, Sheets 76W and 76E (Scotland).

GOWER, P J. 1973.   The middle–upper Dalradian boundary with special reference to the Loch Tay Limestone. Unpublished PhD thesis, University of Liverpool.

— 1977.   The Dalradian rocks of the west coast of the Tayvallich peninsula.   *Scottish Journal of Geology,* Vol. 13, 125–133.

GRAHAM, C M. 1976.   Petrochemistry and tectonic significance of Dalradian metabasaltic rocks of the SW Scottish Highlands.   *Journal of the Geological Society of London,* Vol. 132, 61–84.

— 1986.   The role of the Cruachan Lineament during Dalradian evolution.   *Scottish Journal of Geology,* Vol. 22, 257–270.

— and BORRADAILE, G J. 1984.   The petrology and structure of Dalradian metabasaltic dykes of Jura: implications for early Dalradian evolution.   *Scottish Journal of Geology,* Vol. 20, 257–270.

— and BRADBURY, H J. 1981.   Cambrian and late Precambrian basaltic igneous activity in the Scottish Dalradian: a review.   *Geological Magazine,* Vol. 118, 27–37.

GRAHAM, C M. 1983.   High-pressure greenschist to epidote-amphibolite facies metamorphism of the Dalradian rocks of the SW Scottish Highlands.   *Geological Society Newsletter,* Vol. 12, No. 4, 19.

GRAY, J M, and COXON, P. 1991.   The Loch Lomond Stadial glaciation in Britain and Ireland.   89–105 in *Glacial deposits in Great Britain and Ireland.* EHLERS, J, GIBBARD, P L, and ROSE, J (editors). (Rotterdam: A A Balkema.)

GRAY, J M, and LOWE, J J. 1977 (editors).   *Studies in the Scottish lateglacial environment.* (Oxford: Pergamon.)

GREEN, J F N. 1924.   The structure of the Bowmore–Portsakaig District of Islay.   *Quarterly Journal of the Geological Society of London,* Vol. 89, 72–105.

GRIBBLE, C D. 1967.   The basic intrusive rocks of Caledonian age in the Haddo House and Arnage districts, Aberdeenshire.   *Scottish Journal of Geology,* Vol. 3, 125–136.

— 1968.   The cordierite-bearing rocks of the Haddo House and Arnage districts, Aberdeen.   *Contributions to Mineralogy and Petrology,* Vol. 17, 315–330.

GROOME, D R, and HALL, A. 1974.   The geochemistry of the Devonian lavas of the northern Lorne Plateau, Scotland.   *Mineralogical Magazine,* Vol. 39, 621–640.

GROUT, A, and SMITH, C G. 1989a.   Scottish Highlands and Southern Uplands Mineral Portfolio: limestone and dolomite resources.   *British Geological Survey Technical Report,* WF/89/5.

— — 1989b.   Scottish Highlands and Southern Uplands Mineral Portfolio: talc resources.   *British Geological Survey Technical Report,* WF/89/7.

GUNN, A G, STYLES, M T, STEPHENSON, D, SHAW, M H, and ROLLIN, K E. 1990.   Platinum-group elements in ultramafic rocks of the Upper Deveron Valley, near Huntly, Aberdeenshire.   *Mineral Reconnaissance Report, British Geological Survey,* No. 115.

GUNN, W, CLOUGH, C T, and HILL, J B. 1897.   The geology of Cowal.   *Memoir of the Geological Survey of Scotland,* Sheet 29.

HACKMAN, B D, and KNILL, J L. 1962.   Calcareous algae from the Dalradian of Islay.   *Palaeontology,* Vol. 5, 268–271.

HALL, A, and WALSH, J N. 1972.   Zinnwaldite granite from Glen Gairn, Aberdeenshire.   *Scottish Journal of Geology,* Vol. 8, 265–267.

HALL, A J, BOYCE, A J, and FALLICK, A E. 1994.   A sulphur isotope study of iron sulphides in the late Precambrian Dalradian Ardrishaig Phyllite Formation, Knapdale, Argyll.   *Scottish Journal of Geology,* Vol. 30, 63–71.

— — — and HAMILTON, P J. 1991. Isotope evidence of the depositional environment of Late Proterzoic stratiform barite mineralisation, Aberfeldy, Scotland. *Chemical Geology (Isotope Geoscience Section)*, Vol. 87, 99–114.

HALL, A M (editor) 1984. *Buchan field guide.* (Cambridge: Quaternary Research Association.)

— 1986. Deep weathering patterns in north-east Scotland and their geomorphological significance. *Zeitschrift für Geomorphologie*, NF 30, 407–22.

— 1991. Pre-Quaternary landscape evolution in the Scottish Highlands. *Transactions of the Royal Society of Edinburgh: Earth Sciences*, Vol. 82, 1–26.

— and CONNELL, E R. 1991. The glacial deposits of Buchan, north-east Scotland. 129–136 in *Glacial deposits in Great Britain and Ireland*. EHLERS, J, GIBBARD, P L, and ROSE, J (editors).(Rotterdam: A A Balkema.)

— and JARVIS, J. 1989. A preliminary report on the Late Devensian glaciomarine deposits around St Fergus, Grampian Region. *Quaternary Newsletter*, No. 59, 5–7.

— and SUGDEN, D E. 1987. Limited modification of mid-latitude landscapes by ice sheets. *Earth Surface Processes and Landforms*, Vol. 12, 531–542.

HALLIDAY, A N, AFTALION, M, VAN BREEMEN, O, and JOCELYN, J. 1979. Petrogenetic significance of Rb-Sr and U-Pb isotopic systems in 400 Ma old British Isles granitoids and their hosts. 653–661 *in* The Caledonides of the British Isles—reviewed. HARRIS, A L, HOLLAND, C H and LEAKE, B E (editors). *Special Publication of the Geological Society of London*, No. 8.

— GRAHAM, C M, AFTALION, M, and DYMOKE, P. 1989. The depositional age of the Dalradian Supergroup: U-Pb and Sm-Nd isotopic studies of the Tayvallich Volcanics, Scotland. *Journal of the Geological Society of London*, Vol. 146, 3–6.

HAMIDULLAH, S, and Bowes, D R. 1987. Petrogenesis of the appinite suite of the Appin district, western Scotland. *Acta Universitatis Carolinae—Geologica*, No. 4, 295–396.

HAMBREY, M J, and WADDAMS, P. 1981. Glaciogenic boulder-bearing deposits in the Upper Dalradian Macduff Slates, northeastern Scotland. 571–575 in *Earth's pre-Pleistocene glacial record*. HAMBREY, M J, and HARLAND, W B (editors). (Cambridge: Cambridge University Press.)

HARLAND, W B. 1972. The Ordovician Ice Age. *Geological Magazine*, Vol. 109, 451–456.

HARMON, R S, and HALLIDAY, A N. 1980. Oxygen and strontium isotope relationships in the British late Caledonian granites. *Nature, London*, Vol. 283, 21–25.

— PANKHURST, R J, PLANT J A, and SIMPSON, P R. 1984. Petrogenesis of the Cairngorm granite, east-central Grampian Highlands, Scotland. 72–73 in *Open magmatic systems*. *Proceedings of ISEM field conference*. DUNCAN, M A, GROVE, T L and HILDRETH, W (editors). (Dallas, Texas: Southern Methodist University.)

HARPER, D A T, WILLIAMS, D M, and ARMSTRONG, H A. 1989. Stratigraphical correlations adjacent to the Highland Boundary Fault in the west of Ireland. *Journal of the Geological Society of London*, Vol. 146, 381–384.

HARRIS, A L. 1962. Dalradian geology of the Highland border near Callander. *Bulletin of the Geological Survey of Great Britain*, No. 19, 1–15.

— 1963. Structural investigations in the Dalradian rocks between Pitlochry and Blair Atholl. *Transactions of the Edinburgh Geological Society*, Vol. 19, 256–278.

— 1969. The relationships of the Leny Limestone to the Dalradian. *Scottish Journal of Geology*, Vol. 5, 187–190.

— 1972. The Dalradian rocks at Dunkeld, Perthshire. *Bulletin of the Geological Survey of Great Britain*, No. 38, 1–10.

— 1991.   The growth and structure of Scotland.   1–24 in *Geology of Scotland* (3rd edition). CRAIG, G Y (editor). (London: The Geological Society of London.)

— BALDWIN, C T, BRADBURY, H J, JOHNSON, H D, and SMITH, R A.   1978.   Ensialic basin sedimentation: the Dalradian Supergroup.   115–138 *in* Crustal evolution in northwestern Britain and adjacent regions. BOWES, D R, and LEAKE, B E (editors). *Special Issue of the Geological Journal*, No. 10.

— BRADBURY, H J, and McGONIGAL, N H.   1976.   The evolution and transport of the Tay Nappe.   *Scottish Journal of Geology*, Vol. 12, 103–113.

— and FETTES, D J.   1972.   Stratigraphy and structure of Upper Dalradian rocks at the Highland Border.   *Scottish Journal of Geology*, Vol. 8, 253–264.

— and PITCHER, W S.   1975.   The Dalradian Supergroup.   52–75 *in* A correlation of Precambrian rocks in the British Isles. HARRIS, A L, and others (editors). *Special Report of the Geological Society of London*, No. 6.

HARRIS, M, KAY, E A, WIDNALL, M A, JONES, E M, and STEELE, G B.   1988.   Geology and mineralization of the Lagalochan intrusive complex, western Argyll, Scotland. *Transactions of the Institution of Mining and Metallurgy (Section B: Applied Earth Science)*, Vol. 97, B15–21.

HARRISON, D J.   1985.   Mineralogical and chemical appraisal of Corrycharmaig serpentinite intrusion, Glen Lochay, Perthshire.   *Transactions of the Institution of Mining and Metallurgy (Section B: Applied Earth Science)*, Vol. 94, B147–151.

HARRISON, T N.   1986.   The mode of emplacement of the Cairngorm Granite. *Scottish Journal of Geology*, Vol. 22, 303–314.

— 1987.   The granitoids of eastern Aberdeenshire.   243–250 in *Excursion guide to the geology of the Aberdeen area*. TREWIN, N H, KNELLER, B C, and GILLEN, C (editors). (Edinburgh: Scottish Academic Press for Geological Society of Aberdeen.)

— and HUTCHINSON, J.   1987.   The age and origin of the Eastern Grampians Newer Granites.   *Scottish Journal of Geology*, Vol. 23, 269–282.

HARRY, W T.   1958.   A re-examination of Barrow's Older Granites in Glen Clova. *Transactions of the Royal Society of Edinburgh*, Vol. 53, 393–412.

— 1965.   The form of the Cairngorm Granite Pluton.   *Scottish Journal of Geology*, Vol. 1, 1–8.

HARTE, B.   1979.   The Tarfside succession and the structure and stratigraphy of the eastern Scottish Dalradian rocks.   221–228 *in* The Caledonides of the British Isles— reviewed.   HARRIS, A L, HOLLAND, C H, and LEAKE, B E (editors). *Special Publication of the Geological Society of London*, No. 8, 221–228.

— 1988.   Lower Palaeozoic metamorphism in the Moine–Dalradian belt of the British Isles.   123–134 *in* The Caledonian—Appalachian Orogen. HARRIS, A L, and FETTES, D J (editors). *Special Publication of the Geological Society of London*, No. 38.

— BOOTH, J E, DEMPSTER, J T, FETTES, D J, MENDUM, J R, and WATTS, D.   1984.   Aspects of the post-depositional evolution of Dalradian and Highland Border Complex rocks in the Southern Highlands of Scotland.   *Transactions of the Royal Society of Edinburgh: Earth Sciences*, Vol. 75, 151–163.

— — and FETTES, D J.   1987.   Stonehaven to Findhorn, Dalradian structure and metamorphism.   211–226 in *Excursion guide to the geology of the Aberdeen area*. TREWIN, N H, KNELLER, B C, and GILLEN, C (editors). (Edinburgh: Scottish Academic Press for Geological Society of Aberdeen.)

— and HUDSON, N F C.   1979.   Pelite facies series and the temperatures and pressures of Dalradian metamorphism.   323–337 *in* The Caledonides of the British Isles reviewed. HARRIS, A L, HOLLAND, C H, and LEAKE, B E (editors). *Special Publication of the Geological Society of London*, No. 8.

— and JOHNSON, M R W. 1969. Metamorphic history of Dalradian rocks in Glens Clova, Esk and Lethnot, Angus, Scotland. *Scottish Journal of Geology*, Vol. 5, 54–80.

HASELOCK, P J. 1984. The systematic geochemical variation between two tectonically separate successions in the southern Monadhliaths, Inverness-shire. *Scottish Journal of Geology*, Vol. 20, 191–205.

— and EVANS, R H S. 1990. Discussion on the Grampian Group: a major late Proterozoic clastic sequence in the Central Highlands of Scotland. *Journal of the Geological Society of London*, Vol. 147, 732–734.

— and WINCHESTER, J A. 1982. A note on the stratigraphic relationship of the Leven Schist and Monadhliath Schist in the Central Highlands of Scotland. *Geological Journal*, Vol. 16, 237–241.

— — and WHITTLES, K H. 1982. The stratigraphy and structure of the southern Monadhliath Mountains between Killin and upper Glen Roy. *Scottish Journal of Geology*, Vol. 18, 275–290.

HASLAM, H W. 1968. The crystallisation of intermediate and acid magmas at Ben Nevis, Scotland. *Journal of Petrology*, Vol. 9, 84–104.

— and KIMBELL, G S. 1981. Disseminated copper-molybdenum mineralisation near Ballachulish, Highland Region. *Mineral Reconnaissance Report, British Geological Survey*, No. 43.

HAWSON, C A, and HALL, A J. 1987. Middle Dalradian Corrycharmaig serpentinite, Perthshire, Scotland: an ultramafic intrusion. *Transactions of the Institution of Mining and Metallurgy (Section B: Applied Earth Science)*, Vol. 96, B173–177.

HENDERSON, S M K. 1938. The Dalradian succession of the Southern Highlands. *Report of the Meeting of the British Association for the Advancement of Science, Cambridge, 1938*, 424.

HENDERSON, W G, and FORTEY, N J. 1982. Highland Border rocks of the Aberfoyle district. *Scottish Journal of Geology*, Vol. 18, 227–245.

— and ROBERTSON, A H F. 1982. The Highland Border rocks and their relation to marginal basin development in the Scottish Caledonides. *Journal of the Geological Society of London*, Vol. 139, 433–450.

HICKMAN, A H. 1975. The stratigraphy of late Precambrian metasediments between Glen Roy and Lismore. *Scottish Journal of Geology*, Vol. 11, 117–142.

— 1978. Recumbent folds between Glen Roy and Lismore. *Scottish Journal of Geology*, Vol. 14, 191–212.

— and WRIGHT, A E. 1983. Geochemistry and chemostratigraphical correlation of slates, marbles and quartzites of the Appin Group, Argyll, Scotland. *Transactions of the Royal Society of Edinburgh: Earth Sciences*, Vol. 73, 251–278.

HIGHTON, A J. 1992. The tectonostratigraphical significance of pre-750 Ma metagabbros within the northern Central Highlands, Inverness-shire. *Scottish Journal of Geology*, Vol. 28, 71–76.

— In press. Geology of the country around Aviemore. *Memoir of the British Geological Survey*, Sheet 74E (Scotland).

HILL, J B. 1905. The geology of Mid-Argyll. *Memoir of the Geological Survey, Scotland*, Sheet 37 (Scotland).

HINXMAN, L W. 1896. The geology of West Aberdeenshire and Banffshire with parts of Elgin and Inverness. *Memoir of the Geological Survey, Scotland*, Sheet 75 (Scotland).

— and ANDERSON, E M. 1915. The geology of Mid-Strathspey and Strathdearn, including the country between Kingussie and Grantown. *Memoir of the Geological Survey, Scotland*, Sheet 74 (Scotland).

— CARRUTHERS, R G, and MACGREGOR, M. 1923.   The geology of Corrour and the Moor of Rannoch.   *Memoir of the Geological Survey, Scotland*, Sheet 54 (Scotland).

— and WILSON, J S G. 1902.   The geology of Lower Strathspey.   *Memoir of the Geological Survey, Scotland*, Sheet 85 (Scotland).

HOLE, M J, and MORRISON, M A. 1992.   The differentiated dolerite boss, Cnoc Rhaonastil, Islay: a natural experiment in the low pressure differentiation of an alkali olivine-basalt magma.   *Scottish Journal of Geology*, Vol. 28, 55–69.

HORNE, J. 1923.   The geology of the Lower Findhorn and Lower Strath Nairn, including part of the Black Isle.   *Memoir of the Geological Survey, Scotland*, Sheets 84, part 94 (Scotland).

— and HINXMAN, L W. 1914.   The geology of the country round Beauly and Inverness.   *Memoir of the Geological Survey, Scotland*, Sheet 83 (Scotland).

HUDSON, N F C. 1985.   Conditions of Dalradian metamorphism in the Buchan area. *Journal of the Geological Society of London*, Vol. 142, 63–76.

HUTTON, D H W, and DEWEY, J F. 1986.   Palaeozoic terrane accretion in the western Irish Caledonides.   *Tectonics*, Vol. 5, 1115–1124.

JARVIS, K E. 1987.   The petrogenesis and geochemistry of the diorite complexes south of Balmoral Forest, Angus.   Unpublished PhD thesis, City of London Polytechnic.

JEHU, T J, and CAMPBELL, R. 1917.   The Highland Border rocks of the Aberfoyle district.   *Transactions of the Royal Society of Edinburgh*, Vol. 52, 175–212.

JOHNSON, M R W. 1962.   Relations of movement and metamorphism in the Dalradians of Banffshire.   *Transactions of the Edinburgh Geological Society*, Vol. 19, 29–64.

— 1963.   Some time relations of movement and metamorphism in the Scottish Highlands.   *Geologie en Mijnbouw*, Vol. 42, 121–142.

— and HARRIS, A L. 1967.   Dalradian–?Arenig relations in part of the Highland Border, Scotland and their significance in the chronology of the Caledonian Orogeny. *Scottish Journal of Geology*, Vol. 3, 1–16.

— 1991.   Dalradian.   125–160 in *Geology of Scotland* (3rd edition). CRAIG, G Y (editor). (London: The Geological Society of London.)

— and STEWART, F H. 1960.   On Dalradian structures in north-east Scotland. *Transactions of the Edinburgh Geological Society*, Vol. 18, 94–103.

JOHNSTONE, G S. 1966.   *British regional geology: the Grampian Highlands* (3rd edition). (Edinburgh: HMSO for Geological Survey and Museum.)

— 1975.   The Moine Succession.   30–42 *in* A correlation of Precambrian rocks in the British Isles. HARRIS, A L, and others (editors). *Special Report of the Geological Society of London*, No. 6.

— and SMITH, D I. 1965.   Geological observations concerning the Breadalbane Hydroelectric Project, Perthshire.   *Bulletin of the Geological Survey of Great Britain*, No. 22, 1–52.

— and MYKURA, W. 1989.   *British regional geology: the Northern Highlands of Scotland* (4th edition).   (London: HMSO for British Geological Survey.)

JUDD, J W. 1873.   The Secondary Rocks of Scotland.   *Quarterly Journal of the Geological Society of London*, Vol. 29, 97–197.

KENNEDY, W Q. 1946.   The Great Glen Fault.   *Quarterly Journal of the Geological Society of London*, Vol 102, 41–72.

— 1948.   On the significance of thermal structure in the Scottish Highlands. *Geological Magazine*, Vol. 85, 229–234.

KEY, R M, MAY, F, CLARK, G C, PEACOCK, J D, and CHACKSFIELD, B C. In press. Geology of the country around Glen Roy. *Memoir of the British Geological Survey,* Sheet 63W (Scotland).

— PHILLIPS, E K, and CHACKSFIELD, B C. 1993. Emplacement and thermal metamorphism associated with the post-orogenic Strath Ossian pluton, Grampian Highlands, Scotland. *Geological Magazine,* Vol. 13, 379–390.

KIDSTON, R. 1899. 129–130 in *Summary of progress of the Geological Survey of Great Britain for 1898.* (London: HMSO.)

— and LANG, W H. 1921. On Old Red Sandstone plants showing structure from the Rhynie Chert Bed, Aberdeenshire. Part V. *Transactions of the Royal Society of Edinburgh,* Vol. 52, 855–902.

KILBURN, C, PITCHER, W S, and SHACKLETON, R M. 1965. The stratigraphy and origin of the Portaskaig Boulder Bed Series (Dalradian). *Geological Journal,* Vol. 4, 343–360.

KING, B C, and RAST, N. 1959. Structural geometry of Dalradian rocks at Loch Leven, Scottish Highlands: a discussion. *Journal of Geology,* Vol. 67, 244–246.

KLEIN, G DE V. 1970. Tidal origin of a Precambrian quartzite—the Lower Fine-grained Quartzite (Middle Dalradian) of Islay, Scotland. *Journal of Sedimentary Petrology,* Vol. 40, 973–985.

KNELLER, B C. 1985. Dalradian basin evolution and metamorphism. *Journal of the Geological Society of London,* Vol. 142, 4 (abstract).

— 1987. A geological history of NE Scotland. 1–50 in *Excursion guide to the geology of the Aberdeen area.* TREWIN, H N, KNELLER, B C, and GILLEN, C (editors). (Edinburgh: Scottish Academic Press for Geological Society of Aberdeen.)

— 1988. The geology of part of Buchan. Unpublished PhD thesis, University of Aberdeen.

— and AFTALION, M. 1987. The isotopic and structural age of the Aberdeen Granite. *Journal of the Geological Society of London,* Vol. 144, 717–721.

KNILL, J L. 1959. Palaeocurrents and sedimentary facies of the Dalradian metasediments of the Craignish—Kilmelfort district. *Proceedings of the Geologists' Association,* Vol. 70, 273–284.

— 1963. A sedimentary history of the Dalradian Series. 99–121 in *The British Caledonides.* JOHNSON, M R W, and STEWART, F H (editors). (Edinburgh: Oliver and Boyd.)

KRUHL, J, and VOLL, G. 1975. Large scale pre-metamorphic and pre-cleavage inversion at Loch Leven, Scottish Highlands. *Nues Jahrbuch für Mineralogie,* Vol. 2, 71–78.

KYNASTON, H, and HILL, J B. 1908. The geology of the country near Oban and Dalmally. *Memoir of the Geological Survey, Scotland,* Sheet 45 (Scotland).

LAMBERT, R ST J, and McKERROW, W S. 1976. The Grampian Orogeny. *Scottish Journal of Geology,* Vol. 12, 271–292.

— HOLLAND, J G, and WINCHESTER, J A. 1982. A geochemical comparison of the Dalradian Leven Schists and the Grampian Division Monadhliath Schists of Scotland. *Journal of the Geological Society of London,* Vol. 139, 71–84.

— WINCHESTER, J A, and HOLLAND, J G. 1981. Comparative geochemistry of pelites from the Moinian and Appin Group (Dalradian) of Scotland. *Geological Magazine,* Vol. 118, 477–490.

LEAKE, B E. 1982. Volcanism in the Dalradian. 45–50 in *Igneous rocks of the British Isles.* SUTHERLAND, D S (editor). (Chichester: John Wiley and Sons.)

LEE, G W, and BAILEY, E B. 1925. The pre-Tertiary geology of Mull, Loch Aline and Oban. *Memoir of the Geological Survey, Scotland.* Parts of Sheets 35, 43, 44, 45 and 52 (Scotland).

LEE, M K, and eight others. 1984.   *Investigation of the geothermal potential of the UK–Hot Dry Rocks prospects in Caledonian granites: evaluation of results from the BGS-IC-OU research programme (1981-84)*. (Keyworth, Nottingham: British Geological Survey.)

LESLIE, A G. 1984.   Field relations in the north-eastern part of the Insch mafic igneous mass, Aberdeenshire.   *Scottish Journal of Geology*, Vol. 20, 215–235.

LINDSAY, N G, HASELOCK, P J, and HARRIS, A L. 1989.   The extent of Grampian orogenic activity in the Scottish Highlands.   *Journal of the Geological Society of London*, Vol. 146, 733–735.

LITHERLAND, M. 1975.   Organic remains and traces from the Dalradian of Benderloch, Argyll.   *Scottish Journal of Geology*, Vol. 11, 47–50.

— 1980.   The stratigraphy of the Dalradian rocks around Loch Creran, Argyll.   *Scottish Journal of Geology*, Vol. 16, 105–123.

— 1982.   The structure of the Loch Creran Dalradian and a new model for the SW Highlands.   *Scottish Journal of Geology*, Vol. 18, 205–225.

— 1987.   Sedimentation and tectonics in the Scottish Dalradian: comment.   *Scottish Journal of Geology*, Vol. 23, 315–316.

LONGMAN, C D, BLUCK, B J, and VAN BREEMEN, O. 1979.   Ordovician conglomerates and evolution of the Midland Valley.   *Nature, London,* Vol. 280, 578–581.

LONG, D, SMITH, D E, and DAWSON, A G. 1989.   A Holocene tsunami deposit in eastern Scotland.   *Journal of Quaternary Science*, Vol. 4, 61–66.

LOUDON, T V. 1963.   The sedimentation and structure in the Macduff District of North Banffshire and Aberdeenshire.   Unpublished PhD thesis, University of Edinburgh.

LOWE, J J. 1984.   A critical evaluation of pollen-stratigraphic investigations of pre-Late Devensian sites in Scotland.   *Quaternary Science Reviews*, Vol. 3, 405–32.

— and WALKER, M J C. 1984.   *Reconstructing Quaternary environments*.   (Harlow: Longman.)

McCALLIEN, W J. 1927.   Preliminary account of the post-Dalradian geology of Kintyre.   *Transactions of the Geological Society of Glasgow*, Vol. 18, 40–126.

— 1929.   The metamorphic rocks of Kintyre.   *Transactions of the Royal Society of Edinburgh*, Vol 56, 409–436.

— 1932.   The Kainozoic igneous rocks of Kintyre.   *Geological Magazine*, Vol. 69, 49–61.

— and ANDERSON, R B. 1930.   The Carboniferous sediments of Kintyre.   *Transactions of the Royal Society of Edinburgh*, Vol. 56, 599–619.

MACDONALD, R. 1975.   Petrochemistry of the Early Carboniferous (Dinantian) lavas of Scotland.   *Scottish Journal of Geology*, Vol. 11, 269–314.

— GOTTFRIED, D, FARRINGTON, M J, BROWN, F W, and SKINNER, N G. 1981. Geochemistry of a continental tholeiite suite: late Palaeozoic quartz dolerite dykes of Scotland.   *Transactions of the Royal Society of Edinburgh: Earth Sciences*, Vol. 72, 57–74.

MACGREGOR, A G. 1937.   The Carboniferous and Permian volcanoes of Scotland.   *Bulletin Volcanologique*, Series II Vol. 1, 41–58.

— 1948.   *British regional geology: the Grampian Highlands* (2nd edition).   (Edinburgh: HMSO for Geological Survey and Museum.)

McGREGOR, D M, and WILSON, C D V. 1967.   Gravity and magnetic surveys of the younger gabbros of Aberdeenshire.   *Quarterly Journal of the Geological Society of London*, Vol. 123, 99–123.

MacGregor, S M A, and Roberts, J. 1963. Dalradian pillow lavas, Ardwell Bridge, Banffshire. *Geological Magazine,* Vol. 100, 17–23.

McIntyre, D B. 1951. The tectonics of the area between Grantown and Tomintoul (mid-Strathspey). *Quarterly Journal of the Geological Society of London,* Vol. 107, 1–22.

MacIntyre, R M, McMenamin, T, and Preston, J. 1975. K-Ar results from western Ireland and their bearing on the timing and siting of Thulean magmatism. *Scottish Journal of Geology,* Vol. 11, 227–249.

Mackenzie, D H. 1958. The structure of the Grantown Granite complex, Morayshire. *Geological Magazine,* Vol. 95, 57–70.

Mackie, W. 1897. The sands and sandstones of Eastern Moray. *Transactions of the Edinburgh Geological Society,* Vol. 7, 148.

— 1901a. The occurrence of barium sulphate and calcium fluoride as cementing substances in the Elgin Trias. *Report of the Meeting of the British Association for the Advancement of Science, Glasgow, 1901,* 649–650.

— 1901b. The Pebble Band of the Elgin Trias and its wind-worn pebbles. *Report of the Meeting of the British Association for the Advancement of Science, Glasgow, 1901,* 650.

— 1908. Evidence of contemporaneous volcanic action in the Banffshire Schists. *Transactions of the Edinburgh Geological Society,* Vol. 9, 93–101.

— 1914. The Rock Series of Craigbeg and Ord Hill, Rhynie, Aberdeenshire. *Transactions of the Edinburgh Geological Society,* Vol. 10, 205–236.

— 1923. The principles that regulate the distribution of particles of heavy minerals, etc. *Transactions of the Edinburgh Geological Society,* Vol. 11, 138–164.

McLean, A C, and Deegan, C E. 1978. A synthesis of the solid geology of the Firth of Clyde region. 93–114 *in* The solid geology of the Clyde Sheet (55°N/6°W). McLean, A C, and Deegan, C E (editors). *Report of the Institute of Geological Sciences,* No. 78/9.

McLellan, E L. 1983. Contrasting textures in metamorphic and anatectic migmatites: an example from the Scottish Caledonides. *Journal of Metamorphic Geology,* Vol. 1, 242–262.

— 1985. Metamorphic reactions in the kyanite and sillimanite zones of the Barrovian type area. *Journal of Petrology,* Vol. 26, 789–818.

McQuillin, R, Donato, J A and Tulstrup, J. 1982. Development of basins in the Inner Moray Firth and the North Sea by crustal extension and dextral displacement of the Great Glen Fault. *Earth and Planetary Science Letters,* Vol. 60, 127–139.

Manson, W. 1957. On the occurrence of a marine band in the *Arenicola modiolaris* zone of the Scottish Coal Measures. *Bulletin of the Geological Survey of Great Britain,* No. 12, 66–86.

Marcantonio, F, Dickin, A P, McNutt, R H, and Heaman, L M. 1988. A 1800-million-year-old Proterozoic gneiss terrane in Islay with implications for the crustal evolution of Britain. *Nature, London,* Vol. 335, 62–64.

Marston, R J. 1971. The Foyers granitic complex, Inverness-shire, Scotland. *Quarterly Journal of the Geological Society of London,* Vol. 126, 331–368.

Mason, J, Pattrick, R A D, and Gallagher, M J. 1991. Auriferous structures in the Upper Dalradian near Aberfeldy, Scotland. Abstract in *Exploration and the environment.* Lumb, A J, Brown, M J, and Smith, C G (editors). (Edinburgh, and Keyworth, Nottingham: British Geological Survey.)

Mason, R. 1988. Did the Iapetus Ocean really exist? *Geology,* Vol. 16, 823–826.

Mendum, J R. 1987.   Dalradian of the Collieston coast section. 161–172 in *Excursion guide to the geology of the Aberdeen area.*   Trewin, N H, Kneller, B C, and Gillen, C (editors). (Edinburgh: Scottish Academic Press for Geological Society of Aberdeen.)

— and Fettes, D J. 1985.   The Tay nappe and associated folding in the Ben Ledi– Loch Lomond area.   *Scottish Journal of Geology,* Vol. 21, 41-56.

Merritt, J W. 1992.   A critical review of methods used in the appraisal of onshore sand and gravel resources in Britain.   *Engineering Geology,* Vol. 32, 1–9.

— 1992.   The high-level marine shell-bearing deposits of Clava, Inverness-shire, and their origin as glacial rafts.   *Quaternary Science Reviews,* Vol. 11, 759–779.

Miles, R S. 1968.   The Old Red Sandstone antiarchs of Scotland: Family *Bothriolepidae.   Monograph of the Palaeontographical Society,* No. 552, 1–130.

Miller, J A, and Brown, P E. 1965.   Potassium–argon age studies in Scotland. *Geological Magazine,* Vol. 102, 106-134.

Mohammed, H A. 1987.   Petrography, mineral chemistry, geochemistry and sulphur isotope studies of the Abhainn Strathain copper mineralisation, Meall Mor, south Knapdale, Scotland.   Unpublished PhD thesis, University of Strathclyde.

Moles, N R. 1982.   Sphalerite composition in relation to deposition and metamorphism of the Foss stratiform Ba-Zn-Pb deposit, Aberfeldy, Scotland. *Mineralogical Magazine,* Vol. 47, 487–500.

— 1985.   Metamorphic conditions and uplift history in central Perthshire: evidence from mineral equilibria in the Foss celsian-barite-sulphide deposit.   *Journal of the Geological Society of London,* Vol. 142, 39–52.

Morton, D J. 1979.   Palaeogeographical evolution of the Lower Old Red Sandstone basin in the western Midland Valley.   *Scottish Journal of Geology,* Vol. 15, 97–116.

Mould, D D C P. 1946.   The geology of the Foyers 'granite' and the surrounding country.   *Geological Magazine,* Vol. 83, 249–265.

Muir, R J, Fitches, W R, and Maltman, A J. 1989.   An Early Proterozoic link between Greenland and Scandinavia in the Inner Hebrides of Scotland.   *Terra Abstract,* Vol. 1, 5.

— — — 1992.   Rhinns Complex: a missing link in the Proterozoic basement of the North Atlantic region.   *Geology,* Vol. 20, 1043–1046.

— — — In press.   The Colonsay Group and basement–cover relationships on the Rhinns of Islay, Inner Hebrides.   *Scottish Journal of Geology.*

Munro, M. 1970.   A reassessment of the 'younger' basic igneous rocks between Huntly and Portsoy based on new borehole evidence.   *Scottish Journal of Geology,* Vol. 6, 41–52.

— 1984.   Cumulate relations in the 'Younger Basic' masses of the Huntly–Portsoy area, Grampian Region.   *Scottish Journal of Geology,* Vol. 20, 343–359.

— 1986a.   Mylonite zones in the Insch 'Younger Basic' Mass.   *Scottish Journal of Geology,* Vol. 22, 132–136.

— 1986b. Geology of the country around Aberdeen.   *Memoir of the British Geological Survey,* Sheet 77 (Scotland).

— and Gallagher, J W. 1984.   Disruption of the 'Younger Basic' masses in the Huntly–Portsoy area, Grampian Region.   *Scottish Journal of Geology,* Vol. 20, 361–382.

Murchison, R I. 1859.   On the sandstones of Morayshire containing reptilian remains; and on their relations to the Old Red Sandstone of that country.   *Quarterly Journal of the Geological Society of London,* Vol. 15, 419–423.

Mykura, W. 1982.   Old Red Sandstone east of Loch Ness, Inverness-shire.   *Report of the Institute of Geological Sciences,* No. 82/13.

NAYLOR, H, TURNER, P, VAUGHAN, D J, and FALLICK, A E. 1989. The cherty rock, Elgin: a petrographic and isotopic study of a Permo-Triassic calcrete. *Geological Journal*, Vol. 24, 205–221.

NEALE, J, and FLENLEY, J (editors). 1981. *The Quaternary in Britain*. (Oxford: Pergamon Press.)

NELL, P A R. 1984. The geology of lower Glen Lyon. Unpublished PhD thesis, University of Manchester.

— 1986. Discussion on the Caledonian metamorphic core: an Alpine model. *Journal of the Geological Society of London*, Vol. 143, 723–728.

NEVES, R, and SELLEY, R C. 1975. A review of the Jurassic rocks of north-east Scotland. 1–29 in *Proceedings, Jurassic North Sea Symposium, Stavanger, 1975*, Vol. JNNSS/5. FINSTAD, K G and SELLEY, R C (editors). (Oslo: Norsk Petroleumforening.)

NEWTON, E T. 1894. Reptiles of the Elgin Sandstone, description of two new genera. *Philosophical Transactions of the Royal Society*, Vol. 185B, 573-607.

NICHOLSON, K, and ANDERTON, R. 1989. The Dalradian rocks of the Lecht, NE Scotland: stratigraphy, faulting, geochemistry and mineralisation. *Transactions of the Royal Society of Edinburgh: Earth Sciences*, Vol. 80, 143–157.

NICOL, J. 1852. On the geology of the southern portion of the peninsula of Cantyre, Argyllshire. *Quarterly Journal of the Geological Society of London*, Vol. 8, 406–425.

NOCKOLDS, S R. 1941. The Garabal Hill–Glen Fyne igneous complex. *Quarterly Journal of the Geological Society of London*, Vol. 96, 451–511.

— and MITCHELL, R L. 1948. The geochemistry of some Caledonian plutonic rocks. A study in the relationship between the major and trace elements of igneous rocks and their minerals. *Transactions of the Royal Society of Edinburgh*, Vol. 61, 533–575.

OKONKWO, C T. 1988. The stratigraphy and structure of the metasedimentary rocks of the Loch Laggan–upper Strathspey area, Inverness-shire. *Scottish Journal of Geology*, Vol. 24, 21–34.

OLDERSHAW, W. 1974. The Lochnagar granitic ring complex, Aberdeenshire. *Scottish Journal of Geology*, Vol. 10, 297–309.

O'NIONS, R K, HAMILTON, P J, and HOOKER, P J. 1983. A Nd isotope investigation of sediments related to crustal development in the British Isles. *Earth and Planetary Science Letters*, Vol. 63, 229–240.

PANKHURST, R J. 1970. The geochronology of the basic igneous complexes. *Scottish Journal of Geology*, Vol. 6, 83–107.

— 1974. Rb-Sr whole-rock chronology of Caledonian events in northeast Scotland. *Bulletin of the Geological Society of America*, Vol. 85, 345–350.

— 1979. Isotope and trace element evidence for the origin and evolution of Caledonian granites in the Scottish Highlands. 18–33 in *Origin of granite batholiths: geochemical evidence*. ATHERTON, M P, and TARNEY, J (editors). (Orpington: Shiva Publishing Ltd.)

— and PIDGEON, R T. 1976. Inherited isotope systems and the source region prehistory of the early Caledonian granites in the Dalradian Series of Scotland. *Earth and Planetary Science Letters*, Vol. 31, 58–66.

— and SUTHERLAND, D S. 1982. Caledonian granites and diorites of Scotland and Ireland. 149–190 in *Igneous rocks of the British Isles*. SUTHERLAND, D S (editor). (Chichester: John Wiley and Sons.)

PANTIN, H M. 1956. The petrology of the Ben Vrackie epidiorites and their contact rocks. *Transactions of the Geological Society of Glasgow*, Vol. 22, 48–79.

— 1961.   The stratigraphy and structure of the Blair Atholl–Ben a'Gloe area, Perthshire, Scotland.   *Transactions of the Royal Society of New Zealand*, Vol. 88, 597–622.

PARKER, R T G, CLIFFORD, J A, and MELDRUM, A H.   1989.   The Cononish gold-silver deposit, Perthshire, Scotland.   *Transactions of the Institution of Mining and Metallurgy (Section B: Applied Earth Science)*, Vol. 98, B51–52.

PARNELL, J.   1983.   The Cothall Limestone.   *Scottish Journal of Geology*, Vol. 19, 215–218.

PATERSON, I B, and HALL, I H S.   1986.   Lithostratigraphy of the late Devonian and early Carboniferous rocks in the Midland Valley of Scotland.   *Report of the British Geological Survey*, Vol. 18, No. 3.

— — and STEPHENSON, D.   1990.   Geology of the Greenock district.   *Memoir of the British Geological Survey*, Sheet 30W and part of Sheet 29E (Scotland).

PATTISON, D R M.   1989.   P-T conditions and the influence of phase relations in the Ballachulish aureole, Scotland.   *Journal of Petrology*, Vol. 30, 1219–1244.

— and HARTE, B.   1988.   Evolution of structurally contrasting anatectic migmatites in the 3-kbar Ballachulish aureole, Scotland.   *Journal of Metamorphic Petrology*, Vol. 6, 475–494.

PATTRICK, R A D.   1984.   Sulphide mineralogy of the Tomnadashan copper deposit and the Corrie Buie lead veins, south Loch Tayside, Scotland.   *Mineralogical Magazine*, Vol. 48, 85–91.

— 1985.   Pb-Zn and minor U mineralization at Tyndrum, Scotland.   *Mineralogical Magazine*, Vol. 49, 671–681.

— BOYCE, A, and MACINTYRE, R M.   1988.   Gold-silver vein mineralization at Tyndrum, Scotland.   *Mineralogy and Petrology*, Vol. 38, 61–76.

— CURTIS, S, SCOTT, R A, and TREAGUS, J E.   1991.   Polyphase hydrothermal mineralization in the Tyndrum region.   Abstract in *Exploration and the environment*. (Edinburgh, and Keyworth, Nottingham: British Geological Survey.)

PEACH, B N, and HORNE, J.   1930.   *Chapters on the geology of Scotland*.   (Oxford: Oxford University Press.)

— KYNASTON, H, and MUFF, H B.   1909.   Geology of the seaboard of Mid-Argyll, including the islands of Luing, Scarba, the Garvellachs, and the Lesser Isles, together with the northern part of Jura and a small portion of Mull.   *Memoir of the Geological Survey, Scotland*, Sheet 36 (Scotland).

PEACOCK, J. D.   1966.   Contorted beds in the Permo-Triassic aeolian sandstones of Morayshire.   *Bulletin of the Geological Survey of Great Britain*, No 24, 157–162.

— 1983.   Planning for development: Peterhead project.   *Institute of Geological Science, Technical Report*, No. WA/HI/83/1.

— 1994.   Late Quaternary marine Mollusca as palaeoenvironmental proxies: a compilation and assessment of basic numerical data for NE Atlantic species found in shallow water.   *Quaternary Science Reviews*, Vol. 12, 263–275.

— BERRIDGE, N G, HARRIS, A L, and MAY, F.   1968.   The geology of the Elgin district.   *Memoir of the Geological Survey of Scotland*, Sheet 95 (Scotland).

— and CORNISH, R (editors).   1989.   *Glen Roy area field guide*.   (Cambridge: Quaternary Research Association.)

PHILLIPS, E R, CLARK, G C, and SMITH, D I.   1993.   Mineralogy, petrology, and microfabric analysis of the Eilrig Shear Zone, Fort Augustus, Scotland.   *Scottish Journal of Geology*, Vol. 29, 143–158.

PHILLIPS, F C.   1930.   Some mineralogical and chemical changes induced by progressive metamorphism in the Green Bed group of the Scottish Dalradian.   *Mineralogical Magazine*, Vol. 22, 239–256.

PHILLIPS, W E A, STILLMAN, C J, and MURPHY, T. 1976. A Caledonian plate tectonic model. *Journal of the Geological Society of London*, Vol. 132, 579–609.

PIASECKI, M A J. 1975. Tectonic and metamorphic history of the Upper Findhorn, Inverness-shire, Scotland. *Scottish Journal of Geology*, Vol. 11, 87–115.

— 1980. New light on the Moine rocks of the Central Highlands of Scotland. *Journal of the Geological Society of London*, Vol. 137, 41–59.

— and TEMPERLEY, S. 1988. The Central Highland Division. 46–53 in *Later Proterozoic stratigraphy of the Northern Atlantic regions*. WINCHESTER, J A (editor). (New York: Blackie.)

— and VAN BREEMEN, O. 1979. A Morarian age for the "younger Moines" of central and western Scotland. *Nature, London*, Vol. 278, 734–736.

— — 1979. The Central Highland Granulites: cover-basement tectonics in the Moine. 139–144 *in* The Caledonides of the British Isles—reviewed. HARRIS, A L, HOLLAND, C H, and LEAKE, B E (editors). *Special Publication of the Geological Society of London*, No. 8.

— — 1983. Field and isotopic evidence for a c. 750 Ma tectonothermal event in Moine rocks in the Central Highland region of the Scottish Caledonides. *Transactions of the Royal Society of Edinburgh: Earth Sciences*, Vol. 73, 119-134.

PIDGEON, R T, and AFTALION, M. 1978. Cogenetic vs inherited zircon U-Pb systems in granites: Palaeozoic granites of Scotland and England. 183–220 *in* Crustal evolution of north-west Britain and adjacent regions. BOWES, D E, and LEAKE, B E (editors). *Special Issue of the Geological Journal*, No. 10.

PIPER, J D A. 1985. Continental movements and breakup in late Precambrian–Cambrian times: prelude to Caledonian orogenesis. 19-34 in *The Caledonide Orogen: Scandinavia and related areas*. GEE, D G, and STURT, B A (editors.) (Chichester: John Wiley and Sons.)

PITCHER, W S. 1982. Granite type and tectonic environment. 19–40 in *Mountain building processes*. HSU, K J (editor). (London: Academic Press.)

PLANT, J A. 1986. Models for granites and their mineralising systems in the British and Irish Caledonides. 121–156 in *Geology and genesis of mineral deposits in Ireland*. ANDREW, C J (editor). (Dublin: Irish Association for Economic Geology.)

— COOPER, D C, GREEN, P M, REEDMAN, A J, and SIMPSON, P R. 1991. Regional distribution of As, Sb and Bi in the Grampian Highlands of Scotland and English Lake District: implications for gold metallogeny. *Transactions of the Institution of Mining and Metallurgy (Section B: Applied Earth Science)*, Vol. 100, B135–147.

— WATSON, J V, and GREEN, P M. 1984. Moine–Dalradian relationships and their palaeotectonic significance. *Proceedings of the Royal Society*, Vol. 395a, 185–202.

— HENNEY, P J, and SIMPSON, P R. 1990. The genesis of tin-uranium granites in the Scottish Caledonides: implications for metallogenesis. *Geological Journal*, Vol. 25, 431–442.

PLATTEN, I M. 1991. Zoning and layering in diorites of the Scottish Caledonian Appinite suite. *Geological Journal*, Vol. 26, 329–348.

— and MONEY, M S. 1987. Formation of late Caledonian subvolcanic breccia pipes at Curachan Cruinn, Grampian Highlands, Scotland. *Transactions of the Royal Society of Edinburgh: Earth Sciences*, Vol. 78, 85–103.

POWELL, D. 1974. Stratigraphy and structure of the Western Moine and the problem of Moine orogenesis. *Journal of the Geological Society of London*, Vol. 130, 575–590.

PRICE, R J. 1983. *Scotland's environment during the last 30 000 years*. (Edinburgh: Scottish Academic Press.)

PRINGLE, I R. 1972.   Rb-Sr age determinations on shales associated with the Varanger Ice Age.   *Geological Magazine*, Vol 109, 465–472.

PRINGLE, J. 1941.   On the relationship of the Green Conglomerate to the Margie Grits in the North Esk, near Edzell; and on the probable age of the Margie Limestone. *Transactions of the Geological Society of Glasgow*, Vol. 20, 136–140.

— 1944.   The Carboniferous rocks of Glas Eilean, Sound of Islay, Argyllshire, with an appendix on the petrography by E. B. Bailey.   *Transactions of the Geological Society of Glasgow*, Vol. 20, 249–259.

— 1952.   On the occurrence of Permian rocks in Islay and North Kintyre. *Transactions of the Royal Society of Edinburgh*, Vol. 14, 297–301.

— and MACGREGOR, M. 1940.   The outlier of Carboniferous rocks at Bridge of Awe, Argyllshire.   *Transactions of the Geological Society of Glasgow*, Vol. 20, 73–76.

RAMSAY, D M, and STURT, B A. 1979.   The status of the Banff Nappe.   145–151 *in* The Caledonides of the British Isles—reviewed. HARRIS, A L, HOLLAND, C H and LEAKE, B E (editors). *Special Publication of the Geological Society of London*, No. 8.

RAST, N. 1958.   The tectonics of the Schiehallion Complex.   *Quarterly Journal of the Geological Society of London*, Vol. 114, 25–46.

— 1963.   Structure and metamorphism of the Dalradian rocks of Scotland. 123–142 in *The British Caledonides*. JOHNSON, M R W, and STEWART, F H (editors). (Edinburgh: Oliver and Boyd.)

— and LITHERLAND, M. 1970.   The correlation of the Ballachulish and Perthshire (Islay) successions.   *Geological Magazine*, Vol. 107, 259–272.

READ, H H. 1923.   The geology of the country around Banff, Huntly and Turriff (Lower Banffshire and North-west Aberdeenshire).   *Memoir of the Geological Survey, Scotland*, Sheets 86 and 96 (Scotland).

— 1927.   The igneous and metamorphic history of Cromar, Deeside.   *Transactions of the Royal Society of Edinburgh*, Vol. 55, 317–353.

— 1935.   *British regional geology: the Grampian Highlands* (1st edition).   (Edinburgh: HMSO for Geological Survey and Museum.)

— 1936.   The stratigraphical order of the Dalradian rocks of the Banffshire coast. *Geological Magazine*, Vol. 73, 468–475.

— 1952.   Metamorphism and migmatisation in the Ythan Valley, Aberdeenshire. *Transactions of the Edinburgh Geological Society*, Vol. 15, 265–279.

— 1955.   The Banff Nappe: an interpretation of the structure of the Dalradian rocks of north-east Scotland.   *Proceedings of the Geologists' Association*, Vol. 66, 1–29.

— 1956.   The dislocated south-western margin of the Insch Igneous Mass, Aberdeenshire.   *Proceedings of the Geologists' Association*, Vol. 67, 73–86.

— 1961.   Aspects of the Caledonian magmatism in Britain.   *Proceedings of the Liverpool and Manchester Geological Society*, Vol. 2, 653–683.

— and FARQUHAR, O C. 1956.   The Buchan Anticline of the Banff Nappe of Dalradian rocks in north-east Scotland.   *Quarterly Journal of the Geological Society of London*, Vol. 112, 131–156.

— and HAQ, B T. 1965.   Notes, mainly geochemical, on the granite–diorite complex of the Insch igneous mass, with an addendum on the Aberdeenshire quartz-dolerites. *Proceedings of the Geologists' Association*, Vol. 76, 13–19.

RICE, C M, and TREWIN, N H. 1988.   A Lower Devonian gold bearing hot-spring system, Rhynie, Scotland.   *Transactions of the Institution of Mining and Metallurgy (Section B: Applied Earth Science)*, Vol. 97, 141–144.

RICE, R. 1975. Geochemical exploration in an area of glacial overburden at Arthrath, Aberdeenshire. 82–86 in *Prospecting in areas of glaciated terrain 1975.* (London: Institution of Mining and Metallurgy.)

RICHARDSON, J B. 1967. Some British Lower Devonian spore assemblages and their stratigraphical significance. *Review of Palaeobotany and Palynology,* Vol. 1, 111–129.

RICHARDSON, S W, and POWELL, R. 1976. Thermal causes of the Dalradian metamorphism in the Central Highlands of Scotland. *Scottish Journal of Geology,* Vol. 12, 237–268.

RICHEY, J E. 1939. The dykes of Scotland. *Transactions of the Edinburgh Geological Society,* Vol. 13, 393–435.

— MACGREGOR, A G, and ANDERSON, F W. 1961. *British regional geology. Scotland: the Tertiary Volcanic Districts* (3rd edition). (Edinburgh: HMSO for Institute of Geological Sciences.)

ROBERTS, A M, STRACHAN, R A, HARRIS, A L, BARR, D, and HOLDSWORTH, R E. 1987. The Sgurr Beag Nappe: a reassessment of the stratigraphy and structure of the Northern Highland Moine. *Bulletin of the Geological Society of America,* Vol. 98, 497–506.

ROBERTS, J L. 1966. The emplacement of the Main Glencoe Fault-Intrusion at Stob Mhic Mhartuin. *Geological Magazine,* Vol. 103, 299-316.

— 1966. Sedimentary affiliations and stratigraphic correlation of the Dalradian rocks in the south-west Highlands of Scotland. *Scottish Journal of Geology,* Vol. 2, 200–223.

— 1974a. The structure of the Dalradian rocks in the SW Highlands of Scotland. *Journal of the Geological Society of London,* Vol. 130, 93–124.

— 1974b. The evolution of the Glencoe cauldron. *Scottish Journal of Geology,* Vol. 10, 269–282.

— 1976. The structure of the Dalradian rocks in the north Ballachulish district of Scotland. *Journal of the Geological Society of London,* Vol. 132, 139–154.

— and TREAGUS, J E. 1964. A reinterpretation of the Ben Lui Fold. *Geological Magazine,* Vol. 101, 512–516.

— — 1975. The structure of the Moine and Dalradian rocks in the Dalmally district of Argyllshire, Scotland. *Geological Journal,* Vol. 10, 59–74.

— — 1977a. The Dalradian rocks of the South-west Highlands—introduction. *Scottish Journal of Geology,* Vol. 13, 87–99.

— — 1977b. The Dalradian rocks of the Loch Leven area. *Scottish Journal of Geology,* Vol. 13, 165–184.

— — 1977c. Polyphase generation of nappe structures in the Dalradian rocks of the southwest Highlands of Scotland. *Scottish Journal of Geology,* Vol. 13, 237–254.

— — 1979. Stratigraphical and structural correlation between the Dalradian rocks of the SW and Central Highlands of Scotland. 199–204 *in* The Caledonides of the British Isles—reviewed. HARRIS, A L, HOLLAND, C H, and LEAKE, B E (editors). *Special Publication of the Geological Society of London,* No. 8.

— — 1980. The structural interpretation of the Loch Leven area. *Scottish Journal of Geology,* Vol. 16, 73–75.

— — 1990. Discussion on Dalradian slides and basin development: a radical reinterpretation of stratigraphy and structure in the SW and Central Highlands of Scotland. *Journal of the Geological Society of London,* Vol. 147, 729–731.

ROBERTSON, A H F, and HENDERSON, W G. 1984. Geochemical evidence for the origin of igneous and sedimentary rocks of the Highland Border, Scotland. *Transactions of the Royal Society of Edinburgh: Earth Sciences,* Vol. 75, 135–150.

ROBERTSON, S. 1991.   Older granites in the South-eastern Scottish Highlands. *Scottish Journal of Geology*, Vol. 27, 21–26.

— 1994.   Timing of Barrovian metamorphism and 'Older Granite' emplacement in relation to Dalradian deformation.   *Journal of the Geological Society of London*, Vol. 151, 5–8.

ROCK, N M S. 1983.   The Permo-Carboniferous camptonite-monchiquite dyke-suite of the Scottish Highlands and Islands: distribution, field and petrological aspects. *Report of the Institute of Geological Sciences*, No 82/14.

— 1985.   Value of chronostratigrapical correlation in metamorphic terranes: an illustration from the Colonsay Limestone, Inner Hebrides, Scotland.   *Transactions of the Royal Society of Edinburgh: Earth Sciences*, Vol. 76, 515–517.

— 1984.   Nature and origin of calc-alkaline lamprophyres: minettes, vogesites, kersantites and spessartites.   *Transactions of the Royal Society of Edinburgh: Earth Sciences*, Vol. 74, 193–227.

— 1986.   Chemistry of the Dalradian (Vendian–Cambrian) metalimestones, British Isles.   *Chemical Geology*, Vol. 56, 289–311.

— 1991.   *Lamprophyres.*   (Glasgow and London: Blackie.)

— JEFFREYS, L A, and MACDONALD, R. 1984.   The problem of anomalous local limestone–pelite successions within the Moine outcrop; I: metamorphic limestones of the Great Glen area, from Ardgour to Nigg.   *Scottish Journal of Geology*, Vol. 20, 383–406.

— MACDONALD, R, SZUCS, I, and BOWER, J. 1986.   The comparative geochemistry of some Highland pelites (Anomalous local limestone–pelite successions within the Moine outcrop; II).   *Scottish Journal of Geology*, Vol. 22, 107–126.

ROGERS, D A. 1987.   Devonian correlations, environments and tectonics across the Great Glen Fault.   Unpublished PhD thesis, University of Cambridge.

— MARSHALL, J E A, and ASTIN, T R. 1989.   Devonian and later movements on the Great Glen fault system, Scotland.   *Journal of the Geological Society of London*, Vol. 146, 369–372.

ROGERS, G, DEMPSTER, T J, BLUCK, B J, and TANNER, P W G. 1989.   A high-precision U/Pb age for the Ben Vuirich granite: implications for the evolution of the Scottish Dalradian Supergroup.   *Journal of the Geological Society of London*, Vol. 146, 789–798.

— and DUNNING, G R. 1991.   Geochronology of appinitic and related granitic magmatism in the W Highlands of Scotland: constraints on the timing of transcurrent fault movement.   *Journal of the Geological Society of London*, Vol. 148, 17–27.

— and PANKHURST, R J. 1993.   Unravelling dates through the ages: geochronology of the Scottish metamorphic complexes.   *Journal of the Geological Society of London*, Vol. 150, 447–464.

RUSHTON, A, and PHILLIPS, W E A. 1973.   A *Protospongia* from the Dalradian of Clare Island, Co. Mayo, Ireland.   *Palaeontology*, Vol. 16, 231–237.

RUSSELL, M J, and SMYTHE, D K. 1978.   Evidence for an Early Permian oceanic rift in the Northern North Atlantic.   173–179 in *Petrology and geochemistry of continental rifts*. NEUMANN, E R, and RAMBERG, I B (editors). (Dordrecht: Reidel.)

— — 1983.   Origin of the Oslo Graben in relation to the Hercynian–Alleghenian orogeny and lithospheric rifting in the North Atlantic.   *Tectonophysics*, Vol. 94, 457–472.

SADASHIVAIAH, M S. 1954.   The granite-diorite complex of the Insch Igneous Mass, Aberdeenshire.   *Geological Magazine*, Vol. 91, 286–292.

SAHA, D. 1985.   Clast composition and provenance of psammites in the Colonsay Group and Bowmore Sandstone, SW Argyllshire.   *Scottish Journal of Geology*, Vol. 21, 1–8.

SCOTT, R A. 1987.   Lithostratigraphy, structure and mineralisation of the Argyll Group Dalradian near Tyndrum, Scotland.   Unpublished PhD thesis, University of Manchester.

— POLYA, D A, and PATTRICK, R A D. 1988.   Proximal Cu+Zn exhalites in the Argyll Group Dalradian, Creag Bhocan, Perthshire.   *Scottish Journal of Geology*, Vol. 24, 97–112.

— PATTRICK, R A D, and POLYA, D A. 1991.   Origin of sulphur in metamorphosed stratabound mineralisation from the Argyll Group Dalradian of Scotland.   *Transactions of the Royal Society of Edinburgh: Earth Sciences*, Vol. 82, 91–98.

SCOTTISH COALFIELDS COMMITTEE. 1944.   *Scottish coalfields.*   (Edinburgh: Scottish Home Department.)

SHACKLETON, R M. 1958.   Downward-facing stuctures of the Highland Border.   *Quarterly Journal of the Geological Society of London*, Vol. 113, 361–392.

— 1979.   The British Caledonides: comments and summary.   299–304 *in* The Caledonides of the British Isles—reviewed. HARRIS, A L, HOLLAND, C H, and LEAKE, B E (editors).   *Special Publication of the Geological Society of London*, No. 8.

SHOTTON, F W. 1956.   Some aspects of the New Red desert in Britain.   *Liverpool and Manchester Geological Journal*, Vol. 1, 450–465.

SIMPSON, A, and WEDDEN, D. 1974.   Downward-facing structures in the Dalradian Leny Grits of Bute.   *Scottish Journal of Geology*, Vol. 10, 257–267.

SIMPSON, P R, and six others. 1989.   Gold mineralisation in relation to the extensional volcano-sedimentary basins in the Scottish Dalradian and the Abitibi Belt, Canada.   *Transactions of the Institution of Mining and Metallurgy (Section B: Applied Earth Science)*, Vol. 98, B102–117.

SISSONS, J B. 1967.   *The evolution of Scotland's scenery.*   (Edinburgh: Oliver and Boyd.)

— 1983.   Quaternary.   399–424 in *Geology of Scotland* (2nd edition) CRAIG, G Y (editor). (Edinburgh: Scottish Academic Press.)

SMITH, C G. 1977.   Investigations of stratiform sulphide mineralizations in part of the Dalradian of central Perthshire, Scotland.   *Transactions of the Institution of Mining and Metallurgy (Section B: Applied Earth Science)*, Vol. 86, B50–51.

— 1985.   Recent investigation of manganese mineralization in the Scottish Highlands.   *Transactions of the Institution of Mining and Metallurgy (Section A: Mineral Industry)*, Vol. 94, A159–162.

— 1989a.   Scottish Highlands and Southern Uplands Mineral Portfolio: hard rock aggregate resources.   *British Geological Survey Technical Report*, WF/89/4.

— 1989b.   Scottish Highlands and Southern Uplands Mineral Portfolio: silica sand and silica rock resources.   *British Geological Survey Technical Report*, WF/89/6.

— and nine others. 1977.   Investigation of stratiform sulphide mineralisation at McPhun's Cairn, Argyllshire.   *Mineral Reconnaissance Report, British Geological Survey*, No. 13.

— and six others. 1978.   Investigation of stratiform sulphide mineralisation at Meall Mor, South Knapdale, Argyll.   *Mineral Reconnaissance Report, British Geological Survey*, No 15.

— FORTEY, N J, and COATS, J S. 1991.   Stratabound manganese mineralisation at Lecht, NE Scotland.   Abstract in *Exploration and the environment*. LUMB, A J, BROWN, M J, and SMITH, C G (editors). (Edinburgh, and Keyworth, Nottingham: British Geological Survey.)

— GALLAGHER, M J, COATS, J S, and PARKER, M E. 1984.   Detection and general characteristics of stratabound mineralization in the Dalradian of Scotland.   *Transactions of the Institution of Mining and Metallurgy (Section B: Applied Earth Science)*, Vol. 93, B125–133.

— and eight others. 1988.   Stratabound base-metal mineralisation in Dalradian rocks near Tyndrum, Scotland.   *Mineral Reconnaissance Report, British Geological Survey*, No 93.

SMITH, D E, and DAWSON, A G. 1983.   *Shorelines and isostasy.*   (Glasgow: Academic Press.)

SMITH, D I, and WATSON, J V. 1983.   Scale and timing of movements on the Great Glen Fault, Scotland.   *Geology,* Vol. 11, 523–526.

SMITH, R A, and HARRIS, A L. 1976.   The Ballachulish rocks of the Blair Atholl District.   *Scottish Journal of Geology,* Vol. 12, 153–157.

SMITH, T E. 1968.   Tectonics in Upper Strathspey, Inverness-shire.   *Scottish Journal of Geology,* Vol. 4, 68–84.

SNELLING, N J. 1985.   The chronology of the geology record.   *Memoir of the Geological Society of London,* No. 10. (Oxford: Blackwell Scientific.)

— 1970.   The structural characteristics of the Strathspey Complex, Inverness-shire.   *Geological Magazine,* Vol. 107, 201–215.

SOPER, N J. 1988.   Timing and geometry of collision, terrane accretion and sinistral strike-slip events in the British Caledonides.   481–492 *in* The Caledonian–Appalachian Orogen. HARRIS, A L, and FETTES, D J (editors). *Special Publication of the Geological Society of London,* No. 38.

— and ANDERTON, R. 1984.   Did the Dalradian slides originate as extensional faults? *Nature, London,* Vol. 307, 357–360.

— and HUTTON, D H W. 1984.   Late Caledonian sinistral displacements in Britain: Implications for a three-plate collision model.   *Tectonics,* Vol. 3, 781–794.

SPEIGHT, J M, SKELHORN, R R, SLOAN, T, and KNAAP, R J. 1982.   The dyke swarms of Scotland.   449–460 in *Igneous rocks of the British Isles.* SUTHERLAND, D S (editor). (Chichester: John Wiley and Sons.)

— and MITCHELL, J G. 1979.   The Permo-Carboniferous dyke-swarm of northern Argyll and its bearing on dextral displacement on the Great Glen Fault.   *Journal of the Geological Society of London.* Vol. 136, 3-11.

SPENCER, A M. 1971.   Late Precambrian glaciation in Scotland.   *Memoir of the Geological Society of London,* No. 6.

— 1981. The late Precambrian Port Askaig Tillite in Scotland.   632–636 in *Earth's pre-Pleistocene glacial record.* HAMBREY, M J, and HARLAND, W B (editors). (Cambridge: Cambridge University Press.)

— and PITCHER, W S. 1968.   Occurrence of the Portaskaig Tillite in north-east Scotland.   *Proceedings of the Geological Society of London,* No. 1650, 195–198.

— and SPENCER, M O. 1972.   The late Precambrian/Lower Cambrian Bonahaven Dolomite of Islay and its stromatolites.   *Scottish Journal of Geology,* Vol. 8, 269–282.

STEIGER, R H, and JAEGER, E. 1977.   Subcommission on geochronology: convention on the use of decay constants in geo- and cosmochronology.   *Earth and Planetary Science Letters,* Vol. 36, 359–362.

STEPHENS, W E, and HALLIDAY, A N. 1984.   Geochemical contrasts between late Caledonian plutons of northern, central and southern Scotland.   *Transactions of the Royal Society of Edinburgh: Earth Sciences,* Vol. 75, 259–273.

STEPHENSON, D. 1972.   Middle Old Red Sandstone alluvial fan and talus deposits at Foyers, Inverness-shire. *Scottish Journal of Geology,* Vol. 8, 121–127.

— 1993.   Amphiboles from Dalradian metasedimentary rocks of NE Scotland: environmental inferences and distinction from meta-igneous amphibolites. *Mineralogy and Petrology,* Vol. 49, 45–62.

STEWART, A D. 1962a.   On the Torridonian sediments of Colonsay and their relationship to the main outcrop in north-west Scotland.   *Liverpool and Manchester Geological Journal,* Vol. 3, 121–155.

— 1962b.   Greywacke sedimentation in the Torridonian of Colonsay and Oronsay. *Geological Magazine*, Vol. 99, 399–419.

— 1969.   Torridonian rocks of Scotland reviewed.   595–608 *in* North Atlantic—Geology and Continental Drift, a symposium. KAY, M (editor). *Memoir of the American Association of Petroleum Geology*, Vol. 12.

— 1975.   'Torridonian' rocks of western Scotland.   43–51 *in* A correlation of the Precambrian rocks in the British Isles. HARRIS, A L, and five others (editors). *Special Report of the Geological Society, London*, No. 6.

— and HACKMAN, B D.   1973.   Precambrian sediments of western Islay.   *Scottish Journal of Geology*, Vol. 9, 185–201.

STEWART, F H, and JOHNSON, M R W.   1960.   The structural problem of the younger gabbros of north-east Scotland.   *Transactions of the Edinburgh Geological Society*, Vol. 18, 104–112.

STONE, M.   1957.   The Aberfoyle Anticline, Callander, Perthshire.   *Geological Magazine*, Vol. 94, 265–276.

STRINGER, P.   1957.   Polyphase deformation in the Upper Dalradian rocks of the Southern Highlands of Scotland.   Unpublished PhD thesis, University of Liverpool.

STURT, B A.   1961.   The geological structure of the area south of Loch Tummel. *Quarterly Journal of the Geological Society of London*, Vol. 117, 131–156.

— and HARRIS, A L.   1961.   The metamorphic history of the Loch Tummel area. *Liverpool and Manchester Geological Journal*, Vol. 2, 689-711.

— RAMSAY, D M, PRINGLE, I R, and TEGGIN, D E.   1977.   Precambrian gneisses in the Dalradian sequence of NE Scotland.   *Journal of the Geological Society of London*, Vol. 134, 41–44.

SUMMERHAYES, C P.   1966.   A geochronological and strontium isotope study of the Garabal Hill–Glen Fyne igneous complex, Scotland.   *Geological Magazine*, Vol. 103, 153–165.

SUTHERLAND, D G.   1981.   The high-level marine shell beds of Scotland and the build-up of the last Scottish ice sheet.   *Boreas*, Vol. 10, 247–54.

— 1984.   The Quaternary deposits and landforms of Scotland and the neighbouring shelves: a review.   *Quaternary Science Reviews*, Vol. 3. 157–254.

— 1991.   Late Devensian glacial deposits and glaciation in Scotland and the adjacent offshore region.   53–59 in *Glacial deposits in Great Britain and Ireland*. EHLERS, J, GIBBARD, P L, and ROSE, J (editors). (Rotterdam: A A Balkema.)

SUTTON, J, and WATSON, J V.   1954.   Ice-borne boulders in the Macduff Group of the Dalradian of Banffshire.   *Geological Magazine*, Vol. 91, 391–398.

— — 1955.   The deposition of the Upper Dalradian rocks of the Banffshire coast. *Proceedings of the Geologists' Association*, Vol. 66, 101–133.

— — 1956.   The Boyndie syncline of the Dalradian of the Banffshire coast. *Quarterly Journal of the Geological Society of London*, Vol. 112, 103–130.

SWEET, I P.   1985.   Sedimentology of the Lower Old Red Sandstone near New Aberdour, Grampian Region.   *Scottish Journal of Geology*, Vol. 21, 239–273.

SYNGE, F M.   1956.   The glaciation of north-east Scotland.   *Scottish Geographical Magazine*, Vol. 72, 129–143.

TANNER, P W G, and LESLIE, A G.   1994.   A pre-D2 age for the 590 Ma Ben Vuirich Granite from the Dalradian of Scotland.   *Journal of the Geological Society of London*, Vol. 151, 209–212.

TARLING, D.   1974.   A palaeomagnetic study of the Eocambrian tillites in Scotland. *Journal of the Geological Society of London*, Vol. 130, 163–177.

TARLO, L B H. 1961.   Psammosteids from the Middle and Upper Devonian of Scotland.   *Quarterly Journal of the Geological Society of London*, Vol. 117, 193–213.

TAYLOR, W. 1920.   A new locality for Triassic reptiles, etc.   *Transactions of the Edinburgh Geological Society*, Vol. 11, 11–13.

THIRLWALL, M F. 1981.   Implications for Caledonian plate tectonic models of chemical data from volcanic rocks of the British Old Red Sandstone.   *Journal of the Geological Society of London*, Vol. 138, 123–138.

— 1982.   Systematic variation in chemistry and Nd-Sr isotopes across a Caledonian calc-alkaline volcanic arc: implications for source materials.   *Earth and Planetary Science Letters*, Vol. 58, 27–50.

— 1988.   Geochronology of Late Caledonian magmatism in northern Britain. *Journal of the Geological Society of London*, Vol. 145, 951–967.

THOMAS, C W. 1989.   Application of geochemistry to the stratigraphic correlation of Appin and Argyll Group carbonate rocks from the Dalradian of northeast Scotland. *Journal of the Geological Society of London*, Vol. 146, 631–647.

THOMAS, P R. 1979.   New evidence for a Central Highland Root Zone.   205–211 *in* The Caledonides of the British Isles—reviewed. HARRIS, A L, LEAKE, B E, and HOLLAND, C H (editors). *Special Publication of the Geological Society of London*, No. 8.

— 1980.   The stratigraphy and structure of the Moine rocks north of the Schiehallion Complex, Scotland.   *Journal of the Geological Society of London*, Vol. 137, 469–482.

— 1987.   A9 road section—Blair Atholl to Newtonmore.   39–50 in *An excursion guide to the Moine geology of the Scottish Highlands.* ALLISON, I, MAY, F, and STRACHAN, R A (editors). (Edinburgh: Scottish Academic Press for Edinburgh Geological Society and Geological Society of Glasgow.)

— and TREAGUS, J E. 1968.   The stratigraphy and structure of the Glen Orchy area, Argyllshire, Scotland.   *Scottish Journal of Geology*, Vol. 4, 121–134.

THOMPSON, R N. 1982a.   Geochemistry and magma genesis.   461–477 in *Igneous rocks of the British Isles.* SUTHERLAND, D S (editor). (Chichester: John Wiley and Sons.)

— 1982b.   Magmatism of the British Tertiary volcanic province.   *Scottish Journal of Geology*, Vol. 18, 49–107.

THORP, P W. 1991.   The glaciation and glacial deposits of the western Grampians. 137–149 in *Glacial deposits in Great Britain and Ireland.* EHLERS, P L, GIBBARD, and ROSE, J (editors). (Rotterdam: A A Balkema.)

TILLEY, C E. 1924.   Contact-metamorphism in the Comrie area of the Perthshire Highlands.   *Quarterly Journal of the Geological Society of London*, Vol. 80, 22–71.

— 1925.   A preliminary survey of metamorphic zones in the southern Highlands of Scotland.   *Quarterly Journal of the Geological Society of London*, Vol. 81, 100–112.

TRAQUAIR, R H. 1896.   The extinct vertebrate fauna of the Moray Firth area. 235–285 in *Vertebrate fauna of the Moray Firth.* HARVEY-BROWN, H H, and BUCKLEY, T E (editors). Vol. 2. (Edinburgh.)

— 1897.   Additional notes on the fossil fishes of the Upper Old Red Sandstone of the Moray Firth area.   *Proceedings of the Royal Physical Society of Edinburgh*, Vol. 13, 376–385.

— 1905.   On the fauna of the Upper Old Red Sandstone of the Moray Firth area. *Report of the meeting of the British Association for the Advancement of Science, Cambridge, 1904*, 547.

TREAGUS, J E. 1964.   Notes on the structure of the Ben Lawers Synform.   *Geological Magazine*, Vol. 101, 260–270.

— 1969.   The Kinlochlaggan Boulder Bed.   *Proceedings of the Geological Society of London*, No. 1654, 55–60.

— 1974. A structural cross-section of the Moine and Dalradian rocks of the Kinlochleven area, Scotland. *Journal of the Geological Society of London*, Vol. 130, 525–544.

— 1981. The Lower Dalradian Kinlochlaggan Boulder Bed, Central Scotland. 637–639 in *Earth's pre-Pleistocene glacial record*. HAMBREY, J M, and HARLAND, W B (editors). (Cambridge: Cambridge University Press.)

— 1987. The structural evolution of the Dalradian of the Central Highlands of Scotland. *Transactions of the Royal Society of Edinburgh: Earth Sciences*, Vol. 78, 1–15.

— 1991. Fault displacements in the Dalradian of the Central Highlands. *Scottish Journal of Geology*, Vol. 27, 135–145.

— and KING, G. 1978. A complete Lower Dalradian succession in the Schiehallion district, central Perthshire. *Scottish Journal of Geology*, Vol. 14, 157–166.

— and ROBERTS, J L. 1981. The Boyndie Syncline, a D1 structure in the Dalradian of Scotland. *Geological Journal*, Vol. 16, 125–135.

— TALBOT, C J, and STRINGER, P. 1972. Downward-facing structures in the Birnam Slates, Dunkeld, Perthshire. *Geological Journal*, Vol. 8, 125–128.

— and TREAGUS, S H. 1971. The structures of the Ardsheal peninsula, their age and regional significance. *Geological Journal*, Vol. 7, 335–346.

TREWIN, N H, and RICE, C M. 1992. Stratigraphy and sedimentology of the Devonian Rhynie chert locality. *Scottish Journal of Geology*, Vol. 28, 37–47.

TURNELL, H B. 1985. Palaeomagnetism and Rb-Sr ages of the Ratagan and Comrie intrusions. *Geophysical Journal of the Royal Astronomical Society*, Vol. 83, 363–378.

UPTON, B G J, ASPEN, P, and CHAPMAN, N A. 1983. The upper mantle and deep crust beneath the British Isles: evidence from inclusions in volcanic rocks. *Journal of the Geological Society of London*, Vol. 149, 195–122.

— FITTON, J G, and MACINTYRE, R M. 1987. The Glas Eilean lavas: evidence of a Lower Permian volcano-tectonic basin between Islay and Jura, Inner Hebrides. *Transactions of the Royal Society of Edinburgh: Earth Sciences*, Vol. 77, 289–293.

UPTON, P S. 1986. A structural cross-section of the Moine and Dalradian rocks of the Braemar area. *Report of the British Geological Survey*, Vol. 17, No. 1, 9–19.

URRY, W D, and HOLMES, A. 1941. Age determinations of Carboniferous basic rocks from Shropshire and Colonsay. *Geological Magazine*, Vol. 78, 45–61.

VAN BREEMEN, O, and BOYD, R. 1972. A radiometric age for pegmatite cutting the Belhelvie mafic intrusion, Aberdeenshire. *Scottish Journal of Geology*, Vol. 8, 115–120.

— HALLIDAY, A N, JOHNSON, M R W, and BOWES D R. 1978. Crustal additions in late Precambrian times. 81–106 *in* Crustal evolution in northwestern Britain and adjacent regions. BOWES, D R, and LEAKE, B E (editors). *Special Issue of the Geological Journal*, Vol. 10.

— and PIASECKI, M A J. 1983. The Glen Kyllachy granite and its bearing on the Caledonian orogeny in Scotland. *Journal of the Geological Society of London*, Vol. 140, 47–62.

VAN DE KAMP, P C. 1970. The Green Beds of the Scottish Dalradian Series: geochemistry, origin and metamorphism of mafic sediments. *Journal of Geology*, Vol. 78, 281–303.

VOLL, G. 1964. Deckenbau und fazies im Schottischen Dalradian. *Geologische Rundschau*, Vol. 2, 590–612. [In German].

WADSWORTH, W J. 1986. Silicate mineralogy of the later fractionation stages of the Insch Intrusion, NE Scotland. *Mineralogical Magazine*, Vol. 50, 583–595.

— 1988.   Silicate mineralogy of the Middle Zone cumulates and associated gabbroic rocks from the Insch Intrusion, NE Scotland.   *Mineralogical Magazine,* Vol. 52, 309–322.

— 1991.   Silicate mineralogy of the Belhelvie cumulates, NE Scotland. *Mineralogical Magazine,* Vol. 55, 113–119.

WALKER, A D. 1961.   Triassic reptiles from the Elgin area. *Stagonolepis, Dasygnathus* and their allies.   *Philosophical Transactions of the Royal Society* (B), Vol. 244, 103–204.

— 1964.   Triassic reptiles from the Elgin area: *Ornithosuchus* and the origin of Carnosaurs.   *Philosophical Transactions of the Royal Society* (B), Vol. 248, 53–134.

— 1973.   The age of the Cuttie's Hillock Sandstone (Permo-Triassic) of the Elgin area.   *Scottish Journal of Geology,* Vol. 9, 177–183.

WALKER, F. 1935.   The late Palaeozoic quartz dolerites and tholeiites of Scotland. *Mineralogical Magazine,* Vol. 24, 131–159.

— 1939.   The geology of Maiden Island, Oban.   *Transactions of the Edinburgh Geological Society,* Vol. 13, 475-482.

— 1961.   The Islay–Jura dyke swarm.   *Transactions of the Geological Society of Glasgow.* Vol. 24, 121–137.

— and PATTERSON, E M. 1959.   A differentiated boss of alkali dolerite from Cnoc Rhaonastil, Islay.   *Mineralogical Magazine,* Vol. 32, 140–152.

WALKER, M J C. 1984.   Pollen analysis and Quaternary research in Scotland. *Quaternary Science Reviews,* Vol. 3, 369–404.

— GRAY, J M, and LOWE, J J (editors). 1992.   *The South-west Scottish Highlands: field guide.*   (Cambridge: Quaternary Research Association.)

— MERRITT, J W, AUTON, C A, COOPE, G R, FIELD, M H, HEIJINIS, H, and TAYLOR, B J. 1992.   Allt Odhar and Dalcharn: two pre-Late Devensian (Late Weichselian) sites in northern Scotland.   *Journal of Quaternary Science,* Vol. 7, 69-86.

WARRINGTON, G, and eight others. 1980.   A correlation of Triassic rocks in the British Isles.   *Special Report of the Geological Society of London,* No. 13.

WATKINS, K P. 1983.   Petrogenesis of Dalradian albite porphyroblast schists.   *Journal of the Geological Society of London,* Vol. 140, 601–618.

— 1984.   The structure of the Balquhidder–Crianlarich region of the Scottish Dalradian and its relation to the Barrovian garnet isograd surface.   *Scottish Journal of Geology,* Vol. 20, 53–64.

WATSON, D M S. 1909.   The Trias of Moray.   *Geological Magazine,* Vol. 46, 102–107.

— and HICKLING, G. 1914.   On the Triassic and Permian rocks of Moray.   *Geological Magazine,* Vol. 12, 399–402.

WATSON, J V. 1984.   The ending of the Caledonian Orogeny in Scotland.   *Journal of the Geological Society of London,* Vol. 141, 193–214.

WEBB, P C, and BROWN, G C. 1984.   *Investigations of the geothermal potential of the UK— The Eastern Highlands granites: heat production and related geochemistry.*   (Keyworth, Nottingham: British Geological Survey.)

— TINDLE, A G, and IXER, R A. 1992.   Wo-Sn-Mo-Bi-Ag mineralization associated with zinnwaldite granite from Glen Gairn, Scotland.   *Transactions of the Institution of Mining and Metallurgy, Series B,* Vol. 101, B59–B72.

WEEDON, D S. 1970.   The ultrabasic/basic igneous rocks of the Huntly region. *Scottish Journal of Geology,* Vol. 6, 26–40.

WEISS, L E, and McINTYRE, D B. 1957.   Structural geometry of Dalradian rocks at Loch Leven, Scottish Highlands.   *Journal of Geology,* Vol. 65, 575–602.

WEISS, S, and TROLL, G. 1989. The Ballachulish igneous complex, Scotland: petrography, mineral chemistry and order of crystallization in the monzodiorite–quartz-diorite suite and in the granite. *Journal of Petrology*, Vol. 30, 1069–1115.

WESTOLL, T S. 1951. The vertebrate bearing strata of Scotland. *Report of the XVIIIth International Geological Congress (Great Britain), 1948*, pt XI, 5–21.

— 1977. Northern Britain. 66–93 *in* A correlation of Devonian rocks of the British Isles. HOUSE, M R, and others (editors). *Special Report of the Geological Society of London*, No. 8.

WHITTEN, E H T. 1959. A study of two directions of folding; the structural geology of the Monadhliath and mid-Strathspey. *Journal of Geology*, Vol. 67, 14–47.

WHITTLE, G. 1936. The eastern end of the Insch igneous mass, Aberdeenshire. *Proceedings of the Liverpool Geological Society*, Vol. 17, 64–95.

WILKINSON, S B. 1907. The geology of Islay including Oronsay and portions of Colonsay and Jura. *Memoir of the Geological Survey, Scotland*, Sheets 19 and 27, part 20 (Scotland).

WILLAN, R C R, and COLEMAN, M L. 1983. Sulfur isotope study of the Aberfeldy barite, zinc, lead deposit and minor sulfide mineralization in the Dalradian metamorphic terrain, Scotland. *Economic Geology*, Vol. 78, 1619–1656.

— and HALL, A J. 1980. Sphalerite geobarometry and trace element studies on stratiform sulphide from McPhun's Cairn, Loch Fyne, Argyll, Scotland. *Transactions of the Institution of Mining and Metallurgy (Section B: Applied Earth Science)*, Vol. 88, B31–40.

WILLIAMS, D. 1973. The sedimentology and petrology of the New Red Sandstone of the Elgin Basin, North-East Scotland. Unpublished PhD thesis, University of Hull.

WILLIAMSON, W O. 1935. The composite gneiss and contaminated granodiorite of Glen Shee, Perthshire. *Quarterly Journal of the Geological Society of London*, Vol. 91, 382–422.

WILSON, G V, and FLETT, J S. 1921. The lead, zinc, copper and nickel ores of Scotland. *Memoir of the Geological Survey, Special Report on Mineral Resources*, No. 17.

WILSON, H E, and ROBBIE, J A. 1966. The geology of the country around Ballycastle. *Memoir of the Geological Survey, Northern Ireland*, Sheet 8 (Northern Ireland).

WILSON, J R, AND LEAKE, B E. 1972. The petrochemistry of the epidiorites of the Tayvallich Peninsula, North Knapdale, Argyllshire. *Scottish Journal of Geology*, Vol. 8, 215–252.

WILSON, J S G. 1886. Explanation of Sheet 87, North-east Aberdeenshire and detached portions of Banffshire. *Memoir of the Geological Survey, Scotland*.

WINCHESTER, J A. 1974. The zonal pattern of regional metamorphism in the Scottish Caledonides. *Journal of the Geological Society of London*, Vol. 130, 509–524.

— 1988. Introduction. 1–13 in *Later Proterozoic stratigraphy of the Northern Atlantic regions*. WINCHESTER, J A (editor). (Glasgow and London: Blackie.)

— and GLOVER, B W. 1988. The Grampian Group, Scotland. 146–161 in *Later Proterozoic stratigraphy of the Northern Atlantic regions*. WINCHESTER, J A (editor). (Glasgow and London: Blackie.)

WISEMAN, J D H. 1934. The central and south-west Highland epidiorites: a study in progressive metamorphism. *Quarterly Journal of the Geological Society of London*, Vol. 90, 354–417.

WOODWARD, A S. 1907. On *Scleromochlus taylori* from the Trias of Elgin. *Quarterly Journal of the Geological Society of London*, Vol. 63, 140–144.

WRIGHT, A E. 1988. The Appin Group. 177–199 in *Later Proterozoic stratigraphy of the Northern Atlantic regions*. WINCHESTER, J A (editor). (Glasgow and London: Blackie.)

WRIGHT, W B. 1908.   The two earth-movements of Colonsay.   *Quarterly Journal of the Geological Society of London*, Vol. 64, 297–312.

ZALESKI, E. 1983.   The geology of Strathspey and Lower Findhorn granitoids—a study involving field relations, petrography, mineralogy, geochemistry and geochronology.   Unpublished MSc thesis, University of St Andrews.

— 1985.   Regional and contact metamorphism within the Moy Intrusive Complex. *Contributions to Mineralogy and Petrology*, Vol. 89, 296–306.

Index

Printed in the United Kingdom for HMSO
Dd 301288 11/95 C50 566 59226